Contributions to Management Science

More information about this series at http://www.springer.com/series/1505

Mehdi Mili · Reyes Samaniego Medina
Filippo di Pietro
Editors

New Methods in Fixed Income Modeling

Fixed Income Modeling

Springer

Editors
Mehdi Mili
Department of Economics and Finance,
College of Business Administration
University of Bahrain
Zallaq, Bahrain

Filippo di Pietro
Department of Financial Economics
and Operations Management
University of Seville
Seville, Spain

Reyes Samaniego Medina
Department of Financial Economics
and Accounting
Pablo de Olavide University
Seville, Spain

ISSN 1431-1941 ISSN 2197-716X (electronic)
Contributions to Management Science
ISBN 978-3-030-07008-3 ISBN 978-3-319-95285-7 (eBook)
https://doi.org/10.1007/978-3-319-95285-7

Softcover re-print of the Hardcover 1st edition 2018

This Springer imprint is published by the registered company Springer Nature Switzerland AG
The registered company address is: Gewerbestrasse 11, 6330 Cham, Switzerland

Preface

Over the last 30 years, fixed income markets have undergone a profound change. Among the most salient features of this development are the phenomenal advances in mathematical and stochastic methods applied to finance. The expansion of fixed income markets has been driven by large intermediaries who have a central role in the development of new risk transfer securities. Currently, the pricing of new fixed income securities and the use of derivatives in hedging strategies is a major issue in financial theory for academics as well as for portfolios managers.

Financial theories provide a large number of valuation models for fixed income products. All have been based on the modeling of the term structure of interest rates. These models can be classified into different ways, for example models based on the number of state variables, the deterministic or non-deterministic nature of forward rates, the exogenous or endogenous nature of the initial interest term structure. From a perspective of managing the interest rate risk of a financial institution, most theories are largely silent on the values of the unobserved parameters and where to look for priced sources of risk.

The developments in fixed income derivatives markets have been stimulated by major advances in mathematics and financial modeling. In this context, pricing of financial products is the central object of financial modeling. Modern interest rate theories have invested heavily in the examination of distribution probabilities of the future prices of financial products. In competitive financial markets, the relationships between cash flows generated by financial products and their prices are highlighted through consistent valuation models of financial products, state prices, and risk-neutral probabilities.

Financial models are also involved in the manufacturing and management process of financial products because they provide risk hedging strategies for the sale of these products. Financial modeling steps in as from the product design stage and is often a significant factor in optimizing configurations. It can either help to define the proportion of assets in portfolios, or specify the nature of contingent contracts, particularly options. The creation of new financial products involves residual risks that must be assessed, measured, and possibly hedged in integrated markets. The use of financial models is therefore very useful for measuring and limiting exposure to these residual risks.

This book interestingly addresses two main topics in the field of fixed income modeling. First, contributors make appealing extensions of existing term structure models. Mathematical advances in interest rate asset pricing are illustrated by appropriate implementation to real interest rate markets to illuminate their practical usefulness. Second, we offer insight into fixed income risk management and hedging strategies. Contributors cover a selection of short-term- and long-term-oriented measures as well as their application to the risk management of fixed income portfolios.

This book is written by eminent scholars and international experts to makes valuable contribution to the field of fixed income modeling. Its originality lies in the desire of contributors to mix as harmoniously as possible theory and practice. Throughout the chapters, the authors present interesting advances in modeling interest rate-based products. The application areas are illustrated by real and updated applications on financial markets. Particularly, the topic of this book connects in a significant way with disciplines of research in the field of interest rate modeling and portfolio managers.

This book will appeal to wide readership, from students through researchers to practitioners who want to deepen their knowledge and disclose new advancements in interest rate modeling and valuation of fixed income products. Readers will discover diverse perspectives of the contributing authors such as new rigorous extensions of market and forward models and extensive discussions of updated issues in fixed income markets including new hedging and pricing techniques.

The book is divided into two parts that cover different topics. The first part is an overview of new approaches on term structure modeling. More specifically, in Chapter "Term Structure, Market Expectations of the Short Rate, and Expected Inflation," Jian and Xiaoxia present a model that links the short interest rate and its market expectations, the expectations of the inflation rate, and the dynamics of the term structure of interest rates. In Chapter "A New Approach to CIR Short-Term Rates Modelling," Orlando, Mininni, and Bufalo explain an extension of the CIR framework to fit well to market short interest rate. In Chapter "The Heath-Jarrow-Morton Model with Regime Shifts and Jumps Priced," Elliott and Siu identify the no-arbitrage drift conditions under the general Markovian, regime-switching, Heath–Jarrow–Morton environment which price both the regime-switching and jump risks in the forward rate dynamics. In Chapter "Explicit Computation of the Post-crisis Spot LIBOR in a Jump-Diffusion Framework," Di Persio and Gugole show how to explicitly compute the post-crisis spot LIBOR at different tenors, taking into account the possibility of jumps in the instantaneous spot rate trajectories, representing the so-called OIS short rate. In Chapter "An Overview of Post-crisis Term Structure Models," Martin provides insight into post-crisis modeling of term structures via short-rate models and LIBOR Market models for multiple curves and show how these models are applied in economic scenario generators used for risk management and pricing purposes alike. In Chapter "A Comparison of Estimation Techniques for the Covariance Matrix in a Fixed-Income Framework,"

Neffelli and Resta compare various methodologies to estimate the covariance matrix in a fixed income portfolio, specifically they compare the shrinkage, the nonlinear shrinkage, the minimum covariance determinant, and the minimum regularized covariance determinant estimators against the sample covariance matrix, employed as a benchmark. Finally, in Chapter "The Term Structure Under Non-linearity Assumptions: New Methods in Time Series," Vides, Iglesias, and Golpe summarize an empirical review of the expectations hypothesis of the term structure aiming to establish the adequate procedures for its measurement by using time series. In Chapter "Affine Type Analysis for BESQ and CIR Processes with Applications to Mathematical Finance," Di Persio and Prezioso complement this part, presenting the deep relationships between the Cox-Ingersoll-Ross type-processes, the squared Bessel processes, and the family of affine processes, according to specific dynamics characterizing the dividend structure behind the market scenarios.

The second part of this book is about new advances in fixed Income management. This field is changing deeply since its beginnings. Fixed income management is now at the center of investment management. This book covers the most novel advances in the modeling and management of fixed income and its derivatives. The models presented can be useful in a wide variety of contexts by financial institutions, portfolio managers, and academicians that are focusing their research in this field. In Chapter "Sensitivity Analysis and Hedging in Stochastic String Models," Bueno-Guerrero, Moreno, and Navas analyze certain results on the stochastic string modeling of the term structure of interest rates, and they apply them to study the sensitivities and the hedging of options with payoff functions homogeneous of degree one. In Chapter "Hedging Asian Bond Options with Malliavin Calculus Under Stochastic String Models," Bueno-Guerrero, Moreno, and Navas use some recent hedging results for bond options, obtained with Malliavin calculus in the context of the stochastic string framework, to hedge different types of Asian options. In all the cases, they show that the hedging portfolio has no bank account part. In Chapter "Stochastic Recovery Rate: Impact of Pricing Measure's Choice and Financial Consequences on Single-Name Products," Gambetti, Gauthier, and Vrins compare the default probability curve implied by the International Swap and Derivatives Association (ISDA) model to that obtained from a simple variant accounting for stochastic recovery rate. They show that the former typically leads to underestimating the reference entity's credit risk compared to the latter. In Chapter "Dynamic Linkages Across Country Yield Curves: The Effects of Global and Local Yield Curve Factors on US, UK and German Yields," Coroneo, Garrett and Sanhueza analyze the relationship between the yield curves of the USA, the UK, and Germany using global and local factors. Their focus is on dynamic linkages across and between yield curves and factors. In Chapter "Estimating the No-Negative-Equity Guarantee in Reverse Mortgages: International Sensitivity Analysis," Fuente, Navarro, and Serna perform a sensitivity analysis to show how the value of the no-negative-equity guarantee (NNEG) embedded in reverse mortgage contracts varies with the value of the mortgage roll-up rate, the rental yield rate, as well as the gender and the age of the borrower. In Chapter

"Institutional Versus Retail Investors' Behavior Around Credit Rating News," Abad, Díaz, Escribano, and Robles perform a sensitivity analysis to show how the value of the no-negative-equity guarantee (NNEG) embedded in reverse mortgage contracts varies with the value of the mortgage roll-up rate, the rental yield rate, as well as the gender and the age of the borrower. Finally, in Chapter "The Market and Individual Pricing Kernels Under No Arbitrage Asset Pricing Models," Cosimano and Ma discuss how to use the no-arbitrage asset pricing model to determine the pricing kernel for both the financial markets and an individual with a given degree of constant risk aversion over her terminal wealth. The existence of the market and individual pricing kernel allows them to value any financial contract whose payoff is dependent on the prices from the no-arbitrage asset pricing model.

The book is finally complete. We have to thank all the authors who have contributed generously with their excellent chapters. Without them, obviously, this book would not have been possible.

Zallaq, Bahrain — Mehdi Mili
Seville, Spain — Filippo di Pietro
Seville, Spain — Reyes Samaniego Medina

Scientific Committee

Contents

Part I
New Term Structure Modeling Approaches

Term Structure, Market Expectations of the Short Rate, and Expected Inflation

Jian Luo and Xiaoxia Ye

Abstract Based on the classic Gaussian dynamic term structure model $\mathbb{A}_0(3)$, we rotate the model to a special representation, the so called "Companion Form Realization", in which the state variables comprise the short rate and its related expectations. This unique feature makes the representation very useful in analyzing the response of the yield curve to the shocks in the short rate and its related expectations, and monitoring market expectations. Using the estimated model, we quantify a variety of yield responses to the changes in these important state variables; and also give an "unsurprising" pattern in which changes in state variables have little impact on the long end of the yield curve. Estimated state variables have strong explanatory power for expected inflation. Three case studies of the unconventional monetary policies are presented.

Keywords Term structure of interest rates · Market expectations · Short rate Expected inflation · LSAP · MEP · QE3

JEL Classification E43

1 Introduction

How the yield curve responds to the changes in the market expectations of the short rate is a topic of great concerns to bond traders and monetary policy-makers alike. A good understanding on how the yield curve movement follows the changes in the market expectations can be the key to a successful trading strategy. Policy-makers

J. Luo
Xiamen University and ZhaoDao Technology Co., Shenzhen, China
e-mail: Luo.Jian@zhaodao.info

X. Ye (✉)
University of Liverpool Management School, Liverpool, UK
e-mail: Xiaoxia.Ye@liverpool.ac.uk

M. Mili et al. (eds.), *New Methods in Fixed Income Modeling*, Contributions to Management Science, https://doi.org/10.1007/978-3-319-95285-7_1

monitor the market expectations constantly, a detailed knowledge about the impact of changes in the expectations is crucial for them to set up effective monetary policies.

There have been studies on how to extract the information about market expectations from the interest rate related instruments, for example: Söderlind and Svensson (1997), Brooke et al. (2000), Joyce et al. (2008). It seems that the market expectations of the short rates for the next six months to one year is most informative about the future monetary policy, and relatively free of term premia and liquidity premia. These expectations can be well extracted from the implied forward rate curve. They are closely related to the slope of the implied forward curve (see Appendix 1), therefore tightly linked to the expected inflation, see, e.g., Ang et al. (2008), Frankel and Lown (1994), Mishkin (1990a, b). Besides these, little attention has been paid to the difference between slopes at different maturities on the yield curve. Actually, this quantity measures the expectation of the change in the expectation. For most of the time, this quantity moves in the opposite direction as the expectation of the change in the short rate due to the resilience of the economy system. Exceptions to this rule are very rare. However, once happen, they might be strong signs of substantial movements of the term structure. The expectation of the change in the expectation is also highly related to the expected inflation. Two of these expectation related variables together are able to explain around 60% of the variation in the changes of the long term expected inflation.[1] This is empirically verified in Appendix 2.

There are no previous models coherently link these expectations with the dynamics of the term structure of interest rates. In this chapter, we provide a model to fill in this gap. In this model, the short rate and its related market expectations are naturally constructed as the state variables which follow certain multivariate Gaussian dynamic. Thanks to this unique feature, the model can be employed to effectively infer the information about market expectations from the term structure of interest rates for the purpose of monitoring market movements; more importantly, it can be used to precisely analyze the impacts of the changes in the market expectations, this is the key for successful monetary actions.

In the empirical analysis, using the estimated model, we provide a series of numerical examples to demonstrate how changes in the short rate and its related expectations impact the yield curve, and also show how the impacts evolve over time in the sense of impulse responses. We find firstly that the short end of the term structure can be precisely regarded as a "level factor", given all else being equal, which can uniformly shift the yield curve up to the 10-year maturity; secondly, the market expectation variables are the crucial drivers of the dynamics of the yield curve long end, e.g., 10 bps increase in the market expectation of the short rate change alone can lift up the 10 year zero yield by 16 bps and keep the short end almost intact. Based on these numerical examples, we give an intuitive explanation to a puzzle (see, e.g., Ellingsen and Söderström 2001) that why occasionally the long end of the yield curve moves in a direction opposite to the intent of the policy-makers. Basically, whether

[1]The expected inflation mentioned here is in the sense Haubrich et al. (2012) and Potter (2012). The data is publicly available at http://www.clevelandfed.org/research/data/inflation_expectations/index.cfm.

the policy-makers are able to move the long end of the yield curve according to their intents depends on whether they can control the combined effect of two forces: the "Wicksell effect" and the "Fisher effect". The former is closely related to the short rate and has a "level effect" on the yield curve; the latter is closely related to the expected inflation and has a "slope effect" on the yield curve (more significant impact on the long end than on the short end). Typically, these two forces move in opposite directions. Therefore, the key for the policymakers to lower the long end of yield curve when conducting a monetary easing is to convince the private sector that they will keep the future inflation low so that the "Wicksell effect" on the long end will not be offset by the "Fisher effect"; if they fail to do so, i.e., the private sector may doubt either the ability or the will of the policy-makers to achieve that future inflation, a monetary easing won't be effective as the "Fisher effect" dominates or cancels out the "Wicksell effect", the long end of the yield will likely move in the opposite direction as the short end.

As shown in Kuttner (2001), when monetary policies come in an "unsurprising" manner, the impact would be minor. We therefore quantify an "unsurprising" pattern using parameter estimates and filtered state variables. This pattern can serve as a benchmark to check if a monetary policy is anticipated or not and why so. Specifically, based on the regression (using the expectation states as the dependent variables, and the 1-day lagged short rate as the independent variable) results, when the short rate decreases (increases) by 10 bps, the risk-neutral expected change of the short rate in 1 year is expected to increase (decrease) by about 9 bps. In other words, if a policy induces 10 bps short rate decrease and 9 bps increase in the expected change of the short rate (in 1 year), then this policy will have little (neither short run nor long run) impact on the long end of the yield curve.

In the case studies, we again use the estimated model to analyze the impacts of the announcement effect of the large scale asset purchase program (LSAP), the maturity extension program (MEP), and the third round of Quantitative Easing (QE3). Upon the announcements of the LSAP and MEP, although the forward curve moved in line with the intents of the Fed, the market responded differently to these two announcements: the market responded clearly and positively to the LSAP by lowering the expected inflation at various horizons; however, it responded less so to the MEP by increasing the expected inflation at horizons shorter than 2 years and keeping those at longer horizons intact. Upon the announcements of the QE3, the forward curve moved against the intent of the Fed, i.e., the long-end of the forward curve increased when the QE3 was announced. The expected inflation decreased at short horizons but consistently increased at long horizons. By using the impulse response analysis, we find that the LSAP had a significant impact on lowering the whole yield curve in the long run; but the long run impact of MEP on lowering the yield curve was much less prominent; when it comes to the QE3, the long run impact was actually to push up the whole yield curve. Since the short end has been artificially set in the zero zone, the "Wicksell effect" has been almost no longer in force, the impacts of announcements were mostly attributed to the "Fisher effect", i.e., to the changes in the expectation related state variables.

The rest of the chapter is organized as follows. Section 2 describes the classic model and how the new model is built from the classic model. Section 3 contains the discussion of the empirical results. Section 4 concludes the chapter. Appendices include technical details and supplementary results.

2 A New Usage of the Classic Model

In this section, we apply the modeling framework recently developed by Li et al. (2016a, b) to set up the 3-factor Gaussian dynamic term structure model.

2.1 *Instantaneous Forward Rate*

The modeling framework proposed by Li et al. (2016a, b) (hereafter LYY) is HJM-based, the starting point is the instantaneous forward rate. Following the Musiela parametrization (Brace et al. 1997), the time-t instantaneous forward rate for time-$t + x$ is written as:

$$r(t, x) = r(0, t + x) + \mathbf{\Theta}(t, x) + r_0(t, x)$$

where $\mathbf{\Theta}(t, x) = \int_0^t \sigma(x + t - s) \int_0^{x+t-s} \sigma(v)dvds$, $r_0(t, x) = \int_0^t \sigma(x + t - s)dW_s$, W_t is a 3-dimensional $\mathbb{Q}$-measure Brownian motion, $\sigma(\cdot) \in \mathbb{R}^3$ is the time-invariant volatility function defined in the following SDE:

$$dr(t, x) = \frac{\partial}{\partial x} r(t, x) + \sigma(x) \int_0^x \sigma(x)^\intercal ds + \sigma(x)dW_t.$$

2.2 *Volatility Term Structure and Markov Representation*

The key ingredient under LYY's framework is the volatility function. Here we specify the volatility function to be consistent with the most classic 3-factor Gaussian dynamic model, $AM_0(3)$ due to Dai and Singleton (2000).

$$\sigma(x)_{1\times 3} = \left[e^{-k_1 x},\ e^{-k_2 x},\ e^{-k_3 x} \right] \begin{bmatrix} \Omega_1 & 0 & 0 \\ \Omega_2 & \Omega_4 & 0 \\ \Omega_3 & \Omega_5 & \Omega_6 \end{bmatrix}.$$

Apparently, this volatility function guarantees a Markov representation for the model, as it is consistent with $AM_0(3)$. By the definition, if $\sigma(x)$ is factored as $\mathbf{C}\exp(\mathbf{A}x)\mathbf{B}$, then $r_0(t,x)$ has the following Markov representation:

$$r_0(t,x) = \mathbf{C}(x)Z_t$$
$$dZ_t = \mathbf{A}Z_t dt + \mathbf{B}dW_t$$

where $\mathbf{C}(x) = \mathbf{C}\exp(\mathbf{A}x)$.

Following LYY, we set $r(0,t+x) + \mathbf{\Theta}(t,x)$ to its time-homogeneous counterpart $\varphi + \mathbf{\Theta}^*(x)$, where

$$\mathbf{\Theta}^*(x) = \mathbf{C}(x)\left(\mathbf{A}^{-1}\mathbf{B}\mathbf{B}^\intercal(\mathbf{A}^\intercal)^{-1}\right)\left(\mathbf{C}(0) - \frac{\mathbf{C}(x)^\intercal}{2}\right).$$

Using the modified "essentially affine" (Duffee 2002) market price of risk setting, the measure P dynamic of Z_t is

$$dZ_t = \mathbf{A}^{\mathbb{P}}(Z_t - \mu)dt + \mathbf{B}dW_t^{\mathbb{P}},$$

where $\mathbf{A}^{\mathbb{P}} = \mathbf{A} - \mathbf{B}R_m\mathbf{C}(mV)$, $\mu = \left(\mathbf{A}^{\mathbb{P}}\right)^{-1}\mathbf{B}R_v$ and

$$W_t^{\mathbb{P}} = [R_v + R_m\mathbf{C}(mV)Z_t]t + W_t,$$

mV is a 3-dimensional vector of maturities,[2] $W_t^{\mathbb{P}}$ is a 3-dimensional $\mathbb{P}$-measure Brownian motion.

There are infinitely many representations, as given one realization $\{\mathbf{A}, \mathbf{B}, \mathbf{C}(x)\}$, for any invertible matrix M, $\left\{M\mathbf{A}M^{-1}, M\mathbf{B}, \mathbf{C}(x)M^{-1}\right\}$ is another realization, since

$$\mathbf{C}\exp(\mathbf{A}x)\mathbf{B} = \mathbf{C}M^{-1}\exp\left(M\mathbf{A}M^{-1}x\right)M\mathbf{B}.$$

The realization employed by Dai and Singleton (2000) is only one of them, the details are presented in LYY. Here we present two realizations, the first one is the "Base realization", since this is a Jordan form realization, we use this realization to estimate parameters; the second one is the "Companion form realization" which has some unique features for analyzing the impacts of market expectations on the term structure of interest rates, and will be discussed more in the next section.

- Base realization

$$\mathbf{A}_{Base} = \begin{bmatrix} -k_1 & 0 & 0 \\ 0 & -k_2 & 0 \\ 0 & 0 & -k_3 \end{bmatrix}$$

[2] mV is chosen to be $\left[\text{3 month}\ \ \text{2 year}\ \ \text{10 year}\right]^\intercal$ in the empirical analysis.

$$\mathbf{B}_{Base} = \begin{bmatrix} \Omega_1 & 0 & 0 \\ \Omega_2 & \Omega_4 & 0 \\ \Omega_3 & \Omega_5 & \Omega_6 \end{bmatrix}$$

$$\mathbf{C}(x)_{Base} = \left[e^{-k_1 x}, e^{-k_2 x}, e^{-k_3 x} \right].$$

- Companion form realization

$$\mathbf{A}_{CR} = \begin{bmatrix} 0 & 1 & 0 \\ 0 & 0 & 1 \\ -k_1k_2k_3 & -k_1k_2 - k_1k_3 - k_2k_3 & -k_1 - k_2 - k_3 \end{bmatrix}$$

$$\mathbf{B}_{CR} = \begin{bmatrix} 1 & 1 & 1 \\ -k_1 & -k_2 & -k_3 \\ k_1^2 & k_2^2 & k_3^2 \end{bmatrix} \begin{bmatrix} \Omega_1 & 0 & 0 \\ \Omega_2 & \Omega_4 & 0 \\ \Omega_3 & \Omega_5 & \Omega_6 \end{bmatrix}$$

$$\mathbf{C}(x)_{CR} = \left[e^{-k_1 x}, e^{-k_2 x}, e^{-k_3 x} \right] \begin{bmatrix} 1 & 1 & 1 \\ -k_1 & -k_2 & -k_3 \\ k_1^2 & k_2^2 & k_3^2 \end{bmatrix}^{-1}.$$

2.3 A Closer Look at the Companion Form Realization

The "Companion form realization" are firstly studied by Collin-Dufresne et al. (2008). The first important feature of the Companion form realization is that $Z_{1,t}$ is closely related to the short rate, $r(t, 0)$. Specifically,

$$r(t, 0) = \varphi + \frac{1}{2}\mathbf{C}(0)\left(\mathbf{A}^{-1}\mathbf{B}\mathbf{B}^{\mathsf{T}}\left(\mathbf{A}^{\mathsf{T}}\right)^{-1}\right)\mathbf{C}(0)^{\mathsf{T}} + Z_{1,t},$$

since

$$\mathbf{C}(0)_{CR} = \left[1, 1, 1\right] \begin{bmatrix} 1 & 1 & 1 \\ -k_1 & -k_2 & -k_3 \\ k_1^2 & k_2^2 & k_3^2 \end{bmatrix}^{-1} = \left[1, 0, 0\right].$$

The second important feature is that $Z_{2,t}$ is the risk neutrally expected instantaneous change of $Z_{1,t}$ or the short rate $r(t, 0)$ per unit of time, and $Z_{3,t}$ is the risk neutrally expected instantaneous change of $Z_{2,t}$ per unit of time. This can be easily verified by looking at the conditional expectation dZ_t:

$$\mathbb{E}_t^{\mathbb{Q}}\left(\begin{bmatrix} dZ_{1,t} \\ dZ_{2,t} \\ dZ_{3,t} \end{bmatrix}\right) = \mathbf{A}_{CR}\begin{bmatrix} Z_{1,t} \\ Z_{2,t} \\ Z_{3,t} \end{bmatrix} dt$$

$$= \begin{bmatrix} Z_{2,t} \\ Z_{3,t} \\ -k_1k_2k_3Z_{1,t} - (k_1k_2 + k_1k_3 + k_2k_3)Z_{2,t} - (k_1 + k_2 + k_3)Z_{3,t} \end{bmatrix} dt.$$

The expected instantaneous change of the short rate looks too abstract for the practical use, as mentioned in Söderlind and Svensson (1997) the expected change for a longer horizon, say 6–9 months, is actually more relevant. The last but not least important feature is that the model can be easily rotated to set per unit of time expectations of changes for a longer horizon as the state variables. Specifically, given a horizon h, a new set of state variables X_t^h can be defined as:

$$X_t^h = \begin{bmatrix} Z_{1,t} \\ \mathbb{E}_t^{\mathbb{Q}}\left(\begin{bmatrix} \Delta Z_{1,t+h} \\ \Delta Z_{2,t+h} \end{bmatrix}\right)/h \end{bmatrix} = \begin{bmatrix} Z_{1,t} \\ M_1 \frac{\exp(\mathbf{A}_{CR}h)-I}{h} Z_t \end{bmatrix},$$

where $\Delta Z_{t+h} = Z_{t+h} - Z_t$ and

$$M_1 = \begin{bmatrix} 1\,0\,0 \\ 0\,1\,0 \end{bmatrix}.$$

The first state in X_t^h is still $Z_{1,t}$ which is closely linked to the short rate, the second state is the risk neutrally expected $\Delta Z_{1,t}$ per unit of time, and so on. Therefore $r_0(t, x)$ can be re-written as

$$r_0(t, x) = \mathbf{C}(x) M_{fun}(\mathbf{A}_{CR}, h)^{-1} X_t^h \equiv \mathbf{D}(x, h) X_t^h,$$

where

$$M_{fun}(\mathbf{A}_{CR}, h) = \begin{bmatrix} 1\,0\,0 \\ M_1 \frac{\exp(\mathbf{A}_{CR}h)-I}{h} \end{bmatrix}_{3\times 3}, \quad I = \begin{bmatrix} 1\,0\,0 \\ 0\,1\,0 \\ 0\,0\,1 \end{bmatrix}.$$

These features make the Companion form realization very useful in analyzing the impacts of the monetary policy on the term structure. For example, Kuttner (2001) argues that it is crucial to separate the expected and unexpected components in short rate changes when analyzing how the monetary policy impacts the term structure of interest rates. The Companion form realization gives us a natural separation on the expected and unexpected changes of the short rate in a consistent and non-arbitrage way. More insights on how the Companion form realization can be utilized

Table 1 Parameter estimates

$$\mathbf{A}_{\text{Base}} = \begin{bmatrix} \underset{(0.005)}{-0.1507} & 0 & 0 \\ 0 & \underset{(0.011)}{-0.3204} & 0 \\ 0 & 0 & \underset{(0.049)}{-1.3881} \end{bmatrix}, \mathbf{B}_{\text{Base}} = \begin{bmatrix} \underset{(0.004)}{0.0412} & 0 & 0 \\ \underset{(0.004)}{-0.0410} & \underset{(0.001)}{0.0140} & 0 \\ \underset{(\text{N.A.})}{0} & \underset{(\text{N.A.})}{-0.0140} & \underset{(0.000)}{0.0066} \end{bmatrix}$$

$$R_m = \begin{bmatrix} \underset{(\text{N.A.})}{0} & \underset{(\text{N.A.})}{0} & \underset{(\text{N.A.})}{0} \\ \underset{(44.03)}{124.32} & \underset{(64.22)}{-206.56} & \underset{(65.29)}{217.31} \\ \underset{(50.49)}{76.92} & \underset{(72.68)}{-103.10} & \underset{(68.75)}{149.58} \end{bmatrix}, R_v = \begin{bmatrix} \underset{(\text{N.A.})}{0} & \underset{(0.433)}{1.154} & \underset{(0.484)}{2.486} \end{bmatrix}^{\mathrm{T}}$$

$$\varphi = \underset{(0.000)}{0.0563}$$

This table reports parameter estimates using the Base realization. The standard errors are in parentheses. The sample is daily from January 1994 to October 2012

to analyze the impacts of market expectations on the term structure of interest rates will be discussed in the empirical section.

3 Empirical Results

3.1 Data, Parameter Estimates, and Fitting Performance

The data used in the empirical analysis are the U.S. Treasury constant maturities yield curve rates, downloaded from Federal Reserve Statistical Release, H.15, Selected Interest Rates (Daily).[3] The data-set contains daily observations of the yields with maturities of three, six-month, one, two, three, five, seven, and 10-year. As the Treasury yield curve is considered as the par curve, the par curves are converted to the zero curves by first smoothing par rates then bootstrapping the zero rates from par rates.

The parameters are estimated using Kalman filter in conjunction with QMLE, assuming IID normal measurement errors. Following Duffee (2002), the model is refined by first computing the t-statistics for the parameter estimates of the full model, then setting all the parameters with absolute t-statistics less than one to zeros and re-estimate the model. The refined estimates are presented in Table 1.

In many traditional term structure estimations, yields at certain maturities are assumed to be priced without errors. By this assumption, even one uses a same set

[3]Here is the link: https://www.federalreserve.gov/releases/h15/data.htm.

Table 2 Summary statistics of pricing errors

Maturity	Mean (bp)	Medn (bp)	Std. (bp)	MAE (bp)	Max (%)	Min (%)	VR (%)
3 month	−1.992	−1.561	6.325	4.403	0.236	−0.798	99.912
6 month	2.472	2.126	4.896	4.022	0.503	−0.137	99.948
1 year	0.042	0.314	6.654	4.809	0.419	−0.241	99.903
2 year	0.918	1.164	4.031	3.335	0.144	−0.237	99.963
3 year	−2.120	−1.666	4.060	3.240	0.181	−0.298	99.958
5 year	−1.292	−1.043	4.597	3.767	0.201	−0.178	99.931
7 year	3.204	2.286	5.302	4.545	0.238	−0.121	99.887
10 year	−1.360	−1.359	6.658	5.259	0.239	−0.214	99.760
Ave.	−0.016	0.033	5.315	4.172	0.270	−0.278	99.908

This table reports the summary statistics (sample mean, median, standard deviation, mean absolute error, maximum, and minimum) of the pricing errors over the daily sample from January 1994 to October 2012. The pricing error is defined as the difference between the market observation and the model implied value. The table also reports the variance ratio

of data, the results are different when different maturities are assumed to be priced perfectly. Here we do not make this assumption, all yields are priced with errors. The summary statistics of the pricing errors are shown in Table 2.

Unsurprisingly, this 3-factor Gaussian model does a good job in fitting the dynamic term structure of interest rates. The average MAE (mean absolute error) across maturities is only 4.2 bp, and the average VR (variance ratio) across maturities is as high as 99.9%. This is consistent with the previous literature on the fitting performance of 3-factor Gaussian dynamic term structure models, e.g., Heidari and Wu (2009).

3.2 *State Variables*

To confirm that the first state variable in the Companion form realization represents the short rate, we compare it with the shortest yield, three-month zero yield, used in the estimation. Although the short rate is not the three-month zero yield as the short rate is with zero time-to-maturity, they should look very alike since their maturities are close. The comparison is shown in Fig. 1. By eyeballing, the model implied short rate is very close to the 3-month zero yield.[4] Their correlation is as high as 99.7%. In Piazzesi (2005), the correlation between the short rate implied by her model and LIBOR rate is only 54%. This somewhat confirms the statement that the first state

[4]There are some noticeable exceptions after 2009. These discrepancies might be due to the distortion at the short end of the yield curve which has been artificially set to "zero" since 2009. This distorted nominal short rate can barely reflect the real economic fluctuation. By fitting to the whole yield curve, the model implied short end of the yield curve could be more consistent with the real economy, therefore, gives a better description of the actually nominal short rate.

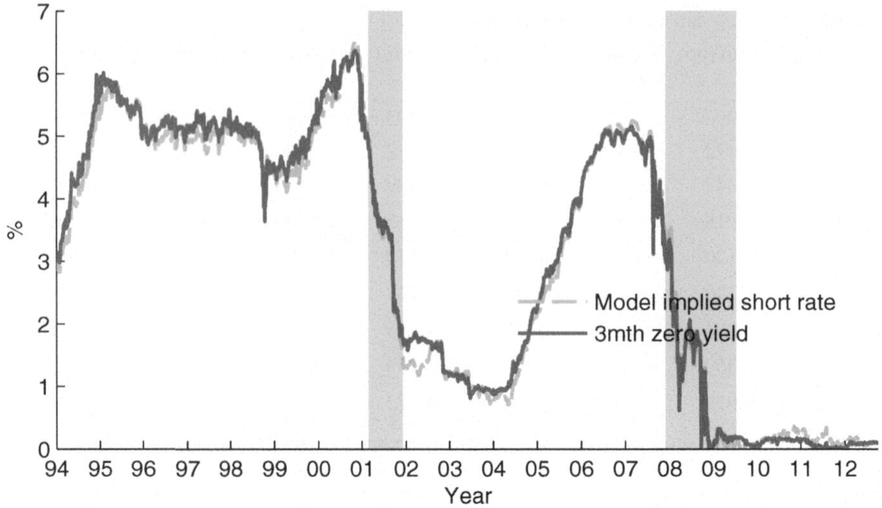

Fig. 1 Model Implied Short Rate versus Three-month Zero Yield. This figure compares the dynamics of the model implied short rate and 3 month zero yield from 1994 to 2012

variable in the Companion form realization can represent the short rate and the model (the realization) used in this chapter is much better in capturing the short rate process than others.

The second and third states are shown in Fig. 2. These two states are significantly negatively correlated. For most of the time, they almost mirror each other on the two sides of the zero line. In order to interpret this, let me borrow some results from Collin-Dufresne et al. (2008): other than the interpretations we mentioned before, these two states also have some other intuitive economic interpretations such as slope and curvature of the yield curve around a maturity of zero. Or in other words, the second and third states represent the first and second order derivatives, respectively, of the term structure function evaluated at the maturity zero. Bearing this understanding in mind, we find that the significant negative-correlation is actually a reflection of the stylized fact that for most of the time, the yield curve near the short end is concave when it is upward-sloping; convex when downward-sloping. Therefore, the shape of the yield curve in fact carries important information about the market expectation of the short rate movements. Another explanation of the negative correlation lies in the resilience of the interest rate system: increasing $Z_{2,t}$ means that the market is expecting the short rate to increase more or decrease less; at the same time, the market also expects a decrease in the speed of this adjustment. This can be very much described as that whenever there is a shock to the short rate, the market will "form" some sort of expectation "cushion" to absorb the effort of the shock. Therefore increasing (decreasing) $Z_{2,t}$ comes with decreasing (increasing) $Z_{3,t}$ in most of the time. The same observations can also be found among the second and third state variables in X_t^h as shown in Fig. 3.

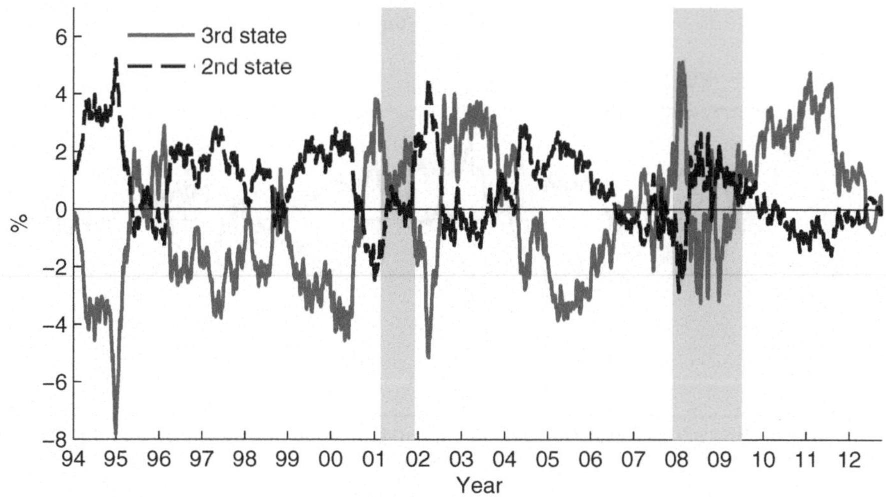

Fig. 2 The 2nd and 3rd state variables in Z_t. This figure shows the dynamic of the 2nd and 3rd state variables in Z_t from 1994 to 2012

3.3 *Yields Responses*

3.3.1 Contemporaneous Responses

Given the forward rate equation in Sect. 2, the zero yield with time-to-maturity of x can be written as:

$$y_t(x) = \varphi + \frac{\int_0^x \mathbf{\Theta}^*(s)ds}{x} + \frac{\int_0^x \mathbf{C}_{CR}(s)ds}{x} Z_t.$$

It can also be written as an affine function of X_t^h:

$$y_t(x) = \varphi + \frac{\int_0^x \mathbf{\Theta}^*(s)ds}{x} + \frac{\int_0^x \mathbf{D}(s)ds}{x} X_t^h.$$

The contemporaneous responses of the zero yield curve to the state variables are measured by the loading coefficients $\frac{\int_0^x \mathbf{C}_{CR}(s)ds}{x}$ and $\frac{\int_0^x \mathbf{D}(s)ds}{x}$. Responses of the zero yields at different maturities are plotted in Fig. 3.

From the figure, first of all we can see the short rate has a relatively flat loading across maturities. This means when other states keep constant, a change in the short rate can even significantly move the long end of the yield curve. Specifically, when there are no changes in other states, 10 bps increase in the short rate can immediately induce 8 bps increase in the 10 year zero yield. Secondly, we find that a change in the second state variable ($Z_{2,t}$ or $X_{2,t}^h$), which is the expectation of the short rate movement, has a small impact on the short end of the yield curve, but a much bigger

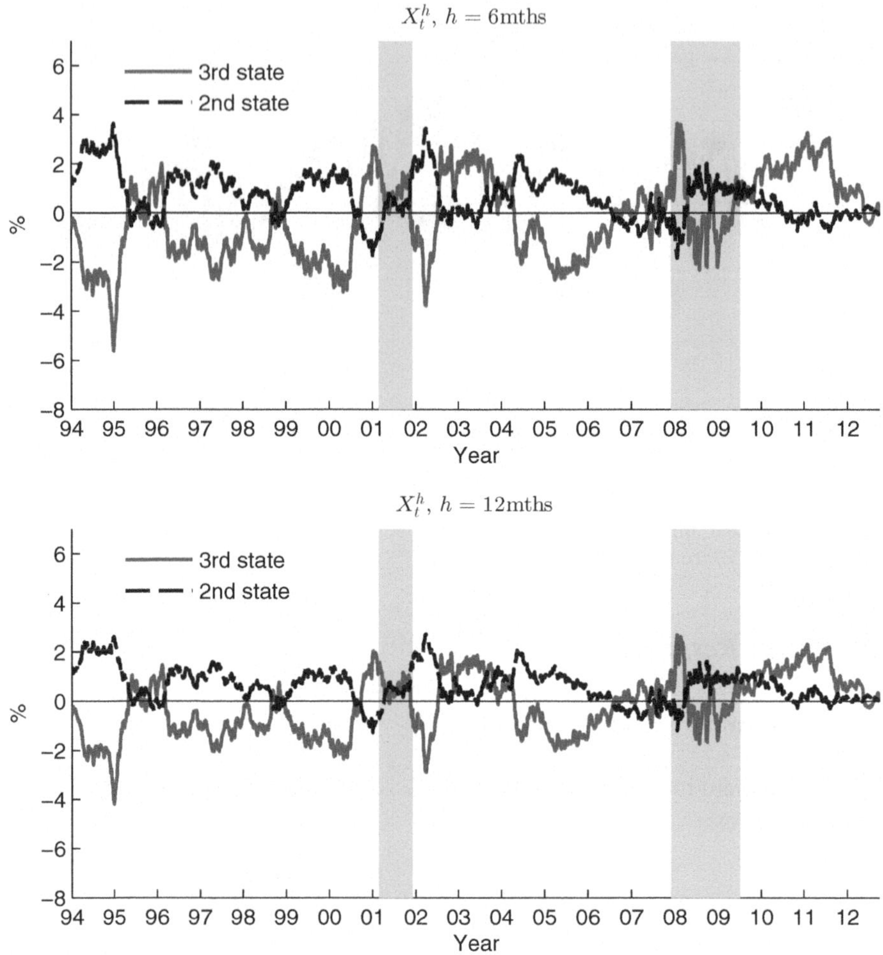

Fig. 3 The 2nd and 3rd state variables in X_t^h. This figure shows the dynamic of the 2nd, 3rd state variables in X_t^h from 1994 to 2012. The upper panel shows the variables with h = six months, and the lower panel shows the variables with h = 12 months

impact on the long end: assuming all else being equal, 10 bps increase in the second state can only increase the three-month zero yield by 1.2 bps, but increase the 10 year zero yield by 16 bps. Notice that for horizons of zero, six months and one year, the contemporaneous impacts of the expectation on the yield curve are of a similar pattern, as we can see that the loadings of the second and third state variables in all three panels in Fig. 4 have similar term structures. This means that even a change in the expectation of the short rate movement in one year can have an immediate impact on the long end of the yield curve. However, as the horizon h increases, the impact of the second state on the long end increases while that of the third state decreases.

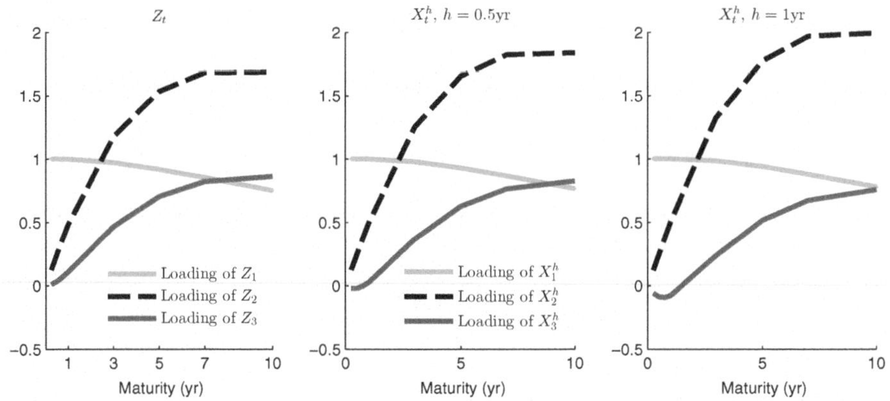

Fig. 4 Contemporaneous Responses of the Zero Yields. This figure shows how the zero yields at different maturities respond to a change in three different sets of state variables (the left panel is for Z_t, the middle panel is for X_t^h when $h = 0.5$ year, the right panel is for X_t^h when $h = 1$ year)

The findings are consistent with the consensus in the literature that only the "surprises" can impact the long end of the yield curve, see, e.g., Kuttner (2001), Gürkaynak et al. (2005), Andersson et al. (2006), and Geiger (Geiger 2011). All state variables can significantly impact the long end in a condition of all else being equal. In other words, they can impact the long end when their changes come in surprise. However, their impacts are of different patterns: when a surprise happens in the short rate, the yield curve will be uniformly shifted; whereas, when a surprise happens in the expectation related states, only the long end of the yield curve will be affected. This again emphasizes that the market expectation is one of the key determinants of the long term interest rates. There are more discussions on this in the next section.

3.3.2 Impulse Responses

The contemporaneous response analysis presented in the last section provides a good sense on how changes in the short rate and the market expectation of the short rate immediately impact the yield curve. This would be very useful for practitioners, as a good understanding on how the yield curve movement follows the changes in the state variables can be the key to a successful trading strategy. However, it is also of a great concern to measure the persistence effect of shocks on the yield curve. Especially from a policy maker's perspective, one would like to know what the expected long-term effect of certain shocks (either endogenous or exogenous) is on the dynamic path of the interest rates. When policy makers are armed with this knowledge, the short interest rate as an instrument can be better deployed to stimulate or restrain the economy. In this section, we conduct an impulse response analysis to further discuss the impact of changes in the short rate and expectations in terms of their persistent effects.

Following Koop et al. (1996), we define the impulse response function as

$$I(n, x, \Delta Z) = \mathbb{E}(y_{t+n}(x)|Z_t = Z + \Delta Z) - \mathbb{E}(y_{t+n}(x)|Z_t = Z),$$

where n is the time horizon concerned, ΔZ is a vector of changes in state variables. Therefore, for the model with Z_t, the impulse response function is

$$I(n, x, \Delta Z) = \frac{\int_0^x \mathbf{C}_{CR}(s)ds}{x} \exp(\mathbf{A}_{CR}n)\Delta Z;$$

for the model with X_t^h,

$$I(n, x, \Delta Z) = \frac{\int_0^x \mathbf{C}_{CR}(s)ds}{x} \exp(\mathbf{A}_{CR}n)M_{fun}(\mathbf{A}_{CR}, h)^{-1}\Delta X.$$

Notice that the contemporaneous response presented before is a special case of the impulse response when $n \to 0$. As mentioned in Koop et al. (1996), the impulse response function can be understood as a multiplier, it captures the properties of the model's propagation mechanism, and compares the value of a yield after the shock has occurred with its benchmark value where the economy has not been subject to any shocks.

Wicksell effect The "Wicksell Effect" used here refers to an "Interest Wicksell Effect" same as that studied by Cwik (2005). When the monetary authority engages in a policy of monetary expansion, e.g., the Fed wants to stimulate the economy, the new money is injected into the monetary system by expanding the supply of investable funds to achieve the targeted Federal funds rate. This will lower the short rate immediately. However, it does not necessary affect the market expectations (e.g., expected inflation) immediately. In this section we show how the Wicksell effect impacts the yield curve dynamically.

We assume that the Wicksell effect only lowers the $Z_{1,t}$ without affecting other variables, i.e.,

$$\Delta Z = \Delta X = \begin{bmatrix} 1, 0, 0 \end{bmatrix}.$$

Since the first state variable is the same for both Z_t and X_t^h, the yield curve responses to this state variable are the same for the three models. We can see from Fig. 5, a shock in the short rate has a uniform impact on the yield curve even in four years of time. One bp decrease in the short rate now is expected to lower the short rate by 0.86 bp in four years; can lower the 10 year zero yield by 0.76 bp now, and is expected to lower it by 0.55 bp in four years. Notice that Wicksell effect can shift the whole curve but barely change the shape of the curve.

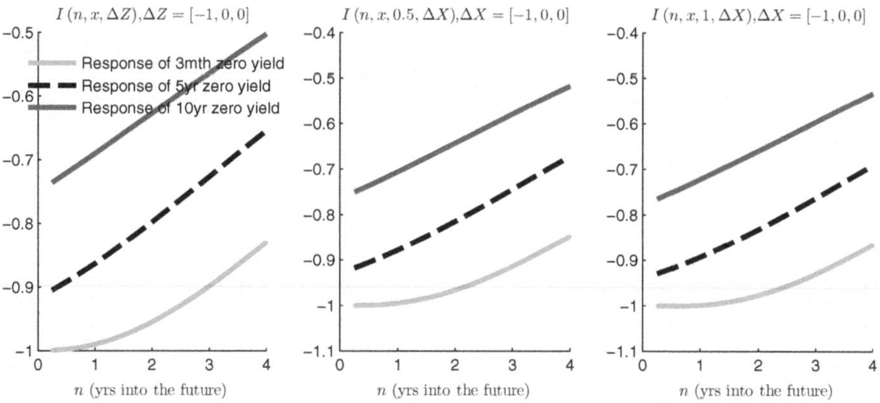

Fig. 5 Impulse Responses of the Zero Yields: Wicksell Effect. This figure shows how the 3 month, 5 year, and 10 year zero yields respond to a shock in the short rate due to the Wicksell effect in 4 year horizon (the left panel is for Z_t, the middel panel is for X_t^h when $h = 0.5$ year, the right panel is for X_t^h when $h = 1$ year)

Fisher effect When the expected inflation rises, interest rates will rise. This result has been named the Fisher effect (Mishkin 2007). Ang et al. (2008) also find that nominal term spreads are primarily driven by changes in the expected inflation. Here we assume a scenario in which the expected inflation has changed and induced a shock on $Z_{2,t}$, $X_{2,t}^{0.5}$, and $X_{2,t}^{1}$ but no direct impact on the short rate and other state variables, i.e.,

$$\Delta Z = \Delta X = \left[0,\ 1,\ 0\right].$$

The impulse responses for a 4 year horizon are shown in Fig. 6. It is found that if the shock only happens in the expectation of the short rate change, the impact on the long end of the yield curve is significant, instant, and persistent. Specifically, one bp increase in $Z_{2,t}$ can raise the 10 year zero yield by 1.6 bp and the impact stays there for at least four years; the impact of an increase in $X_{2,t}$ has a similar pattern but with an even higher magnitude. The Fisher effect can also have impacts on the short rate, however, it comes in a lagged and slowly increasing manner (from 0.4 bp to two bps in four years of time). The persistent impact of a shock in $X_{2,t}^{0.5}$ is larger than that of $Z_{2,t}$, and that of $X_{2,t}^{1}$ is even larger.

Therefore the Fisher effect in this scenario has indeed an impressive impact on the yield curve. However, we argue that this scenario is actually unreasonable and misleading. If we take a look at the dynamics of $Z_{2,t}$, $Z_{3,t}$ and $X_{2,t}^h$, $X_{3,t}^h$ in Figs. 2 and 3, we can quickly find that the absolute correlations between their changes are significantly high. The correlation coefficient of $\Delta Z_{2,t}$ and $\Delta Z_{3,t}$ is -0.9714, it is -0.956 and -0.9374, for $\Delta X_{2,t}^{0.5}$ versus $\Delta X_{3,t}^{0.5}$ and $\Delta X_{2,t}^{1}$ versus $\Delta X_{3,t}^{1}$, respectively. This means that it will be very rare that a shock only happens in $Z_{2,t}$ and $X_{2,t}^h$ without

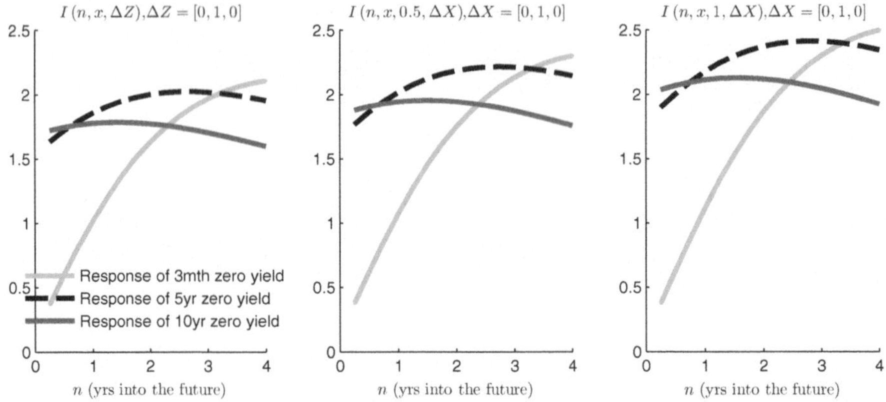

Fig. 6 Impulse Responses of the Zero Yields: Fisher Effect I. This figure shows how the 3 month, 5 year, and 10 year zero yields respond to a shock in the expectation of the short rate change in a 4 year horizon (the left panel is for Z_t, the middel panel is for X_t^h when $h = 0.5$ year, the right panel is for X_t^h when $h = 1$ year)

influencing $Z_{3,t}$ and $X_{3,t}^h$. Therefore, a more sensible scenario is a change in the expected inflation accompanies shocks on $Z_{2,t}$, $Z_{3,t}$ and $X_{2,t}^h$, $X_{3,t}^h$. We now set ΔZ and ΔX as

$$\Delta Z = \left[0,\ 1,\ -1.6\right],\quad \Delta X = \left[0,\ 1,\ -1.4\right],$$

where -1.6 is the sample median of $-\left|\frac{Z_{3,t}}{Z_{2,t}}\right|$; -1.4 is that for $-\left|\frac{X_{3,t}^1}{X_{2,t}^1}\right|$. The impulse response analysis of the Fisher effect in this scenario is presented in Fig. 7. Surprisingly, we find that in this scenario, the impacts of the Fisher effect drop substantially. Now 1 bp increase in $Z_{2,t}$ can only lift the long end by about 0.3 bp in 4 years of time due to 1.6 bp decrease in $Z_{3,t}$. Similarly, 1 bp increase in $X_{2,t}^{0.5}$ ($X_{2,t}^1$) can only induce 0.5 bp (0.7 bp) increase in the long end. Since the second and third states represent the shape of the forward curve (see Appendix 1), the main consequence of Fisher effect on the curve is the change of shape. In other words, a change of the market expectation is typically associated with some change in the shape of the forward curve.

Although this finding is somewhat surprising, it is quite consistent with the consensus that only the "surprises" can significant impact the long end of the yield curve. The shock in the second scenario are less surprising than that in the first one, because the first one is really rare given the historical observations. Therefore the Fisher effect in the first "unusual" scenario has a much more prominent impact on the long end. And this finding should also remind us not to ignore the expectation of the expectation changes.

Wicksell and Fisher effects combined In reality, the Wicksell and Fisher effects often come together. Intuitively, with a policy of monetary expansion, the Wicksell

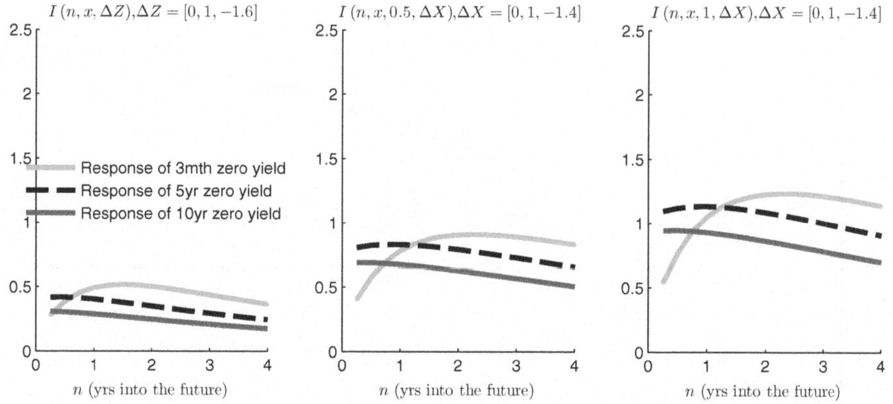

Fig. 7 Impulse Responses of the Zero Yields: Fisher Effect II. This figure shows how the 3 month, 5 year, and 10 year zero yields respond to shocks in the expectation related state variables in a 4 year horizon (the left panel is for Z_t, the middel panel is for X_t^h when $h = 0.5$ year, the right panel is for X_t^h when $h = 1$ year)

effect will lower the whole yield curve with a more significant impact on the short end; at the same time, the expected inflation will likely increase, so the Fisher effect will lift the long end to a certain extent. This combined effect is illustrated in Fig. 8 which is a variant of Fig. 1 in Cwik (2005). Although Fig. 8 illustrates a policy easing, a similar analysis with everything in the opposite direction applies to a policy tightening. Asking how the long end of the yield curve will be affected by the combined effect leads to an open question which has been heavily discussed in the literature: Cook and Hahn (1989) report long-term interest rates move in the same direction as monetary policy actions based on a relatively old data-set; Gürkaynak et al. (2005) argue that the movement of the forward curve is the other way around based on an updated data set; and Ellingsen and Söderström (2001) build a theoretical model to prove that depending on whether the shocks are endogenous or exogenous both scenarios are possible (endogenous shocks lead to the gay line, exogenous ones lead to the dash line). Here we argue that which direction the long end will move is just a matter of domination. If the Wicksell effect dominates the Fisher effect, the long end of the yield curve will fall, and will rise the other way around. We quantitatively illustrate this point by presenting a few numerical examples.

For the sake of a clearer comparison, we only consider $I(n, x, 1, \Delta X)$, the results for three different scenarios are shown in Fig. 9. The left panel in Fig. 9 shows a scenario in which the Fisher effect dominates the Wicksell effect, i.e., if $X_{1,t}^1$ decreases by 1 bp and $X_{2,t}^1$ increases by 1.3 bp, the impact of the combined effect will tilt the yield curve counterclockwise in the short run and even lift the whole yield curve in the long run (say, in the course of four years); the middle panel shows a scenario in which the combined effect only transitorily impact the short end and has no impact on the long end at all, i.e., the Fisher effect cancels out the Wicksell effect at the long end. For this to happen, when $X_{1,t}^1$ increases by 1 bp, $X_{2,t}^1$ should decrease

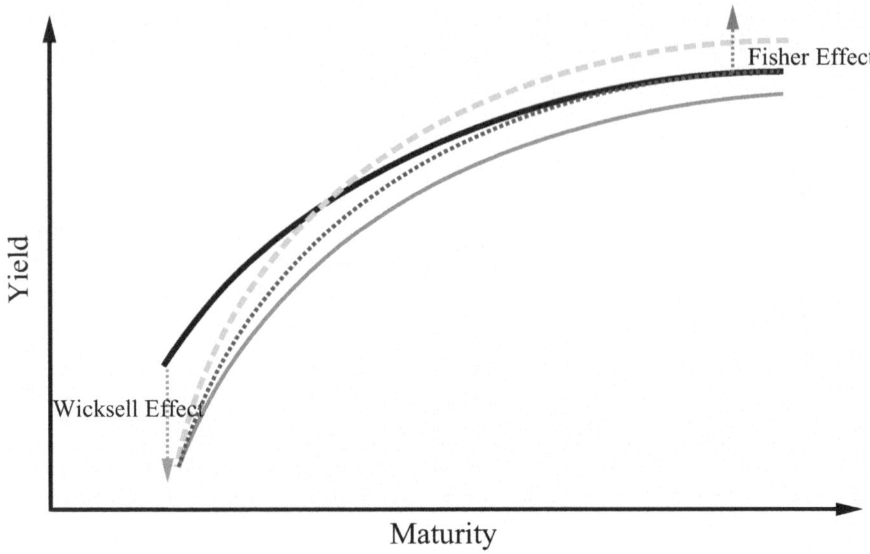

Fig. 8 Wicksell and Fisher Effects Combined. This figure shows how the Wicksell and Fisher effects combined can move the yield curve. The black line is the initial curve, the gay line is the resulting curve when the Wicksell effect dominates the Fisher effect, the dash line is the resulting curve when the Fisher effect dominates the Wicksell effect, and the dot line is the resulting curve when the Wicksell effect and the Fisher effect cancel out at the long end of the curve

0.83 bp; the right panel shows a scenario where the Wicksell effect dominates the Fisher effect, so the combined effect shifts the whole yield curve downward. This happens when $X^1_{2,t}$ increases relatively less (about 1/3) comparing to the amount by that $X^1_{1,t}$ decreases.

These observations convey an intuitive idea about why sometimes the monetary actions are ineffective and when they are so. If the central banks can convince the private sector that they will keep the future inflation low when conducting a monetary easing,[5] then this easing will be considered as a sign of a positive economic outlook, so it will typically take effect by inducing the Wicksell effect to dominate the Fisher effect, and lowering the whole yield curve; however, if they fail to do so, i.e., the private sector may doubt either the ability or the will of the central banks to achieve that future inflation, a monetary easing can only be interpreted as a further accelerant of the inflation, so it won't be effective as the Fisher effect dominates or cancels out the Wicksell effect, the long end of the yield will likely move in the opposite direction as the short end. Therefore, accurately measuring and monitoring the expected inflation, and the credibility of the central banks are the keys to build up the channels of monetary transmission.

[5]Here we assume the easing is conducted via a conventional monetary policy, e.g., setting the Fed funds rate's targets.

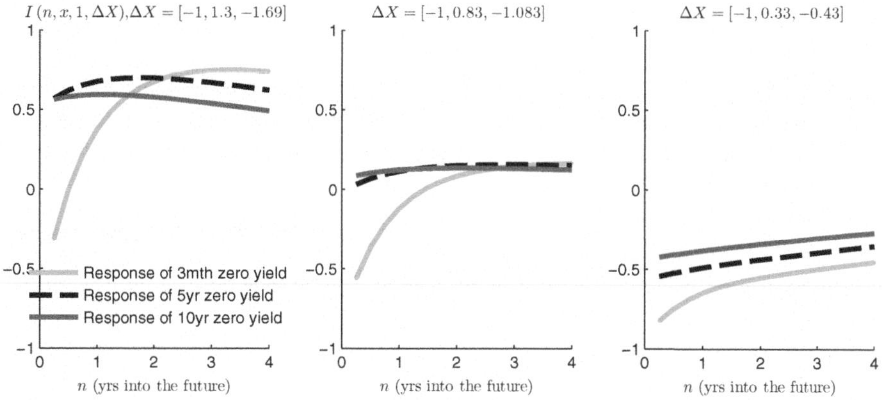

Fig. 9 Responses of the Zero Yields: Wicksell and Fisher Effects Combined. This figure shows how the 3 month, 5 year, and 10 year zero yields respond to the combined Wicksell and Fisher effect in three different ways (in the left panel the Fisher effect dominates the Wicksell effect, in the middle panel the Wicksell effect and the Fisher effect cancel out, in the right panel the Wicksell effect dominates the Fisher effect)

Before finishing up this section, let's take a look at how the combined effect impacts the yield curve under an "unsurprising" circumstance. According to Kuttner (2001), anticipated monetary moves have little impact on the long end of the yield curve. Therefore, under an "unsurprising" circumstance the Wicksell effect and the Fisher effect should cancel out as the scenario in the middle panel of Fig. 9. Now the question is: what would be a sensible "unsurprising" circumstance? A natural answer is a representative of the most often encountered circumstances. To find this representative, we estimate the following three regressions (again we take the model with X_t^1):

$$\begin{aligned}
\Delta X_{1,t}^1 &= 0.00 + \beta_1(= 0.476^{**})\Delta X_{1,t-1}^1 + \epsilon_{1,t},\ R^2 = 0.23 \\
\Delta X_{2,t}^1 &= -0.00 + \beta_2(= -0.452^{**})\Delta X_{1,t-1}^1 + \epsilon_{2,t},\ R^2 = 0.09 \\
\Delta X_{3,t}^1 &= 0.00 + \beta_3(= 0.639^{**})\Delta X_{1,t-1}^1 + \epsilon_{3,t},\ R^2 = 0.08.
\end{aligned}$$

** indicates the 5% confidence level. All estimates of β s are significant at 5%. Generally speaking, 1 pb decrease in $X_{1,t}^1$ accompanies $-\beta_2/\beta_1$ bp increase in $X_{2,t}^1$, and β_3/β_1 bp decrease in $X_{3,t}^1$. Therefore, a representative $\Delta X_{1,t}^1$ would be

$$\Delta X = \left[-1,\ -\tfrac{\beta_2}{\beta_1}\ -\tfrac{\beta_3}{\beta_1}\right].$$

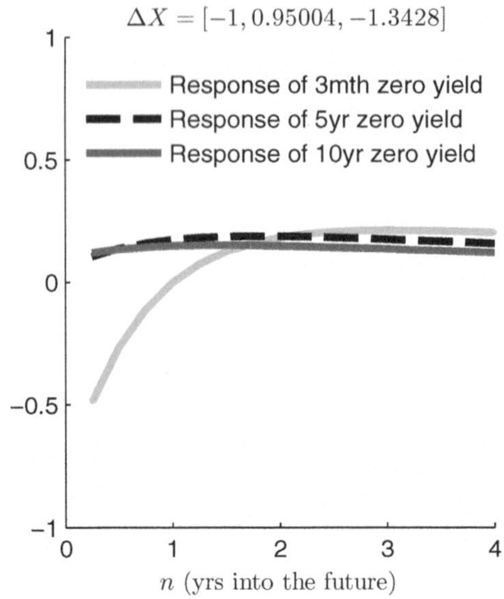

Fig. 10 The Unsurprising Circumstance. This figure shows the responses of 3 month, 5 year, and 10 year zero yields to the combined Wicksell and Fisher effect in an unsurprising circumstance

The response function $I(n, x, 1, \Delta X)$ shown in Fig. 10 is very similar to the middle panel in Fig. 9. This just confirms that the model presented in the chapter is able to infer useful information about the market expectations. By using the model to monitor the information, one might be able to easily isolate the impact of shocks by categorizing them as "surprises" and "unsurprises".

3.4 Case Studies: Impacts of LSAP, MEP, and QE3 Announcements

In December 2008, the Federal Reserve reduced its target for the federal funds rate—the traditional tool of U.S. monetary policy—to a range of 0–25 basis points, essentially the lower bound of zero. This means that despite the severity of the recession, the conventional option of reducing the Funds rate was no longer available. Concerned that economic conditions would deteriorate, the FOMC chose to pursue unconventional monetary policies since there was no scope for further cuts in short-term interest rates.

On November 25, 2008, the Federal Reserve announced that it would purchase up to $100 billion in agency debt, and up to $500 billion in agency MBS, and later on a series of largescale asset purchase programs (LSAP) were implemented on those debts and long-term Treasury securities as well. LSAP is also referred to as "Quantitative Easing" (QE). On September 21, 2011, the FOMC announced the Maturity Extension Program (MEP), under which the FOMC will purchase, by the end of June 2012, $400 billion of Treasury securities with remaining maturities of 6–30 years while simultaneously selling an equal amount of Treasuries with remaining maturities of 3 years or less. MEP is also referred to as "operation twist". On September 12 and 13, 2012, the third round of QE, "QE3" was announced, which entails buying $40 billion in mortgage-backed securities per month. And no specific end date was mentioned.

On November 25, 2008,

$$\Delta X_t^1 = \left[-1.717, \ -4.736 \ -7.404 \right] \text{bps};$$

on September 21, 2011,

$$\Delta X_t^1 = \left[-0.377, \ 3.450 \ -11.600 \right] \text{bps};$$

from September 11 to 13, 2012,

$$\Delta X_t^1 = \left[-0.457, \ -2.509 \ 10.024 \right] \text{bps}.$$

Comparing these three ΔX_t^1s with the one presented in Fig. 10, we can see LSAP, MEP, and QE3 were apparently surprises to the economy system. Now let's see how these surprises impacted the yield curve.

For the LSAP and MEP, from the implied forward curve changes shown in Figs. 11a and 13a, we find that the contemporaneous responses of market were consistent with the intents of the Fed: the LSAP was targeting the long-term yields, in Fig. 11a it shows that the forward curve was tilted clockwise with the short end being intact; in Fig. 13a the forward curve rose in the short end but fell in the long end, this is a response to the "twist" effect of the MEP. We also find that the MEP had a significantly smaller impact on the forward curve, this is somewhat expected, as the MEP can actually be considered as a part of the LSAP.

However, for the QE3, from Fig. 15a, it is found that the contemporaneous responses of the forward curve did not go as what the Fed intended: although just as the LSAP, the QE3 was also targeting the long-term yields, in Fig. 15a it shows that the forward curve is tilted counter clockwise with the short end being a bit flatter. In other words, upon the announcement of the QE3, the long-term forward rates actually increased while the short-term forward rates decreased.

It is interesting to see how the expected change of the short rate responded to the LSAP announcement, this expected change is associated with the slope of the implied forward curve and often interpreted as the expected inflation. From the panels titled "X_2^h" and "ΔX_2^h" in Fig. 11b, we find that the market seemed (at least on the day of the LSAP announcement) to have a strong believe that the LSAP would have a positive influence on the recovery of the economy instead of just boosting up the inflation without stimulating the recovery: upon the LSAP announcement, the expected inflation for horizons up to six years fell by different extents. Looking at the shape change of the forward curves, we find that the LSAP announcement decreased the slope of the forward curve at two-year most, i.e., increasing the short

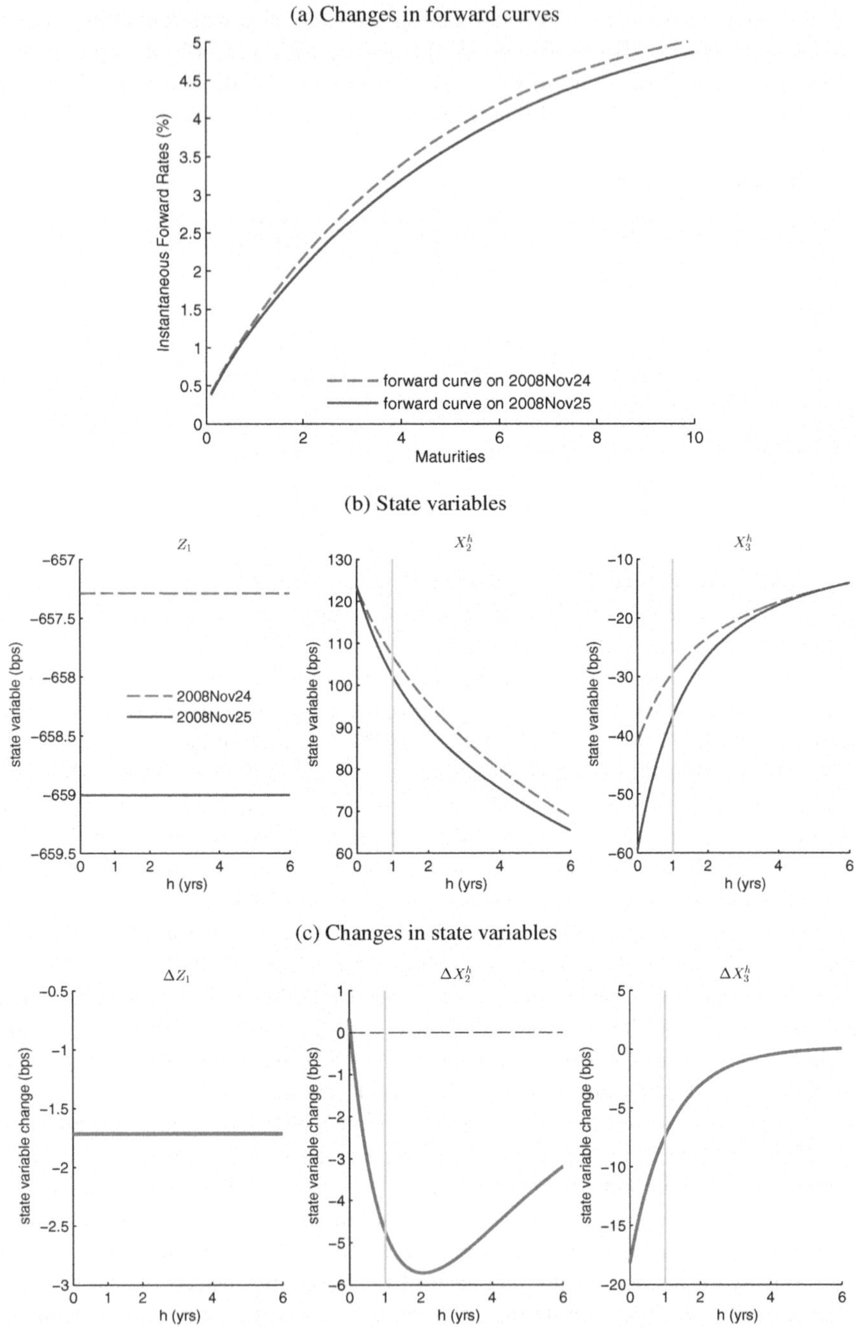

Fig. 11 Changes upon the LSAP announcement. This figure shows changes of the forward curve and state variables happened on November 25, 2008 when the Fed announced the LSAP

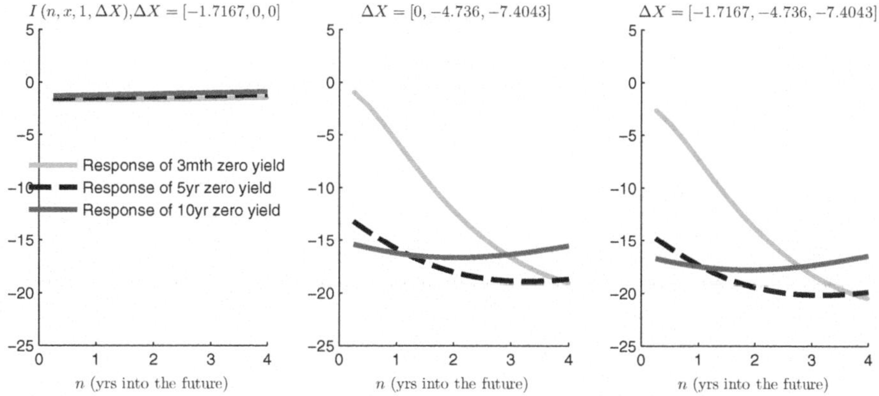

Fig. 12 Responses of the Zero Yields upon the LSAP announcement. This figure shows the responses of the 3 month, 5 year, and 10 year zero yields to the shocks happened on November 25, 2008 when the Fed announced the LSAP

end curvature of the forward curve. The reaction of the yield curve upon the LSAP announcement can be understood as a consequence of a new version of Wicksell and Fisher effects combined in which both effects push the yield curve to a same direction (downwards).

Figure 12 shows the expected impacts of the LSAP announcement on the yield curve four years into the future. Normally, X_2^h and X_3^h move in opposite directions, however, we find from the panels titled "X_2^h" and "X_2^h " in Fig. 11b that X_3^h moved in the same direction as X_2^h. This might indicate that the market was expecting a further decrease of X_2^h, and the LSAP announcement was indeed a big surprise to the market. Therefore the expected long run impacts of the LSAP announcement should be significant, and this is confirmed in Fig. 12: the LSAP announcement had a persistent impact on the long end of the yield curve and a gradually increasing one on the short end, specifically, the LSAP announcement lowered the long end by 16 bps instantly and the impact would stay there for at least four years; although it only contemporaneously lowered the short end by about three bps, the impact would be gradually increasing, it is expected to lower the short end by 20 bps in the course of four years.[6] The impact of the LSAP announcement is mostly attributed to the changes in the expectation related state variables.

The MEP announcement had a much smaller impact on the yield curve, and it impacted the yield curve quite differently. As we can see in the panel titled "ΔX_2^h" in Fig. 13b, the announcement of MEP increased X_2^h for h < 2 year, but had no impact on X_2^h for h > 2 year. This means that the announcement increased the short term expected inflation while kept the long term ones constant. Comparing the one day responses of the forward curves upon the announcements of the LSAP and the MEP, it seems the market remained relatively calm upon the MEP announcement, and believed that the additional purchase of $400 billion of long term Treasury securities

[6]The downward impact on the short end would be bounded by the zero lower bound.

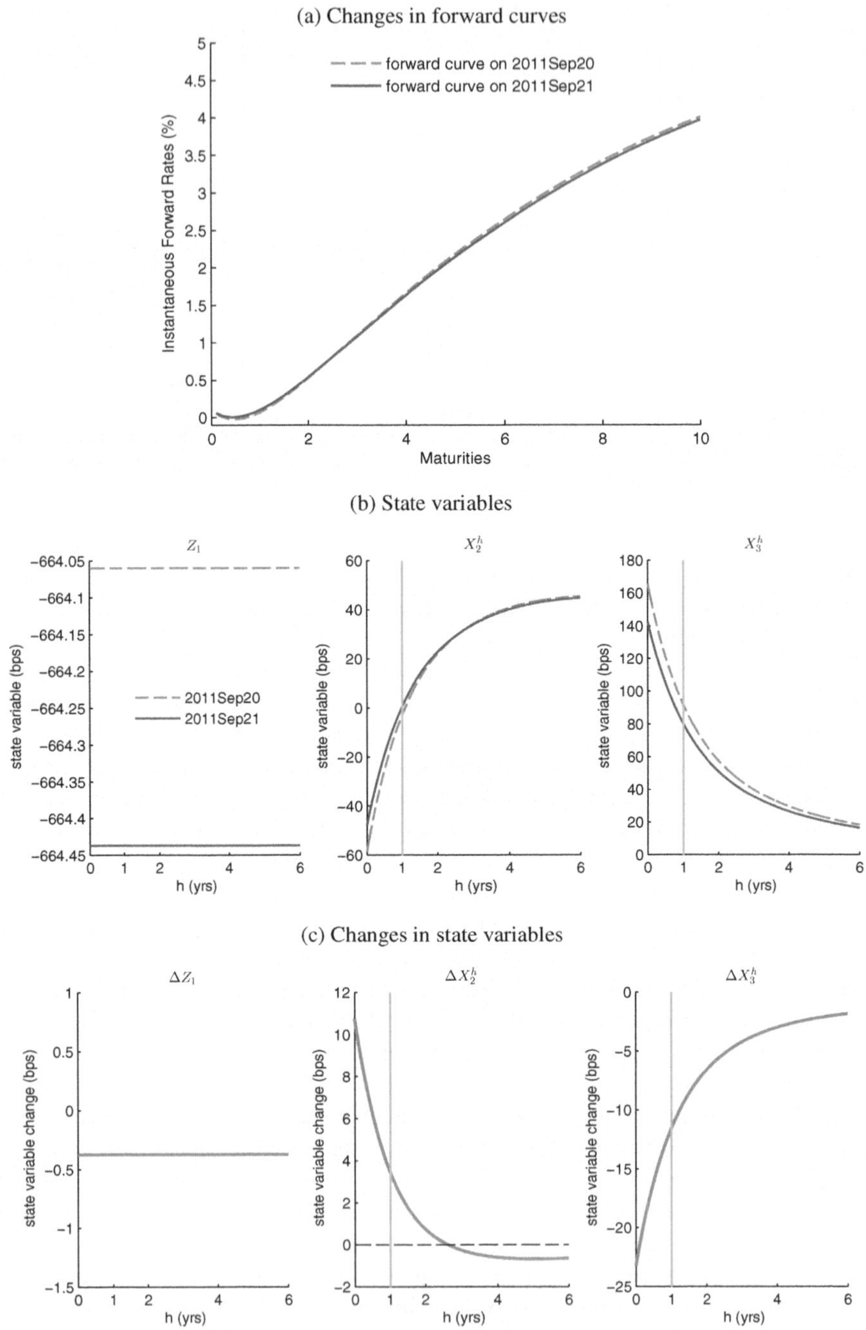

Fig. 13 Changes upon the MEP announcement. This figure shows changes of the forward curve and state variables happened on September 21, 2011 when the Fed announced the MEP

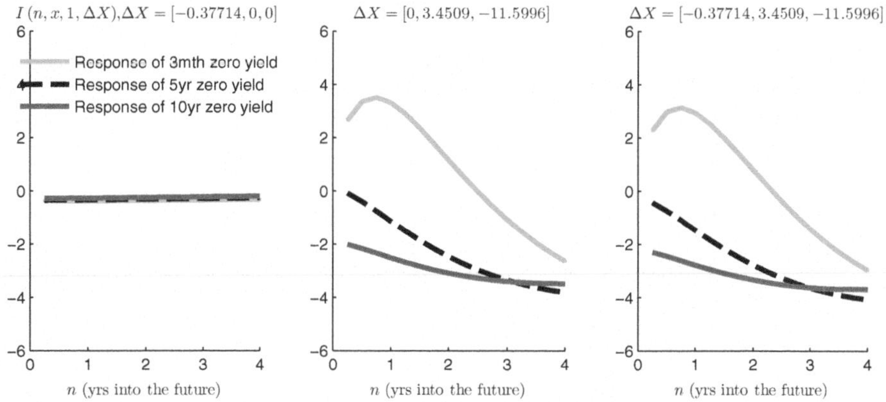

Fig. 14 Responses of the Zero Yields upon the MEP announcement. This figure shows the responses of the 3 month, 5 year, and 10 year zero yields to the shocks happened September 21, 2011 when the Fed announced the MEP

financed by selling a same amount of the short term securities would actually boost up the inflation in the short run rather than having any further influence on the recovery. So under this scenario, the Fisher effect made the slope at the short end less negative, and slightly push down the long end of the yield curve. The contribution of the Wicksell effect on pushing down the long end was just minor; unsurprisingly, the expected long run impact of the MEP announcement was much less notable. This is confirmed in Fig. 14, specifically, the announcement was expected to lower the yield curve by about four bps in the course of four years, which was only 1/4 of the impact of the LSAP announcement. Again the impact was mostly attributed to the changes in the expectation related state variables. However, one observation that needs to be mentioned is that, unlike the LSAP, the MEP announcement was expected to increase the short rate in about two years before it eventually pushing down the short rate. This is consistent with the increase of the expected inflation at short horizons.

As we can see in the panel titled "ΔX_2^h" in Fig. 15b, the announcement of QE3 decreased X_2^h for h $<$ 2 year, but consistently increased X_2^h for h $>$ 2 year. This means that the market seemed to believe the QE3 will decrease the short term expected inflation but increase the long term expected inflation. Under this scenario, even the Wicksell effect was positive on the yield curve, as the short rate increased by half bp upon the announcement; the Fisher effect combined with the Wicksell effect move the long end of the yield curve to the direction which is against the intent of the Fed when the QE3 was announced. In the long run, the announcement was expected to push up the yield curve even further as shown in Fig. 16: the whole yield curve was expected to increase by four bps or so in four years of time, although in the short run (within two years), the short end was expected to slightly decrease (by up to two bps). Same as previous two cases, the impact was mostly attributed to the changes in the expectation related state variables.

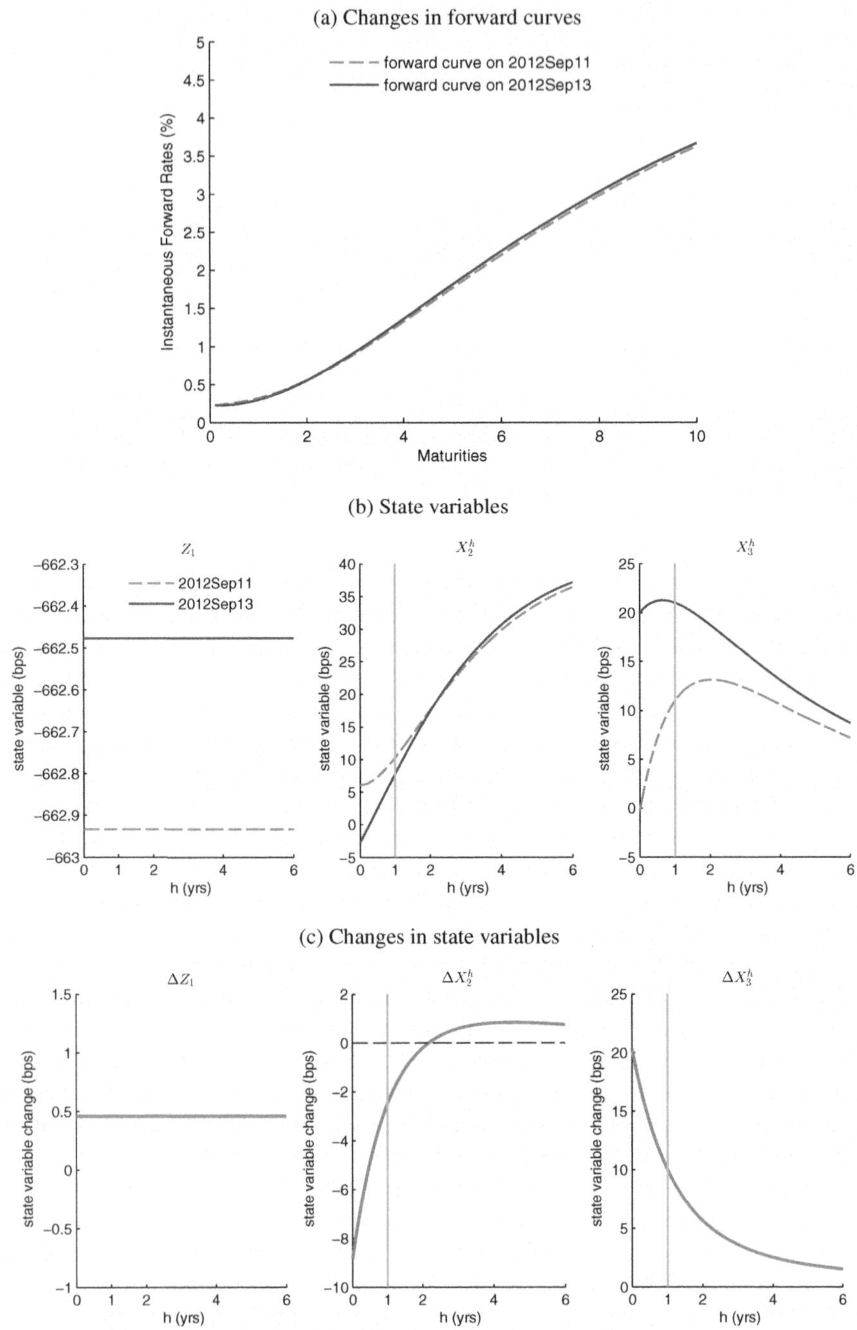

Fig. 15 Changes upon the QE3 announcement. This figure shows changes of the forward curve and state variables happened on September 12 and 13, 2012 when the Fed announced the QE3

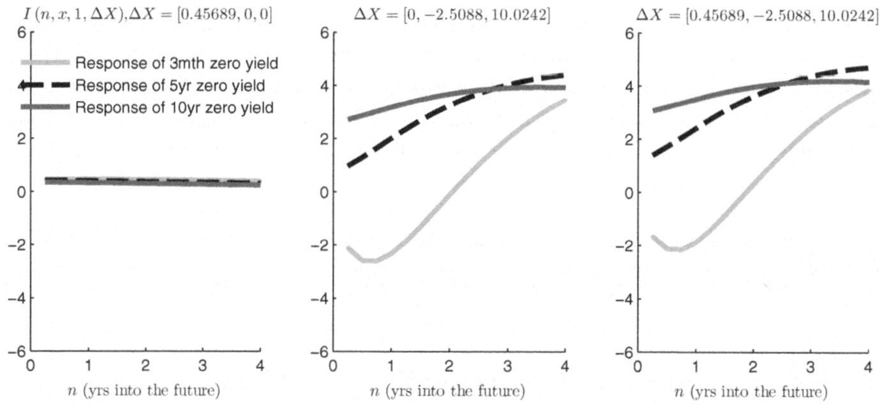

Fig. 16 Responses of the Zero Yields upon the QE3 announcement. This figure shows the responses of the 3 month, 5 year, and 10 year zero yields to the shocks happened on September 12 and 13, 2012 when the Fed announced the QE3

Based on these three case studies: from the LSAP in 2008, the MEP in 2011, then to the QE3 in 2012, it seems that the announcements of the unconventional monetary policies has started to lose their effectiveness and to have unintended consequences. More solid assessment and justification of the unconventional monetary policies are definitely needed before more unconventionals are implemented. Although using a different approach, the case studies conducted here reach similar conclusions as those in Ye (2015).

4 Conclusions

In this chapter, based on the classic $\mathbb{A}_0(3)$, we develop a new model in which the state variables comprise the short rate and its related expectations. This unique feature makes the model especially useful in analyzing the responses of the yield curve to the market expectations, and monitoring the changes of the expectations. The model is estimated using about 18 years of daily zero yield data, then based on the estimated model, various numerical examples are provided to help understand the mechanism behind the monetary transmission.

The main observations include: the short rate has a direct and uniform impact on the yield curve (the Wicksell Effect) but has no direct impact on the shape of the yield curve, while the market expectation of the short rate change, Z_2 or X_2^h, along with the market expectation of the expectation change, Z_3 or X_3^h, are directly related to the shape of the forward curve and have an immediate impact on the long end of the yield curve(the Fisher Effect); how a monetary policy would impact the long end of the yield curve is a matter of domination, e.g., an easing policy would lower the long end of the yield curve when the expected inflation is low and the market is confident in the monetary authority, i.e., the Wicksell Effect dominates the Fisher Effect, however, it would also lift up the long end when the situation is the other way around; expected movements (unsurprising pattern) in the state variables can barely affect the long end of the yield curve; three case studies show that the unconventional monetary policies has started to lose their effectiveness and to have unintended consequences.

Acknowledgements We thank the participants at WB/BIS Joint Fourth Public Investors' Conference for their helpful comments. All errors are ours.

Appendix 1: State Variables as the Forward Curve Characteristics

This section shows that the state variables Z_t and X_t^h can precisely be interpreted as some characteristics of the forward curve.

In what follows, Δt denotes the limit of a time interval which approaches to zero infinitely. For convenience, the notation of lim is dropped.

$$Z_{2,t} = \frac{\mathbb{E}_t^{\mathbb{Q}}(Z_{1,t+\Delta t}) - Z_{1,t}}{\Delta t} = \frac{r(t,\Delta t) - r(t,0)}{\Delta t} + \frac{\Theta^*(0) - \Theta^*(\Delta t)}{\Delta t} \approx \frac{r(t,\Delta t) - r(t,0)}{\Delta t},$$

$$\begin{aligned} Z_{3,t} &= \frac{\mathbb{E}_t^{\mathbb{Q}}(Z_{2,t+\Delta t}) - Z_{2,t}}{\Delta t} \\ &= \frac{\frac{r(t,2\Delta t)-r(t,\Delta t)}{\Delta t} - \frac{r(t,\Delta t)-r(t,0)}{\Delta t}}{\Delta t} + \frac{\frac{\Theta^*(\Delta t)-\Theta^*(2\Delta t)}{\Delta t} - \frac{\Theta^*(0)-\Theta^*(\Delta t)}{\Delta t}}{\Delta t} \\ &= \frac{r(t,2\Delta t) - 2r(t,\Delta t) + r(t,0)}{(\Delta t)^2} + \frac{-\Theta^*(0) + 2\Theta^*(\Delta t) - \Theta^*(2\Delta t)}{(\Delta t)^2} \\ &\approx \frac{r(t,2\Delta t) - 2r(t,\Delta t) + r(t,0)}{(\Delta t)^2}. \end{aligned}$$

Given the parameter estimates, we have:

$$\frac{\Theta^*(0) - \Theta^*(1/12)}{1/12} = 0.016\,\text{bp}, \quad \frac{-\Theta^*(0) + 2\Theta^*(1/12) - \Theta^*(1/6)}{(1/12)^2} = 0.337\,\text{bp}.$$

Comparing to the magnitudes of $Z_{2,t}$ and $Z_{3,t}$, these error terms are negligible. Therefore it is clear that $Z_{2,t}$ and $Z_{3,t}$ represent the slope and curvature of the forward curve $r(t,x)$ at $x = 0$, respectively.

$$X_{2,t}^h = \frac{\mathbb{E}_t^{\mathbb{Q}}(Z_{1,t+h}) - Z_{1,t}}{h} = \frac{r(t,h) - r(t,0)}{h} + \frac{\Theta^*(0) - \Theta^*(h)}{h} \approx \frac{r(t,h) - r(t,0)}{h},$$

$$\begin{aligned} X_{3,t}^h &= \frac{\mathbb{E}_t^{\mathbb{Q}}(Z_{2,t+h}) - Z_{2,t}}{h} = \frac{\mathbb{E}_t^{\mathbb{Q}}\left(\frac{r(t+h,\Delta t)-r(t+h,0)}{\Delta t}\right) - \frac{r(t,\Delta t)-r(t,0)}{\Delta t}}{h} \\ &= \frac{\frac{r(t,\Delta t+h)-r(t,h)}{\Delta t} - \frac{r(t,\Delta t)-r(t,0)}{\Delta t}}{h}. \end{aligned}$$

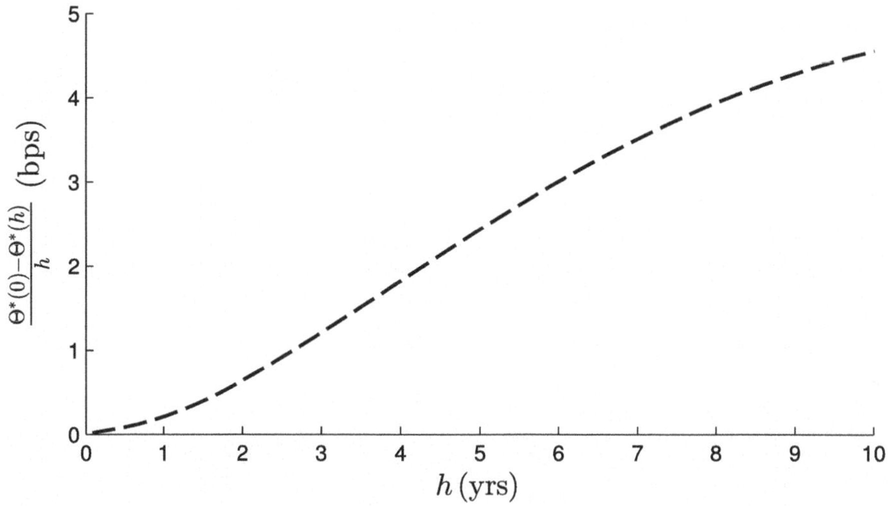

Fig. 17 Size of the error term: $\frac{\Theta^*(0)-\Theta^*(h)}{h}$

As shown in Fig. 17, $\frac{\Theta^*(0)-\Theta^*(h)}{h}$ (evaluated at the parameter estimates) is no more than five bps even when $h = 10$ year; for $h < 2$ year, this error is less than one bp. So, $\frac{\Theta^*(0)-\Theta^*(h)}{h}$ is negligible too.

Therefore, $X^h_{2,t}$ represents the slope between $r(t, h)$ and $r(t; 0)$ and $X^h_{3,t}$ represents the difference between slopes at $r(t, h)$ and $r(t, 0)$.

Appendix 2: Expected Inflation and Short Rate Expectations

Although the strong linkage between the expected inflation and the slope of the yield curve or forward curve or term spreads has been mentioned in many of previous studies, see, e.g., Ang et al. (2008) and Söderlind and Svensson (1997); in this appendix, we again empirically verify the strong relationship between the long term expected inflation and short rate expectations to support the points about the expected inflation made in this article. The 10-year expected inflation data used here are estimated by the Federal Reserve Bank of Cleveland according to a model proposed in Haubrich et al. (2012) and Potter (2012),[7] and publicly available at https://www.clevelandfed.org/our-research/indicators-and-data/inflation-expectations.aspx In what follows, the expected inflation is denoted by π^e.

[7]Since the data used to estimate the expected inflation provided by the Federal Reserve Bank of Cleveland includes the TIPS data, and the 10-year TIPS are most liquid, the 10-year expected inflation is relatively free of estimation errors.

Table 3 Expected inflation versus state variables regression results

	$\widehat{\alpha}$	$\hat{\beta}_1$	$\hat{\beta}_2$	R^2 (%)
$\Delta\pi_t^e = \alpha + \beta_1 \Delta Z_{1,t} + \varepsilon_t$	0	0.02	N.A.	0.22
$\Delta\pi_t^e = \alpha + \beta_1 \Delta X_{2,t}^1 + \varepsilon_t$	0	0.23**	N.A.	48.62
$\Delta\pi_t^e = \alpha + \beta_1 \Delta X_{3,t}^1 + \varepsilon_t$	0	−0.09**	N.A.	25.8
$\Delta\pi_t^e = \alpha + \beta_1 \Delta X_{2,t}^1 + \beta_2 \xi_t + \varepsilon_t$; $\Delta X_{3,t}^1 = \theta + \lambda \Delta X_{2,t}^1 + e_t$; $\xi_t = \Delta X_{3,t}^1 - \hat{\theta} - \hat{\lambda}\Delta X_{2,t}^1$;	0	0.23**	0.13**	57.27
$\Delta\pi_t^e = \alpha + \beta_1 \Delta X_{3,t}^1 + \beta_2 \xi_t + \varepsilon_t$; $\Delta X_{2,t}^1 = \theta + \lambda \Delta X_{3,t}^1 + e_t$; $\xi_t = \Delta X_{2,t}^1 - \hat{\theta} - \hat{\lambda}\Delta X_{3,t}^1$;	0	−0.09**	0.43**	57.27
$\pi_t^e = \alpha + \beta_1 Z_{1,t} + \varepsilon_t$	0.03**	0.2**	N.A.	65.87
$\pi_t^e = \alpha + \beta_1 X_{2,t}^1 + \varepsilon_t$	0.02**	0.32	N.A.	19.1
$\pi_t^e = \alpha + \beta_1 X_{3,t}^1 + \varepsilon_t$	0.03**	−0.28	N.A.	41.81

**Indicates 5% significance, "N.A." represents "Not Available"

Form Table 3, we can see $\Delta X_{2,t}^1(\Delta X_{3,t}^1)$ alone can explain about 50% (26%) of the variation in $\Delta\pi^e$; $\Delta X_{2,t}^1$ and $\Delta X_{3,t}^1$ together can explain about 60% of the variation.[8] and as expected $\Delta\pi^e$ is positively related to $\Delta X_{2,t}^1$, i.e., the slop of the forward curve (the coefficient in front of $\Delta X_{2,t}^1$ is 0.23 with 5% significance). We can also see that the change of the short rate ($\Delta Z_{1,t}$) can barely explain any variation in $\Delta\pi^e$, as the R^2 of the regression is only 0.2% and the coefficient is insignificant. However, interestingly, it is noted that the short rate ($Z_{1,t}$) per se is able to explain the level of the expected inflation (π^e) up to 66%. These observations tell us that the change of the expected inflation from one period to another is primarily driven by the changes of the market expectations, and in the long run the level of the expected inflation is mainly affected by the short rate. These observations are also presented in Fig. 18.

[8]Notice that $\Delta X_{2,t}^1$ and $\Delta X_{3,t}^1$ are not directly included in the regressions, instead residual ξ_t is used in the regressions to avoid the multicollinearity issue and have meaningful interpretations for coefficients.

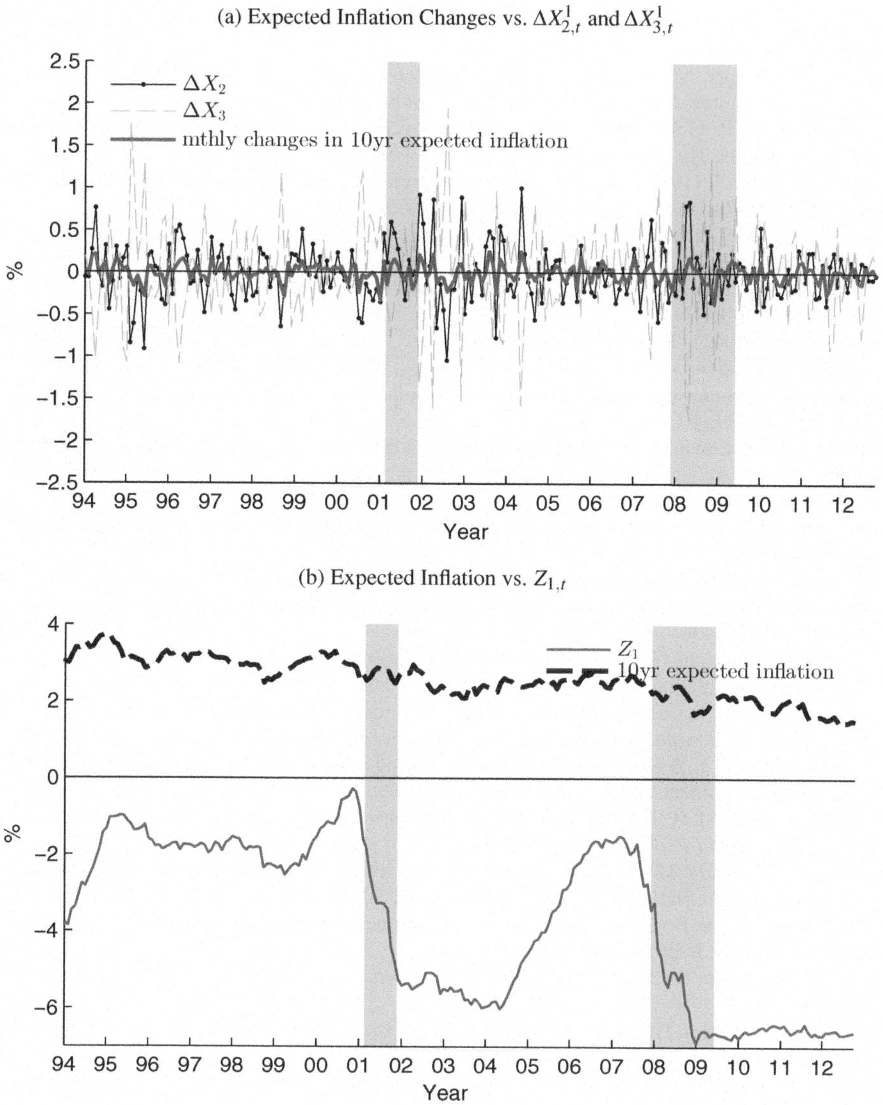

Fig. 18 Dynamics of expected inflation and its changes versus state variables

References

Andersson, M., Dillén, H., & Sellin, P. (2006). Monetary policy signaling and movements in the term structure of interest rates. *Journal of Monetary Economics, 53,* 1815–1855.

Ang, A., Bekaert, G., & Wei, M. (2008). The term structure of real rates and expected inflation. *The Journal of Finance, 63,* 797–849.

Brace, A., Gatarek, D., & Musiela, M. (1997). The market model of interest rate dynamics. *Mathematical Finance, 7,* 127–147.
Brooke, M., Cooper, N., & Scholtes, C. (2000). Inferring market interest rate expectations from money market rates. *Bank of England Quarterly Bulletin, November 2000.*
Collin-Dufresne, P., Goldstein, R. S., & Jones, C. S. (2008). Identification of maximal affine term structure models. *The Journal of Finance, 63,* 743–795.
Cook, T., & Hahn, T. (1989). The effect of changes in the federal funds rate target on market interest rates in the 1970s. *Journal of Monetary Economics, 24,* 331–351.
Cwik, P. F. (2005). The inverted yield curve and the economic downturn. *New Perspectives on Political Economy, 1,* 1–37.
Dai, Q., & Singleton, K. J. (2000). Specification analysis of affine term structure models. *Journal of Finance, 55,* 1943–1978.
Duffee, G. R. (2002). Term premia and interest rate forecasts in affine models. *The Journal of Finance, 57,* 405–443.
Ellingsen, T., & Söderström, U. (2001). Monetary policy and market interest rates. *The American Economic Review, 91,* 1594–1607.
Frankel, J. A., & Lown, C. S. (1994). An indicator of future inflation extracted from the steepness of the interest rate yield curve along its entire length. *The Quarterly Journal of Economics, 109,* 517–530.
Geiger, F. (2011). *The yield curve and financial risk premia: Implications for monetary policy.* Berlin: Springer.
Gürkaynak, R. S., Sack, B., & Swanson, E. (2005). The sensitivity of long-term interest rates to economic news: Evidence and implications for macroeconomic models. *The American Economic Review, 95,* 425–436.
Haubrich, J., Pennacchi, G., & Ritchken, P. (2012). Inflation expectations, real rates, and risk premia: Evidence from inflation swaps. *Review of Financial Studies, 25,* 1588–1629.
Heidari, M., & Wu, L. (2009). A joint framework for consistently pricing interest rates and interest rate derivatives. *Journal of Financial and Quantitative Analysis, 44,* 517–550.
Joyce, M., Relleen, J., & Sorensen, S. (2008). Measuring monetary policy expectations from financial market instruments. *Bank of England Working Papers.*
Koop, G., Pesaran, M. H., & Potter, S. M. (1996). Impulse response analysis in nonlinear multivariate models. *Journal of Econometrics, 74,* 119–147.
Kuttner, K. N. (2001). Monetary policy surprises and interest rates: Evidence from the Fed funds futures market. *Journal of Monetary Economics, 47,* 523–544.
Li, H., Ye, X., & Yu, F. (2016a). A unified HJM approach to non-markov gaussian dynamic term structure models: International evidence.
Li, H., Ye, X., & Yu, F. (2016b). Unifying Gaussian dynamic term structure models from an HJM perspective.
Mishkin, F. S. (1990a). The information in the longer maturity term structure about future inflation. *The Quarterly Journal of Economics, 105,* 815–828.
Mishkin, F. S. (1990b). What does the term structure tell us about future inflation? *Journal of Monetary Economics, 25,* 77–95.
Mishkin, F. S. (2007). *The economics of money, banking, and financial markets.* Pearson Education.
Piazzesi, M. (2005). Bond yields and the Federal Reserve. *Journal of Political Economy, 113,* 311–344.
Potter, S. M. (2012). Improving the measurement of inflation expectations. *Federal Reserve Bank of New York Speech.*
Söderlind, P., & Svensson, L. (1997). New techniques to extract market expectations from financial instruments. *Journal of Monetary Economics, 40,* 383–429.
Ye, X. (2015). A new approach to measuring market expectations and term premia. *Journal of Fixed Income, 24,* 22–46.

A New Approach to CIR Short-Term Rates Modelling

Giuseppe Orlando, Rosa Maria Mininni and Michele Bufalo

Abstract It is well known that the CIR model, as introduced in 1985, is inadequate for modelling the current market environment with negative short rates, $r(t)$. Moreover, in the CIR model, the stochastic part goes to zero with the rates, neither volatility nor long term mean change with time, or fit with skewed (fat tails) distribution of $r(t)$, etc. To overcome the limitations of the CIR, several different approaches have been proposed to date: multi-factor models such as the Hull and White or the Chen models to the CIR++ by Brigo and Mercurio. Here, we explain how our extension of the CIR framework may fit well to market short interest rates.

Keywords CIR Model · Short Interest Rates · Forecasting and Simulation

JEL Classification G12 · E43 · E47
2010 MSC 91G30 · 91B84 · 91G70

Preface

The CIR model was first introduced in 1985 by Cox et al. (1985) as an extension of the Vasicek model (1977). The purpose of the CIR model is to describe the evolution

G. Orlando (✉)
Department of Economics and Finance, Università degli Studi di Bari Aldo Moro,
Via C. Rosalba 53, 70124 Bari, Italy
e-mail: giuseppe.orlando@uniba.it; giuseppe.orlando@unicam.it

G. Orlando
School of Science and Technologies, Università degli Studi di Camerino,
Via M. delle Carceri 9, 62032 Camerino, Italy

R. M. Mininni
Department of Mathematics, Università degli Studi di Bari Aldo Moro,
Via Orabona 4, 70125 Bari, Italy
e-mail: rosamaria.mininni@uniba.it

M. Bufalo
Department of Methods and Models for Economics,
Territory and Finance, Università degli Studi di Roma "La Sapienza",
Via del Castro Laurenziano 9, 00185 Roma, Italy

M. Mili et al. (eds.), *New Methods in Fixed Income Modeling*, Contributions to Management Science, https://doi.org/10.1007/978-3-319-95285-7_2

of short-term interest rates, r, by the following stochastic differential equation (SDE)

$$dr(t) = [k(\theta - r(t)) - \lambda(r(t))]dt + \sigma\sqrt{r(t)}dW(t), \tag{1}$$

with initial condition $r(0) = r_0 > 0$. $W(t)$ denotes a standard Brownian motion modelling a random risk factor. The CIR model is classified as a one-factor, time-homogeneous model, because: the parameters σ, k and θ in (1) are time-independent, and the interest rate dynamics are driven only by the market risk factor $\lambda(r(t)) := \lambda r(t)$, where λ is a constant. Therefore, the model is composed of two parts: the drift component, $k[\theta - r(t)]$, to ensure mean reversion, at speed k, towards the long run value θ and the random component $W(t)$, which is scaled by the standard deviation $\sigma\sqrt{r(t)}$.

The diffusion process r solution to (1) is non-negative; it can be shown that, if $2k\theta > \sigma^2$, $r(t)$ is strictly positive, and, for small $r(t)$, the process rebounds as the random perturbation dampens with $r(t) \to 0$.

The CIR model has become very popular in finance among practitioners because of its relatively handy implementation, tractability and because it was perceived as an improvement over the Vasicek model. Namely, the latter, did not allow for negative rates (an undesirable feature under pre-2007 financial crisis assumptions) and introduced a rate-dependent volatility. Apart from simulating short-term interest rate, other applications include modelling mortality intensities, see Dahl (2004), as well as default intensities in credit risk, see Duffie (2005). However, after some years, arose the need for more sophisticated models, which could accommodate multiple sources of risk, as well as the need of shocks and/or structural changes of the market. This burgeoning need has led to the development of a number of models extending the CIR framework (as described in Sect. 1). These extensions to the CIR model notwithstanding, they not cope with the low to negative interest rates of recent years. In fact, the financial crisis of 2007 and the ensuing quantitative-easing policies, not only brought down interest rates as a consequence of reduced growth of developed economies but, also, accustomized markets to unprecedented negative interest regimes under the so-called "new normal" (Engelen 2015). Therefore, while the CIR model was still a handy tool in modelling interest rates it shows some limitations that financial institutions find difficult to manage in the current market setting. These limitations within the CIR framework are as follows: (1) interest rates rebound as they approach zero; (2) volatility dampens when rates are low; (3) the instantaneous volatility σ is constant with time; (4) there are no jumps; (5) the risk premium $\lambda(r(t))$ is linear with $r(t)$; (6) the only risk factor for interest rates is the market risk.

The aim of this work is to show an alternative approach to overcoming the aforementioned issues. This paper is organized as follows: the first section gives an account of the literature on the CIR model, the second describes the proposed CIR# model along with the results of our simulations over market data. A final section draws some concluding remarks and provides some direction for future research.

1 Literature Review

As mentioned in the Preface, CIR is a one-factor, time-homogeneous model (as well as Vasicek, Dothan, and the Exponential Vasicek model) widespread in the financial sector because of its simplicity and ability to model other phenomena (Dahl 2004; Duffie 2005), etc. However, as explained in the Preface, after some years the need emerged for mathematical models with an exact fit of the currently observed yield curve, which could take into account multiple risk factors and market shocks. Therefore, financial innovation pushed through a range of solutions including, the following best-known models: the Hull-White (1990) model which considers time-dependent coefficients, the Chen (1996) three-factor model, the Black-Karasinski model which considers log-normal interest rates (Black and Karasinski 1991). Among those that tried to innovate the CIR framework are Brigo et al. Some of the advancements to the CIR framework include the one-factor CIR++ model (Brigo and Mercurio 2000), where short rates are shifted by a deterministic function, and the two-factor models CIR2 and CIR2++ (Brigo and Mercurio 2006) by Brigo and Mercurio, the JCIR++ jump model by Brigo and El-Bachir (2006), where jumps are described by a time-homogeneous Poisson process. In addition to the shortcomings of the CIR model we have listed, Brigo & El-Bachir paper mentions that the CIR model cannot "reproduce some typical shapes, like that of an inverted yield curve" (Brigo and El-Bachir 2006). Cox, Ingersoll and Ross, instead, stated explicitly that the model produces"only normal, inverse or humped shapes" (Cox et al. 1985). The apparent contradiction lies in the practical implementation as explained by Carmona and Tehranchi (Carmona and Tehranchi 2006): "tweaking the parameters can produce yield curves with one hump or one dip, but it is very difficult (if not impossible) to calibrate the parameters so that the hump/dip sits where desired. On a different path, Keller-Ressel and Steiner (2008) adopted the definition of one-dimensional, conservative, affine short rate process by Duffie et al. (2003), which implies exponentially affine structure of zero-coupon bond prices and, thus, affine term structure of yields and forward rates. Moreover, the mentioned definition of affine processes includes, also, processes with jumps (even when the jumps' intensity depends in an affine way on the state of the process itself). They proved (Keller-Ressel and Steiner 2008, Theorem 3.9) that in any time-homogeneous, affine one-factor model, the attainable yield curves are either inverse, normal or humped.

Last but not least, Zhu (2014), in order to incorporate the default clustering effects, proposed a CIR model with jumps that are modelled by the Hawkes process (a point process with self-exciting property and the desired clustering effect). Moreno et al. (2015) introduced a cyclical square-root model, where the long-run mean and the volatility parameters are driven by harmonic oscillators, and Najafi et al. (2017), Najafi and Mehrdoust (2017) proposed some extensions of the CIR model where a mixed fractional Brownian motion is added to display the random part of the model.

2 Numerical and Empirical Results

Our dataset records monthly EUR and USD interest rates (spanning from December 31, 2010 to July 29, 2016) on 1/360A-50YA maturity [i.e. 1-Day (Overnight), 1-Month, 2-Months, 3-Months,..., 1-Year, 2-Years,..., 50-Years] (IBA). A qualitative analysis of the dataset showed that the most challenging task is to fit short-term interest rates on 1/360A, 30/360A, 60/360A, 90/360A, 120/360A,..., 360/360A (i.e., 1-Day, 1-Month, 2-Months, 3-Months, 4-Months,..., 1-Year) maturity due to the presence of next-to-zero and/or negative spot rate values. For this reason, in this report we have started to examine samples of interest rates with maturity up to 1-Year. In particular, we shall limit ourselves to simulating just the expected interest rates in the fixed time horizon. To fit the short interest rates adequately, we will implement in a future work a novel numerical procedure, which we call the CIR# model.

In the sequel we will consider the data sample of 68 monthly observed EUR interest rates with maturity T = 30-Years. All computations have been executed using MATLAB® R2012b.

First of all, we have to move spot rates to positive values to eliminate negative/near-zero values. The translation is done by adding a deterministic positive quantity α as follows:

$$r_{new}(t) = r_{old}(t) + \alpha. \tag{2}$$

The translation leaves unchanged the stochastic dynamics of short interest rates, i.e. $dr_{new}(t) = dr_{old}(t)$. There are many values that could be assigned to α, but the most appropriate choice is to determine a quantity depending on the empirical interest rates probability distribution. We decided to set α equal to that value corresponding to the 99th-percentile of the analysed data sample's empirical distribution. This translation guarantees that if the previous translation is not sufficient to move old interest rates to new positive values (e.g., when negative interest rates are greater than the 99th-percentile), we can set α equal to the 1st-percentile of the empirical distribution. In this case, (2) becomes $r_{new}(t) = r_{old}(t) - \alpha$. Alternatively, other transformations may be chosen such as, of the type

$$r_{new}(t) = \sigma\, r_{old}(t) + \mu, \tag{3}$$

where μ and σ are respectively the mean and standard deviation of the sample $r_{old}(t), t \in [0, T]$ for a fixed maturity T. In practice, (3) does not represent the best choice due to some possible complications (e.g., persistence of negative values, worse fitting, changes in short interest rates dynamics).

For the numerical simulation, we consider the well-known, easy-to-implement, Euler scheme. For the CIR model, it reads as follows

$$r_{i+1} = r_i + k(\theta - r_i)\Delta + \sigma\sqrt{r_i \Delta} W_i, \tag{4}$$

where Δ is a time-step and W_i's are i.i.d. $N(0, 1)$. To obtain more accurate strong Taylor schemes, Kloeden and Platen (1992, Section 10.4) suggested a more accurate approximation of diffusion processes by truncating the Itô-Taylor expansion for integrating SDEs to the third order term. Thus, for the CIR model this strong Itô-Taylor scheme consists in summing to (4) the following terms up to the third order approximation

$$\frac{\sigma^2}{4}((\sqrt{\Delta}\,W_i)^2 - \Delta) - k\,\sigma\sqrt{r(t_i)}\left[\frac{\Delta^{3/2}}{2}\left(W_i + \frac{\widetilde{W}_i}{\sqrt{3}}\right)\right] - \frac{\Delta^2 k^2 (\theta - r(t_i))}{2}$$
$$+ \frac{\sigma}{2\sqrt{r(t_i)}}\left(k\,(\theta - r(t_i)) - \frac{\sigma^2}{4}\right)\left[\Delta\sqrt{\Delta}\,W_i - \frac{\Delta^{3/2}}{2}\left(W_i + \frac{\widetilde{W}_i}{\sqrt{3}}\right)\right], \quad (5)$$

where $\widetilde{W}_i$ are i.i.d. $N(0, 1)$ uncorrelated with W_i.

For the calibration of the parameters k, θ and σ in (1), we compared the ML estimation method by Kladìviko (2007) to the Bayesian inference technique proposed in Bibby et al. (2010, Example 5.4). The latter method has turned out to be very useful in obtaining, improving and studying estimators for discretely sampled diffusion-type models, wherein the likelihood function is usually not explicitly known. For the CIR model, the estimating function obtained in Bibby et al. (2010) gives the following explicit estimators of the three parameters

$$\hat{k} = -\ln\left(\frac{(n-1)\sum_{i=2}^{n} r_i/r_{i-1} - (\sum_{i=2}^{n} r_i)(\sum_{i=2}^{n} r_{i-1}^{-1})}{(n-1)^2 - (\sum_{i=2}^{n} r_{i-1})(\sum_{i=2}^{n} r_{i-1}^{-1})}\right),$$

$$\hat{\theta} = \frac{1}{(n-1)}\sum_{i=2}^{n} r_i + \frac{e^{-\hat{k}_n}}{(n-1)(1-e^{-\hat{k}_n})}(r_n - r_1),$$

$$\hat{\sigma} = \frac{\sum_{i=2}^{n} r_{i-1}^{-1}(r_i - r_{i-1}e^{-\hat{k}_n} - \hat{\theta}_n((1-e^{-\hat{k}_n}))^2}{\sum_{i=2}^{n} r_{i-1}^{-1}((\hat{\theta}_n/2 - r_{i-1})e^{-2\hat{k}_n} - (\hat{\theta}_n - r_{i-1})e^{-\hat{k}_n} + \hat{\theta}_n)/\hat{k}_n}, \quad (6)$$

where n is the sample size (in our case $n = 68$). It is worth noting that statistical inference methods for diffusion-type models available in literature are applicable to the CIR model only when the sampled interest rates are non-negative. For this reason, the parameters estimation has been done after translation of the sampled interest rates to positive values by (2). Table 1 lists the estimates computed by the two methods mentioned above.

To test the goodness of fit of the fitted CIR interest rates, $\{\hat{r}_i; i = 1, \ldots, n\}$ simulated by using the scheme (4)–(5) with the parameters estimated by (6), to the real data $\{r_{real,i}\,; i = 1, \ldots, n\}$, we will compute the coefficient R^2 [see Kvalseth (1985), formula (4)], and the mean square error (MSE)

Table 1 Estimates of the parameters k, θ, σ

	PRIBOR 3M		Estimating functions
	Initial values	Optimal values	Optimal values
$\hat{k}$	0.3300	0.3217	0.0278
$\hat{\theta}$	4.5799	4.5201	4.5799
$\hat{\sigma}$	0.2832	0.2881	0.0830

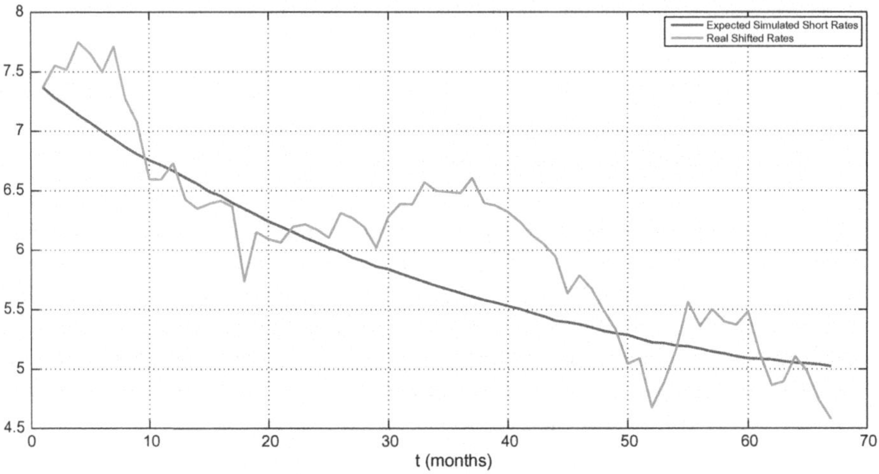

Fig. 1 Expected simulated interest rates versus real shifted rates

$$\varepsilon = \frac{1}{n}\sum_{i=1}^{n}(r_{real,i} - \hat{r}_i)^2. \tag{7}$$

Figure 1 shows the expected (simulated) interest rates compared with the real rates shifted to positive values. In particular, $R^2 = 0.7738$ and $\varepsilon = 0.1928$.

To improve the aforementioned results for each sample in the dataset with maturity 1/360A,..., 360/360A, we can split the entire sample into suitable sub-samples chosen according to the empirical probability distribution, which is unknown and clearly different from the Chi-square conditional distribution or the Gamma stationary distribution of the CIR interest rate process. We hypothesized the empirical distribution to be a mixture of normal distributions, given the presence of negative values. Therefore, our idea is to identify each sub-sample with a different normal distribution. To do this, we implemented an algorithm including the Lilliefors test and the Johnson's transformation (Johnson 1949) (applied to the real shifted rates to ensure that sub-samples are normally distributed). Finally, we applied the procedure above described to each sub-sample. A mixture of Gamma distributions may be alternatively used. To partition the original sample into sub-samples with a Gamma distribution, we first need to translate the real interest rates to positive values.

Table 2 MSE and the statistics R^2 for the sample data with maturity $T = 30/360A$ after segmentation

	Normal distribution					Gamma distribution	
	Sub. 1	Sub. 2	Sub. 3	Sub. 4	Sub. 5	Sub. 1	Sub. 2
ε^k	0.2015	1.85×10^{-5}	1.22×10^{-4}	3.92×10^{-4}	8.23×10^{-4}	0.1510	0.0140
R^2	0.3855	0.5983	0.8501	0.7483	0.9638	0.4741	0.9248
ε	2.1969					2.5655	
R^2	0.6527					0.7095	

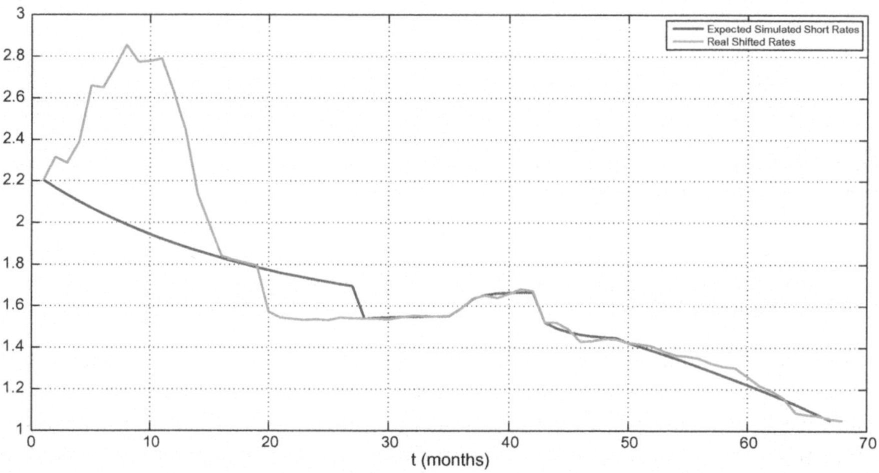

Fig. 2 Expected simulated interest rates (blue line) versus real shifted rates (green line) after segmentation with Normal Distribution (Color figure online)

As a numerical example, we considered the sample from the dataset consisting of $n = 68$ real interest rates with maturity $T = 30/360A$ and applied the aforementioned segmentation by assuming Normal or Gamma distribution. The sample was partitioned, respectively, into $k = 5$ normally distributed sub-samples and $k = 2$ sub-samples with Gamma distribution. Table 2 lists, for both the partitions, the MSE, ε^k, and the statistics R^2 computed for each sub-sample. The total MSE ε and the total R^2 have been computed over the whole sample as a weighted mean of the corresponding values for each sub-group. Figures 2 and 3 compare the expected simulated interest rates with the real shifted rates after the corresponding segmentation. The jump appearing in both the figures occurs because of the translation (2).

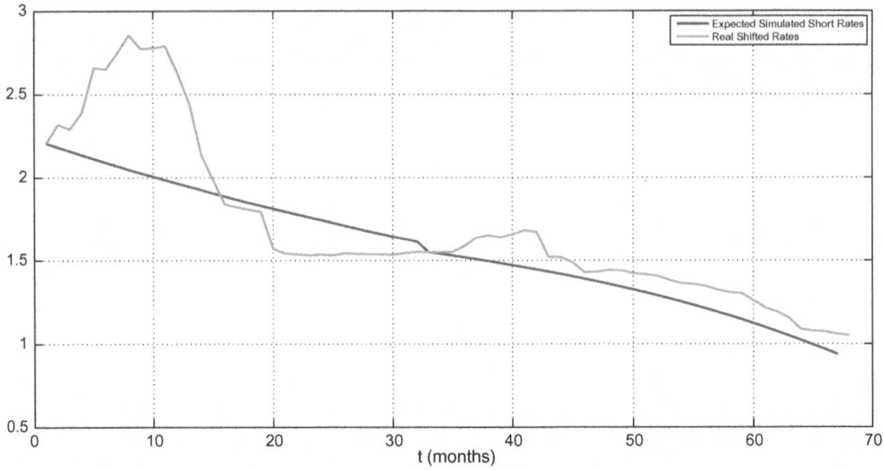

Fig. 3 Expected simulated interest rates (blue line) versus real shifted rates (green line) after segmentation with Gamma Distribution (Color figure online)

3 Conclusions

For the reasons explained in the Preface, the classic CIR model on short-term interest rates is not adequate to describe the real interest rates dynamics. Therefore we have, first, introduced opportune transformations (i.e., translations) to solve the problem of negative values. Second, we have calibrated the model's parameters, and we have shown two different numerical and statistical (and equally valid) approaches for doing that. Third, we have displayed shifted rates versus their corresponding simulated values. From the results as provided in Table 2, we note that the rates with shorter maturity have worse fitting. The reason for this fact is the presence of variability for volatility, a phenomenon that is not well explained by the original CIR model (i.e., the volatility parameter σ, the diffusion term decreases as the interest rate decreases, etc.). So, by a statistical test, we split the entire observed data sample into a number of sub-samples such that each sub-sample belongs to the same probability distribution. The choice for these distributions was either Gaussian or Gamma. The first for its well-known properties and the second because the conditional distribution of the original CIR process is a Gamma. With this modification to the classic CIR model, we improved the goodness of fit as well as the forecasting. However, it must be said that the values of the mean square error and the statistic R^2 are worse on the first sub-samples, where the volatility variability is greater. Moreover, the adjusted model proposed does not present conditional heteroscedasticity, but suffers from autocorrelation. Future research can concentrate on further improvements to the proposed CIR# model able to handle issues such as changes in volatility, autocorrelation, etc.

References

Bibby, B. M., Jacobsen, M., & Sørensen, M. (2010). Estimating functions for discretely sampled diffusion type models. In: Y. Aït-Sahalia & L. P. Hansen (Eds.), *Handbook of financial econometrics: Tools and Techniques*, *1*, 203–268.

Black, F., & Karasinski, P. (1991). Bond and option pricing when short rates are lognormal. *Financial Analysts Journal*, *47*(4), 52–59.

Brigo, D., & El-Bachir, N. (2006). Credit derivatives pricing with a smile-extended jump stochastic intensity model. Available at SSRN 950208.

Brigo, D., & Mercurio, F. (2000). The CIR++ model and other deterministic-shift extensions of short rate models. In *Proceedings of the 4th Columbia-JAFEE conference for mathematical finance and financial engineering* (pp. 563–584).

Brigo, D., & Mercurio, F. (2006). *Interest rate models-theory and practice: With smile, inflation and credit*. Berlin: Springer.

Carmona, R. A., & Tehranchi, M. R. (2006). *Interest rate models: An infinite dimensional stochastic analysis perspective*. Berlin: Springer.

Chen, L. (1996). *Stochastic mean and stochastic volatility: A three-factor model of the term structure of interest rates and its applications in derivatives pricing and risk management*. Blackwell Publishers.

Cox, J. C., Ingersoll, J. E., & Ross, S. A. (1985). A theory of the term structure of interest rates. *Econometrica: Journal of the Econometric Society*, 385–407.

Dahl, M. (2004). Stochastic mortality in life insurance: Market reserves and mortality-linked insurance contracts. *Insurance: Mathematics and Economics*, *35*(1), 113–136.

Duffie, D. (2005). Credit risk modeling with affine processes. *Journal of Banking & Finance*, *29*(11), 2751–2802.

Duffie, D., Filipović, D., & Schachermayer, W. (2003). Affine processes and applications in finance. *Annals of Applied Probability*, *13*(3), 984–1053.

Engelen, K. C. (2015). The unthinkable as the new normal. *The International Economy*, 29 (Summer 2015)(3):30.

Hull, J., & White, A. (1990). Pricing interest-rate-derivative securities. *Review of Financial Studies*, *3*(4), 573–592.

IBA. Data vendor codes. ICE Benchmark Administration (IBA). https://www.theice.com/publicdocs/IBA_Quote_Vendor_Codes.xlsx.

Johnson, N. L. (1949). Systems of frequency curves generated by methods of translation. *Biometrika*, *36*(1/2), 149–176.

Keller-Ressel, M., & Steiner, T. (2008). Yield curve shapes and the asymptotic short rate distribution in affine one-factor models. *Finance and Stochastics*, *12*(2), 149–172.

Kladìvko, K. (2007). Maximum likelihood estimation of the Cox-Ingersoll-Ross process: The Matlab implementation. *Technical Computing Prague*.

Kloeden, P. E., & Platen, E. (1992). *Numerical solution of stochastic differential equations*. Berlin: Springer.

Kvalseth, T. O. (1985). Cautionary note about R^2. *The American Statistician*, *39*(4), 279–285.

Moreno, M., & Platania, F. (2015). A cyclical square-root model for the term structure of interest rates. *European Journal of Operational Research*, *1*, 109–121.

Najafi, A. R., & Mehrdoust, F. (2017). Bond pricing under mixed generalized CIR model with mixed Wishart volatility process. *Journal of Computational and Applied Mathematics*, *319*(C), 108–116.

Najafi, A. R., Mehrdoust, F., & Shima, S. (2017). Pricing American put option on zero-coupon bond under fractional CIR model with transaction cost. *Communications in Statistics - Simulation and Computation*, 1–7.

Vasicek, O. (1977). An equilibrium characterization of the term structure. *Journal of Financial Economics*, *5*(2), 177–188.

Zhu, L. (2014). Limit theorems for a Cox-Ingersoll-Ross process with Hawkes jumps. *Journal of Applied Probability*, *51*(3), 699–712.

The Heath-Jarrow-Morton Model with Regime Shifts and Jumps Priced

Robert J. Elliott and Tak Kuen Siu

Abstract The Heath-Jarrow-Morton model is an important tool for describing the term structure of interest rates. A regime switching version was considered by Elliott and Siu (Quant Finance 16(12):1791–1800, 2016). It is of interest to price the risk due to the regime switching and this was discussed in Elliott and Siu (Quant Finance 16(12):1791–1800, 2016). In this paper, an extended Heath-Jarrow-Morton model for stochastic forward rates, incorporating both regime shifts and jumps is considered, where jumps in the forward rate dynamics are directly triggered by the regime switches. No-arbitrage drift conditions, which take into account the pricing of both the regime-switching and jump risks, are derived in two situations. The first situation starts with a risk-neutral measure while the second situation starts with the real-world measure.

Keywords Heath-Jarrow-Morton model · Regime shifts · Jumps · Forward rate processes · No-arbitrage drift conditions

1 Introduction

The Heath-Jarrow-Morton (HJM) model is an important tool which has made significant impacts to both academic research and market practice. It was pioneered by the seminal work of Heath, Jarrow and Morton (1992). Rather than modelling the stochastic behavior of short-term interest rates, the HJM model directly describes the evolution of the forward rate process over time. Essentially, it aims to describe

R. J. Elliott
School of Commerce, University of South Australia, Adelaide, Australia
e-mail: relliott@ucalgary.ca

R. J. Elliott
Haskayne School of Business, University of Calgary, Calgary, Alberta, Canada

T. K. Siu (✉)
Department of Actuarial Studies and Business Analytics, Faculty of Business and Economics, Macquarie University, Sydney, NSW 2109, Australia
e-mail: Ken.Siu@mq.edu.au; ktksiu2005@gmail.com

M. Mili et al. (eds.), *New Methods in Fixed Income Modeling*, Contributions to Management Science, https://doi.org/10.1007/978-3-319-95285-7_3

the stochastic evolution of the whole yield curve over time. The HJM model may be thought of as a theoretical version of the LIBOR market model pioneered by Brace et al. (1997), Jamshidian (1997) and Miltersen et al. (1997), etc, where the latter describes the London Interbank Offer Rate (LIBOR) directly and is popular among market practitioners. There are various extensions to the HJM model in the literature. One direction of extensions for the HJM model and its practical version, namely the LIBOR market model, is to incorporate the impact of structural changes in economic conditions. This has received some attention. See, for example, Valchev (2004), Elliott and Valchev (2004), Elhouar (2008), Chiarella et al. (2010), Steinruecke et al. (2014, 2015), Elliott and Siu (2016). Particularly, Steinruecke et al. (2014, 2015) extends the LIBOR market model by incorporating both regime switches and jumps using Markovian regime-switching jump-diffusion models. Elliott and Siu (2016) consider the pricing of regime switching risk in a Markovian regime-switching HJM model. See Elliott and Siu (2016) for related discussions and some relevant literature.

This paper also discusses the pricing of regime-switching risk in the Markovian, regime-switching, Heath-Jarrow-Morton environment of Elliott and Siu (2016). However, in addition to assuming that the model coefficients, such as the drift and volatility of the forward rate process, depend on the state of an economy, a more general situation is considered where jumps in the forward rate dynamics are triggered by transitions in the economic regime. As is usual, the evolution of the state of the economy over time is described by a continuous-time, finite-state, Markov chain which is assumed to be observable. We then identify the no-arbitrage drift conditions in this general Markovian, regime-switching, Heath-Jarrow-Morton (MRS-HJM) environment to price both the regime-switching and jump risks in the forward rate dynamics. Two situations are considered. In the first situation, we start with a risk-neutral probability measure and derive the corresponding no-arbitrage drift condition. In the second situation, we start with a real-world probability measure. Then as in Elliott and Siu (2016), a general spot martingale measure, which prices both the regime-switching and jump risks, is specified using a product of two density processes for a measure change. One density process is for a measure change for the Brownian motion. The other less well known density process due to Dufour and Elliott (1999) provides the Girsanov-type measure change for the Markov chain. A no-arbitrage HJM-type drift condition for the second situation is also obtained. A particular specification is introduced for the risk-neutral transition rates of the Markov chain which can specify a particular "direction" for the measure change of the Markov chain. This specification gives rise to a restriction on defining a spot martingale measure for pricing both the regime-switching and jump risks in the general MRS-HJM model.

This theoretical contribution is based on the previous paper of Elliott and Siu (2016). However, the theoretical approach discussed in this paper may clarify the arguments and consolidate the modeling structure for pricing regime switching risk under the HJM-type modeling framework for term structures of interest rate. It is our belief that the incorporation and pricing of regime switching risk due to changes in the economic condition in modeling term structures of interest rate are issues. It is hoped that the results from this paper may provide some theoretical insights into consolidating our understanding of these issues. Empirical investigation of the model and its potentially testable implications will be conducted in future work.

The rest of this paper is organised as follows. The next section presents the general MRS-HJM model. Section 3 derives the no-arbitrage drift condition under the first situation. In Sect. 4, the no-arbitrage drift condition under the second situation is derived. The final section gives a summary. An appendix, which gives the proofs of the theorems presented in this paper, is placed before the reference section.

2 The MRS-HJM Model with Jumps

A continuous-time economy is considered, where the time parameter set $\mathcal{T} := [0, T]$. As is usual, uncertainties are modeled by a complete probability space $(\Omega, \mathcal{F}, \mathbb{P})$, where $\mathbb{P}$ is a risk-neutral probability measure (or a real-world probability measure) in the first (or second) situation to be considered below.

The evolution of the underlying state of an economy, (i.e., the economic regime), is modelled by a continuous-time, finite-state, Markov chain $\{\mathbf{X}(t)|t \in \mathcal{T}\}$ on $(\Omega, \mathcal{F}, \mathbb{P})$. To simplify our discussion, it is supposed that the chain is observable. For mathematical convenience, it is assumed that the state space of the chain $\mathcal{S}$ is identified with a set of unit basis vectors $\{\mathbf{e}_1, \mathbf{e}_2, \ldots, \mathbf{e}_N\}$ in $\Re^N$, where the jth component of $\mathbf{e}_i$ is the Kronecker delta δ_{ij}, for each $i, j = 1, 2, \ldots, N$. See, for example, Elliott et al. (1995). The probability laws of the chain under the measure $\mathbb{P}$ are specified by a family of rate matrices, say $\{\mathbf{A}(t)|t \in \mathcal{T}\}$, where $\mathbf{A}(t) = [a_{ij}(t)]_{i,j=1,2,\ldots,N}$ and $a_{ji}(t)$ is the transition rate of the chain from state $\mathbf{e}_i$ to state $\mathbf{e}_j$ at time t. It is well-known that $a_{ji}(t) \geq 0$, for $j \neq i$, and $\sum_{j=1}^{N} a_{ji}(t) = 0$. To avoid trivalities, we assume that $a_{ji}(t) > 0$, for $j \neq i$, so that $a_{ii}(t) < 0$. The following semimartingale dynamics for the chain are well-known, see, for example, Elliott et al. (1995):

$$\mathbf{X}(t) = \mathbf{X}(0) + \int_0^t \mathbf{A}(u)\mathbf{X}(u)du + \mathbf{M}(t). \tag{2.1}$$

Here $\mathbf{M} := \{\mathbf{M}(t)|t \in \mathcal{T}\}$ is an $\Re^N$-valued $(\mathbb{F}^{\mathbf{X}}, \mathcal{P})$-martingale and $\mathbb{F}^{\mathbf{X}} := \{\mathcal{F}^{\mathbf{X}}(t)|t \in \mathcal{T}\}$ is the right-continuous, $\mathbb{P}$-completed, natural filtration generated by the chain.

In the sequel, the general Markovian regime-switching Heath-Jarrow-Morton (MRS-HJM) model for the forward rate dynamics is described. Suppose $\{W(t)|t \in \mathcal{T}\}$ is a standard Brownian motion on $(\Omega, \mathcal{F}, \mathbb{P})$ and $\mathbb{F}^W$ is the $\mathbb{P}$-augmentation of the natural filtration generated by the Brownian motion. To simplify the analysis, as in, for example, Elliott and Siu (2016), it is supposed that the chain and the Brownian motion are independent under $\mathbb{P}$.

For each $t \in \mathcal{T}$, let $\alpha(t, T, \mathbf{X}(t))$, $\sigma(t, T, \mathbf{X}(t))$ and $\gamma(t, T, \mathbf{X}(t))$ be the drift, the volatility and the jump size of a forward rate process at time t. It is assumed here that these model coefficients depend on the state of the economy $\mathbf{X}(t)$. Mathematically,

$$\alpha(t, T, \mathbf{X}(t)) := \langle \boldsymbol{\alpha}(t, T), \mathbf{X}(t) \rangle,$$
$$\sigma(t, T, \mathbf{X}(t)) := \langle \boldsymbol{\beta}(t, T), \mathbf{X}(t) \rangle,$$
$$\gamma(t, T, \mathbf{X}(t)) := \langle \boldsymbol{\gamma}(t, T), \mathbf{X}(t) \rangle.$$

Here $\langle \cdot, \cdot \rangle$ is the scalar product in $\Re^N$. The vectors $\boldsymbol{\alpha}(t, T)$, $\boldsymbol{\sigma}(t, T)$ and $\boldsymbol{\gamma}(t, T)$ are, respectively, defined as follows:

$$\boldsymbol{\alpha}(t, T) := (\alpha_1(t, T), \alpha_2(t, T), \ldots, \alpha_N(t, T))' \in \Re^N,$$
$$\boldsymbol{\sigma}(t, T) := (\sigma_1(t, T), \sigma_2(t, T), \ldots, \sigma_N(t, T))' \in \Re^N,$$
$$\boldsymbol{\gamma}(t, T) := (\gamma_1(t, T), \gamma_2(t, T), \ldots, \gamma_N(t, T))' \in \Re^N,$$

where, for each $i = 1, 2, \ldots, N$, $\{\alpha_i(t, T) | t \in \mathcal{T}\}$, $\{\beta_i(t, T) | t \in \mathcal{T}\}$ and $\{\gamma_i(t, T) | t \in \mathcal{T}\}$ are $\mathbb{F}^W$-progressively measurable processes.

Then we suppose that the forward rate process $\{f(t, T) | t \in \mathcal{T}\}$ is governed by the following general MRS-HJM model.

$$df(t, T) = \alpha(t, T, \mathbf{X}(t))dt + \sigma(t, T, \mathbf{X}(t))dW(t) + \langle \boldsymbol{\gamma}(t, T), d\mathbf{X}(t) \rangle. \quad (2.2)$$

Under these dynamics, a transition in the state of the economy described by the Markov chain not only leads to a structural change in the model coefficients, but also triggers a jump in the forward rate dynamics. This is different from the Markovian, regime-switching, HJM model considered in Elliott and Siu (2016), where a transition in the chain does not trigger a jump in the forward rate dynamics. Due to the presence of regime switching and jump risks, the market described by the general MRS-HJM model is incomplete. In other words, there is more than one pricing kernel for valuing interest-rate contingent claims under the general MRS-HJM model.

To complete the specification of the information structure of the general MRS-HJM model, we define an enlarged filtration $\mathbb{G} := \{\mathcal{G}(t) | t \in \mathcal{T}\}$ as follows. For each $t \in \mathcal{T}$, let

$$\mathcal{G}(t) := \mathcal{F}^W(t) \vee \mathcal{F}^{\mathbf{X}}(t).$$

Here, for example, $\mathcal{A}_1 \vee \mathcal{A}_2$ is the minimal σ-algebra containing both the σ-algebras $\mathcal{A}_1$ and $\mathcal{A}_2$. It is then obvious that the forward rate process $\{f(t, T) | t \in \mathcal{T}\}$ in Eq. (2.2) is $\mathbb{G}$-adapted.

3 No-Arbitrage Drift Condition I

In this section we start with a given risk-neutral probability measure, say $\mathbb{P}$. Then following the techniques in, for example, Elliott and Kopp (2005), Chap. 9, Sect. 9.7, the no-arbitrage drift condition for the general MRS-HJM model is derived. This extends the no-arbitrage drift condition for the standard HJM.

From Eq. (2.2), the forward rate process described by the general MRS-HJM model under the given risk-neutral measure $\mathbb{P}$ is given by:

$$f(t,T) = f(0,T) + \int_0^t \alpha(t_1, T, \mathbf{X}(t_1))dt_1 + \int_0^t \sigma(t_1, T, \mathbf{X}(t_1))dW(t_1) + \int_0^t \langle \gamma(t_1, T), d\mathbf{X}(t_1)\rangle. \quad (3.1)$$

Write, for each $t \in \mathcal{T}$,

$$Z(t) := -\int_t^T f(t,u)du. \quad (3.2)$$

Since $\{f(t,u)|u \in [t,T]\}$ is $\mathcal{G}(t)$-measurable, the process $\{Z(t)|t \in \mathcal{T}\}$ is $\mathbb{G}$-adapted.

It is well-known that the price process $\{B(t,T)|t \in \mathcal{T}\}$ of a zero-coupon bond maturing at time T is given by:

$$B(t,T) = \exp(Z(t)), \quad B(T,T) = 1. \quad (3.3)$$

See, for example, Elliott and Kopp (2005), Page 279.

We can then prove the following no-arbitrage drift condition.

Theorem 1 *Suppose $\mathbb{P}$ is a given risk-neutral probability measure and the forward rate process under $\mathbb{P}$ is governed by Eq.* (2.2)*. Then the no-arbitrage drift condition for the general MRS-HJM model in Eq.* (2.2) *is given by:*

$$\alpha(t, T, \mathbf{X}(t)) = \sigma(t, T, \mathbf{X}(t)) \int_t^T \sigma(t, u, \mathbf{X}(t))du - \left\langle \frac{\partial \mathbf{G}(t,T)}{\partial T}, \mathbf{A}(t)\mathbf{X}(t) \right\rangle, \quad (3.4)$$

where

$$\frac{\partial \mathbf{G}(t,T)}{\partial T} = (e^{\gamma_1^*(t,T)}\gamma_1(t,T), e^{\gamma_2^*(t,T)}\gamma_2(t,T), \ldots, e^{\gamma_N^*(t,T)}\gamma_N(t,T))' \in \Re^N. \quad (3.5)$$

Under the risk-neutral measure $\mathbb{P}$, the semimartingale representation for the chain is given by Eq. (2.1) *and $\gamma_i^*(t,T) := \int_t^T \gamma_i(t,u)du$.*

Proof See the Appendix. □

Remark 1 When the jump term triggered by transitions in the chain in the forward rate process in Eq. (2.2) is absent, (i.e., when $\gamma_i(t,u) = 0$ for each $i = 1, 2, \ldots, N$ and each $u \in [t,T]$), the no-arbitrage drift condition in Eq. (3.4) becomes:

$$\alpha(t, T, \mathbf{X}(t)) = \sigma(t, T, \mathbf{X}(t)) \int_t^T \sigma(t, u, \mathbf{X}(t))du. \quad (3.6)$$

It is not difficult to check that the no-arbitrage drift condition in Eq. (3.6) is the same as the no-arbitrage drift condition in the Markovian regime-switching HJM model considered in Elliott and Siu (2016) if we start with a given risk-neutral probability measure.

Remark 2 Due to the presence of the last term "$\left\langle \frac{\partial \mathbf{G}(t,T)}{\partial T}, \mathbf{A}(t)\mathbf{X}(t) \right\rangle$" in Eq. (3.4), the no-arbitrage drift condition in Eq. (3.4) in Theorem 1 prices both the regime-switching risk and the jump risk triggered by regime switches. When the underlying chain is absent, the no-arbitrage drift condition in Eq. (3.6) becomes:

$$\alpha(t, T) = \sigma(t, T) \int_t^T \sigma(t, u) du. \tag{3.7}$$

This is the no-arbitrage drift condition for the standard HJM model in, for example, Elliott and Kopp (2005), Page 280. The no-arbitrage drift condition in Eq. (3.7) does not price either the regime-switching risk or the jump risk.

4 No-Arbitrage Drift Condition II

In this section, we start with a real-world probability measure. Here with a slight abuse of notation, we use $\mathbb{P}$ to denote the real-world probability measure. As in, for example, Elliott and Siu (2016), we consider a product of two density processes, one for a measure change for the Brownian motion and another one for the chain. These give an equivalent martingale measure in the Markovian regime-switching HJM model. For the measure change for the chain, we use the density process in, for example, Dufour and Elliott (1999), which has been used in Elliott and Siu (2016) for a measure change for the modulating Markov chain of a Markovian regime-switching HJM model. The measure change for the Brownian motion is obtained using the standard Girsanov theorem for Brownian motion. The results for the measure changes are briefly presented below. For the details and the proofs of the measure change for the chain, see, for example, Dufour and Elliott (1999).

Let $\{\theta(t)|t \in \mathcal{T}\}$ be a $\mathbb{G}$-optional process such that

$$\theta(t) = \langle \boldsymbol{\theta}(t), \mathbf{X}(t) \rangle, \tag{4.1}$$

where $\boldsymbol{\theta}(t) := (\theta_1(t), \theta_2(t), \ldots, \theta_N(t))' \in \Re^N$; for each $i = 1, 2, \ldots, N$, $\{\theta_i(t)|t \in \mathcal{T}\}$ is an $\mathbb{F}^W$-optional process satisfying a linear growth condition.

Consider the following $\mathbb{G}$-adapted process $\{\Lambda_1^\theta(t)|t \in \mathcal{T}\}$:

$$\Lambda_1^\theta(t) := \exp\left(-\int_0^t \theta(u) dW(u) - \frac{1}{2} \int_0^t \theta^2(u) du \right). \tag{4.2}$$

Since $\{\theta_i(t)|t \in \mathcal{T}\}$ satisfies a linear growth condition, for each $i = 1, 2, \ldots, N$, an application of Lemma 13.37 in Elliott (1982) gives:

$$\mathrm{E}[\Lambda_1^{\theta}(T)] = 1.$$

The process $\{\Lambda_1^{\theta}(t)|t \in \mathcal{T}\}$ is used as the density process for the measure change for the Brownian motion.

To specify the density process for the measure change for the chain, we need to specify a family of rate matrices of the chain under a new equivalent probability measure. In Elliott and Siu (2016), this family was specified by employing the concept of rapidly fluctuating Markov chains as in Yin and Zhang (2007). Here we aim to specify one particular equivalent martingale measure based on the product of the two density processes using a parametrization of the family of rate matrices which depends on the optional process $\{\theta(t)|t \in \mathcal{T}\}$. This parametrization is described in the sequel.

Let $\mathbf{A}^{\theta} := \{\mathbf{A}^{\theta}(t)|t \in \mathcal{T}\}$, where $\mathbf{A}^{\theta}(t) := [a_{ij}^{\theta}(t)]_{i,j=1,2,\ldots,N}$ and $a_{ji}^{\theta}(t)$ is the transition rate of the chain from $\mathbf{e}_i$ to $\mathbf{e}_j$ at time t. We suppose that for each $i, j = 1, 2, \ldots, N$ with $i \neq j$,

$$a_{ij}^{\theta}(t) := |\theta(t)|a_{ij}(t), \tag{4.3}$$

where $|\theta(t)|$ is the absolute value of $\theta(t)$.

The reason for using the absolute value of $\theta(t)$ in the above specification is that it preserves the sign of the transition rates $a_{ij}(t)$. Consequently, $a_{ij}^{\theta}(t) \geq 0$ for each $i, j = 1, 2, \ldots, N$ with $i \neq j$. Furthermore,

$$\sum_{j=1}^{N} a_{ji}^{\theta}(t) = |\theta(t)| \sum_{j=1}^{N} a_{ji}(t) = 0, \tag{4.4}$$

so $a_{ii}^{\theta}(t) \leq 0$ for each $i = 1, 2, \ldots, N$.

The condition that $a_{ij}^{\theta}(t) \geq 0$ for each $i, j = 1, 2, \ldots, N$ with $i \neq j$ and the condition in Eq. (4.4) imply that the parametrization in Eq. (4.3) gives a family of rate matrices. Using this parametrization for the family of rate matrices for the chain, the density process for the measure change of the chain in Dufour and Elliott (1999), which was adopted in Elliott and Siu (2016), may be simplified.

Indeed, under this parametrization, the family of "ratio" rate matrices of the chain, denoted by $\mathbf{\Theta} := \{\mathbf{\Theta}(t)|t \in \mathcal{T}\}$ with the (i, j)-element given by the ratio $\frac{a_{ij}^{\theta}(t)}{a_{ij}(t)}$ is given by:

$$\mathbf{\Theta}(t) := \left[\frac{a_{ij}^{\theta}(t)}{a_{ij}(t)}\right]_{i,j=1,2,\ldots,N} = [|\theta(t)|]_{i,j=1,2,\ldots,N}. \tag{4.5}$$

That is, $\mathbf{\Theta}(t)$ is an $\Re^N \otimes \Re^N$-matrix with all elements being $|\theta(t)|$.

Write $\mathbf{\Theta}_0(t)$ for the $\Re^N \otimes \Re^N$-matrix $\mathbf{\Theta}(t)$, but with all elements in the main diagonal being zero. That is,

$$\mathbf{\Theta}_0(t) := \mathbf{\Theta}(t) - \mathbf{diag}(|\theta(t)|, |\theta(t)|, \ldots, |\theta(t)|).$$

(Here $\mathbf{diag}(|\theta(t)|, |\theta(t)|, \ldots, |\theta(t)|)$ is a diagonal matrix with all elements in the main diagonal being $|\theta(t)|$.)

Similarly, the $\Re^N \otimes \Re^N$-matrix $\mathbf{A}_0(t)$ is defined by:

$$\mathbf{A}_0(t) := \mathbf{A}(t) - \mathbf{diag}(a_{11}(t), a_{22}(t), \ldots, a_{NN}(t)).$$

Suppose $\mathbf{J} := \{\mathbf{J}(t)|t \in \mathcal{T}\}$ denotes a vector-valued counting process on $(\Omega, \mathcal{F}, \mathbb{P})$ such that

1. $\mathbf{J}(t) := (J_1(t), J_2(t), \ldots, J_N(t))' \in \Re^N$;
2. for each $i = 1, 2, \ldots, N$, $J^i(t)$ represents the number of times the chain $\mathbf{X}$ jumps to the state $\mathbf{e}_i$ from any other state during the time interval $[0, t]$.

It is shown in Dufour and Elliott (1999), (see also Elliott and Siu (2016)), that $\widetilde{\mathbf{J}} := \{\widetilde{\mathbf{J}}(t)|t \in \mathcal{T}\}$ defined by:

$$\widetilde{\mathbf{J}}(t) := \mathbf{J}(t) - \int_0^t \mathbf{A}_0(u)\mathbf{X}(u)du,$$

is a vector martingale associated with the vector counting process $\mathbf{J}$.

The martingale $\widetilde{\mathbf{J}}$ will be used as a "random shock" in the density process $\{\Lambda_2^\theta(t)|t \in \mathcal{T}\}$ for the measure change for the chain. This is defined as follows:

$$\Lambda_2^\theta(t) = 1 + \int_0^t \Lambda_2^\theta(u-)[\mathbf{\Theta}(u)\mathbf{X}(u-) - \mathbf{1}]'d\widetilde{\mathbf{J}}(u),$$

where $\mathbf{1} := (1, 1, \ldots, 1)' \in \Re^N$. It is obvious that $\{\Lambda_2^\theta(t)|t \in \mathcal{T}\}$ is a $(\mathbb{G}, \mathbb{P})$-(local)-martingale. However, we assume that $|\theta(t)|$ is bounded so that $\{\Lambda_2^\theta(t)|t \in \mathcal{T}\}$ is a $(\mathbb{G}, \mathbb{P})$-martingale.

The product of the two density processes for the measure changes $\{\Lambda^\theta(t)|t \in \mathcal{T}\}$ is now defined as:

$$\Lambda^\theta(t) := \Lambda_1^\theta(t) \cdot \Lambda_2^\theta(t).$$

Then a new probability measure $\mathbb{P}^\theta \sim \mathbb{P}$ on $\mathcal{G}(T)$ can be defined by putting:

$$\left.\frac{d\mathbb{P}^\theta}{d\mathbb{P}}\right|_{\mathcal{G}(T)} := \Lambda^\theta(T). \tag{4.6}$$

It is well-known by the standard Girsanov's theorem for Brownian motions, that the process $\{W^{\theta}(t)|t \in \mathcal{T}\}$ defined by:

$$W^{\theta}(t) := W(t) + \int_0^t \theta(u)du, \tag{4.7}$$

is a $(\mathbb{G}, \mathbb{P}^{\theta})$-standard Brownian motion.

Furthermore, the Girsanov transform for Markov chains in, for example, Dufour and Elliott (1999) gives that under the measure $\mathbb{P}^{\theta}$ the chain $\mathbf{X}$ has the family of rate matrices $\mathbf{A}^{\theta}$. Consequently, it has the following semimartingale dynamics under $\mathbb{P}^{\theta}$.

$$\mathbf{X}(t) = \mathbf{X}(0) + \int_0^t \mathbf{A}^{\theta}(u)\mathbf{X}(u)du + \mathbf{M}^{\theta}(t). \tag{4.8}$$

Here $\mathbf{M}^{\theta} := \{\mathbf{M}^{\theta}(t)|t \in \mathcal{T}\}$ is a $(\mathbb{G}, \mathbb{P}^{\theta})$-martingale. See also Elliott and Siu (2016) for some related discussions.

The following theorem gives the necessary and sufficient condition for the absence of arbitrage opportunities in the generalized MRS-HJM model of Eq. (2.2)

Theorem 2 *The general Markovian, regime-switching, HJM model in Eq.* (2.2) *is arbitrage-free if and only if there exists a $\mathbb{G}$-optional process $\{\theta(t)|t \in \mathcal{T}\}$ such that for each $(t, \mathbf{x}) \in \mathcal{T} \times \mathcal{E}$,*

$$\alpha(t, T, \mathbf{x}) = \sigma(t, T, \mathbf{x}) \int_t^T \sigma(t, u, \mathbf{x})du + \theta(t)\sigma(t, T, \mathbf{x}) - |\theta(t)| \left\langle \frac{\partial \mathbf{G}(t, T)}{\partial T}, \mathbf{A}(t)\mathbf{x} \right\rangle, \tag{4.9}$$

where $\theta(t) = \theta(t, \mathbf{x})$ and the partial derivative $\frac{\partial \mathbf{G}(t,T)}{\partial T}$ is given by Eq. (3.5).

Proof See the Appendix. □

Again due to the presence of the last term "$|\theta(t)| \left\langle \frac{\partial \mathbf{G}(t,T)}{\partial T}, \mathbf{A}(t)\mathbf{x} \right\rangle$" in Eq. (4.9), the no-arbitrage drift condition in Eq. (4.9) prices both the jump risk and regime-switching risk. When the jump term triggered by regime switching in the general MRS-HJM model in Eq. (2.2) vanishes, the necessary and sufficient condition for the absence of arbitrage in the general MRS-HJM model in Eq. (2.2) in Theorem 2 becomes the one in Elliott and Siu (2016). Furthermore, when the modulating Markov chain is absent, (i.e., there are no regime switches), the no-arbitrage condition in Theorem 2 becomes the one in the original HJM model of Heath, Jarrow and Morton (1992), (see also Elliott and Kopp (2005), Theorem 9.7.3).

5 Conclusion

The pricing of regime switching risk in a Markovian regime-switching HJM model in Elliott and Siu (2016) is re-visited and a more general Markovian regime-switching HJM model is considered. This incorporates both regime switches in the model coefficients and jumps in the forward rate process triggered by regime switches. The no-arbitrage drift condition for the general Markovian regime-switching HJM model is derived when a risk-neutral probability measure is given in advance. Then a second situation is considered where a risk-neutral probability measure is defined by a product of two density processes for measure changes and a parametrization for the family of rate matrices of the underlying chain. The parametrization simplifies the measure change for the chain. The no-arbitrage drift condition for the general Markovian regime-switching HJM model is derived under this model. The main results of this paper may provide some theoretical insights in gaining a deeper understanding of pricing regime-switching risk in a general Markovian regime-switching HJM model.

Acknowledgements The authors would like to thank the Australian Research Council for continued support.

Appendix

Proof of Theorem 1:

Proof Differentiating $Z(t)$ in Eq. (3.2) gives:

$$dZ(t) = f(t,t)dt - \int_t^T \alpha(t,u,\mathbf{X}(t))dtdu - \int_t^T \sigma(t,u,\mathbf{X}(t))dW(t)du \\ - \int_t^T \langle \gamma(t,u), d\mathbf{X}(t) \rangle du. \tag{A.1}$$

Write, for each $t \in \mathcal{T}$,

$$\alpha^*(t,T,\mathbf{X}(t)) := \int_t^T \alpha(t,u,\mathbf{X}(t))du,$$

$$\sigma^*(t,T,\mathbf{X}(t)) := \int_t^T \sigma(t,u,\mathbf{X}(t))du,$$

$$\gamma^*(t,T) := \int_t^T \gamma(t,u)du.$$

Note that for each $t \in \mathcal{T}$,

$$f(t,t) = r(t).$$

Then the dynamics for the process $\{Z(t)|t \in \mathcal{T}\}$ in Eq. (3.2) can be written as:

$$dZ(t) = r(t)dt - \alpha^*(t, T, \mathbf{X}(t))dt - \sigma^*(t, T, \mathbf{X}(t))dW(t) - \langle \gamma^*(t, T), d\mathbf{X}(t) \rangle. \tag{A.2}$$

Applying Itô's differentiation rule for semimartingales, (see, for example, Elliott (1982), Chap. 12) to $B(t, T) = \exp(Z(t))$ and using Eq. (A.2) gives:

$$\begin{aligned}
B(t, T) = B(0, T) &+ \int_0^t e^{Z(t_1)}\left(r(t_1) - \alpha^*(t_1, T, \mathbf{X}(t_1)) + \frac{1}{2}\sigma^*(t_1, T, \mathbf{X}(t_1))^2\right)dt_1 \\
&- \int_0^t e^{Z(t_1)}\sigma^*(t_1, T, \mathbf{X}(t_1))dW(t_1) - \int_0^t e^{Z(t_1-)}\langle \gamma^*(t_1, T), d\mathbf{X}(t_1)\rangle \\
&- \sum_{0<t_1\le t}\left(e^{Z(t_1)} - e^{Z(t_1-)} - e^{Z(t_1-)}\Delta Z(t_1)\right) \\
= B(0, T) &+ \int_0^t e^{Z(t_1)}\left(r(t_1) - \alpha^*(t_1, T, \mathbf{X}(t_1)) + \frac{1}{2}\sigma^*(t_1, T, \mathbf{X}(t_1))^2\right)dt_1 \\
&- \int_0^t e^{Z(t_1)}\sigma^*(t_1, T, \mathbf{X}(t_1))dW(t_1) - \sum_{0<t_1\le t} e^{Z(t_1-)}\langle \gamma^*(t_1, T), \Delta\mathbf{X}(t_1)\rangle \\
&- \sum_{0<t_1\le t}\left(e^{Z(t_1)} - e^{Z(t_1-)} - e^{Z(t_1-)}\langle \gamma^*(t_1, T), \Delta\mathbf{X}(t_1)\rangle\right) \\
= B(0, T) &+ \int_0^t e^{Z(t_1)}\left(r(t_1) - \alpha^*(t_1, T, \mathbf{X}(t_1)) + \frac{1}{2}\sigma^*(t_1, T, \mathbf{X}(t_1))^2\right)dt_1 \\
&- \int_0^t e^{Z(t_1)}\sigma^*(t_1, T, \mathbf{X}(t_1))dW(t_1) - \sum_{0<t_1\le t} e^{Z(t_1-)}\left(e^{\langle \gamma^*(t_1-, T), \Delta\mathbf{X}(t_1)\rangle} - 1\right).
\end{aligned} \tag{A.3}$$

Recall

$$\gamma^*(t, T) := \int_t^T \gamma(t, u)du,$$

where $\gamma(t, u) := (\gamma_1(t, u), \gamma_2(t, u), \ldots, \gamma_N(t, u))' \in \Re^N$.

Consequently, we can write:

$$\gamma^*(t, T) := (\gamma_1^*(t, T), \gamma_2^*(t, T), \ldots, \gamma_N^*(t, T))' \in \Re^N,$$

so that for each $i = 1, 2, \ldots, N$,

$$\gamma_i^*(t, T) := \int_t^T \gamma_i(t, u)du.$$

Define the following vector in $\Re^N$:

$$\mathbf{G}(t,T) := (G_1(t,T), G_2(t,T), \ldots, G_N(t,T))' \in \Re^N, \tag{A.4}$$

where $G_i(t,T) := \exp(\gamma_i^*(t,T))$ for each $i = 1, 2, \ldots, N$.

Write $\mathbf{1} := (1, 1, \ldots, 1)' \in \Re^N$. Then

$$\begin{aligned}
&\sum_{0<t_1\leq t} e^{Z(t_1-)}\left(e^{\langle \gamma^*(t_1-,T), \Delta\mathbf{X}(t_1)\rangle} - 1\right)\\
&= \sum_{0<t_1\leq t} e^{Z(t_1-)}\langle \mathbf{G}(t_1-,T) - \mathbf{1}, \Delta\mathbf{X}(t_1)\rangle\\
&= \int_0^t e^{Z(t_1-)}\langle \mathbf{G}(t_1-,T) - \mathbf{1}, d\mathbf{X}(t_1)\rangle\\
&= \int_0^t e^{Z(t_1-)}\langle \mathbf{G}(t_1-,T) - \mathbf{1}, \mathbf{A}(t_1)\mathbf{X}(t_1)\rangle dt_1 + \int_0^t e^{Z(t_1-)}\langle \mathbf{G}(t_1-,T) - \mathbf{1}, d\mathbf{M}(t_1)\rangle.
\end{aligned} \tag{A.5}$$

Let $\{\widetilde{B}(t,T) | t \in \mathcal{T}\}$ be the discounted bond price process, where

$$\widetilde{B}(t,T) := \exp\left(-\int_0^t r(t_1)dt_1\right)B(t,T).$$

Then combining Eq. (A.3) and Eq. (A.5) gives the following dynamics for the discounted bond price process $\{\widetilde{B}(t,T) | t \in \mathcal{T}\}$:

$$\begin{aligned}
\widetilde{B}(t,T) &= \widetilde{B}(0,T) + \int_0^t \widetilde{B}(t_1,T)\left(-\alpha^*(t_1,T,\mathbf{X}(t_1)) + \frac{1}{2}\sigma^*(t_1,T,\mathbf{X}(t_1))^2\right)dt_1\\
&\quad - \int_0^t \widetilde{B}(t_1-,T)\langle \mathbf{G}(t_1-,T) - \mathbf{1}, \mathbf{A}(t_1)\mathbf{X}(t_1)\rangle dt_1\\
&\quad - \int_0^t \widetilde{B}(t_1,T)\sigma^*(t_1,T,\mathbf{X}(t_1))dW(t_1)\\
&\quad - \int_0^t \widetilde{B}(t_1-,T)\langle \mathbf{G}(t_1-,T) - \mathbf{1}, d\mathbf{M}(t_1)\rangle\\
&= \widetilde{B}(0,T) + \int_0^t \widetilde{B}(t_1,T)\Big(-\alpha^*(t_1,T,\mathbf{X}(t_1)) + \frac{1}{2}\sigma^*(t_1,T,\mathbf{X}(t_1))^2\\
&\quad - \langle \mathbf{G}(t_1-,T) - \mathbf{1}, \mathbf{A}(t_1)\mathbf{X}(t_1)\rangle\Big)dt_1 - \int_0^t \widetilde{B}(t_1,T)\sigma^*(t_1,T,\mathbf{X}(t_1))dW(t_1)\\
&\quad - \int_0^t \widetilde{B}(t_1-,T)\langle \mathbf{G}(t_1-,T) - \mathbf{1}, d\mathbf{M}(t_1)\rangle.
\end{aligned} \tag{A.6}$$

This is a $\mathbb{G}$-special semimartingale. Note that to preclude arbitrage opportunities, the discounted bond price process $\{\widetilde{B}(t, T)|t \in \mathcal{T}\}$ is a $(\mathbb{G}, \mathbb{P})$-martingale since $\mathbb{P}$ is the given risk-neutral probability measure. See, for example, Harrison and Pliska (1981, 1983). By the unique decomposition of a special semimartingale, the bounded variation term in the representation (A.6) of the discounted bond price process must be indistinguishable from the zero process. This then gives:

$$\alpha^*(t, T, \mathbf{X}(t)) = \frac{1}{2}\sigma^*(t, T, \mathbf{X}(t))^2 - \langle \mathbf{G}(t, T) - \mathbf{1}, \mathbf{A}(t)\mathbf{X}(t)\rangle. \tag{A.7}$$

Differentiating Eq. (A.7) with respect to T gives:

$$\alpha(t, T, \mathbf{X}(t)) = \sigma(t, T, \mathbf{X}(t)) \int_t^T \sigma(t, u, \mathbf{X}(t))du - \langle \frac{\partial \mathbf{G}(t, T)}{\partial T}, \mathbf{A}(t)\mathbf{X}(t)\rangle.$$

□

Proof of Theorem 2:

Proof From Eqs. (A.6), (4.7) and (4.8), the discounted bond price process $\{\widetilde{B}(t, T)|t \in \mathcal{T}\}$ under the measure $\mathbb{P}^\theta$ is:

$$\begin{aligned}\widetilde{B}(t, T) = \widetilde{B}(0, T) + \int_0^t \widetilde{B}(t_1, T)\Big(&- \alpha^*(t_1, T, \mathbf{X}(t_1)) + \frac{1}{2}\sigma^*(t_1, T, \mathbf{X}(t_1))^2 \\ &+\theta(t_1)\sigma^*(t_1, T, \mathbf{X}(t_1)) - \langle \mathbf{G}(t_1-, T) - \mathbf{1}, \mathbf{A}^\theta(t_1)\mathbf{X}(t_1)\rangle \Big)dt_1 \\ &- \int_0^t \widetilde{B}(t_1, T)\sigma^*(t_1, T, \mathbf{X}(t_1))dW^\theta(t_1) \\ &- \int_0^t \widetilde{B}(t_1-, T)\langle \mathbf{G}(t_1-, T) - \mathbf{1}, d\mathbf{M}^\theta(t_1)\rangle. \end{aligned} \tag{A.8}$$

Note that $\mathbf{A}^\theta(t) = |\theta(t)|\mathbf{A}(t)$, so

$$\langle \mathbf{G}(t-, T) - \mathbf{1}, \mathbf{A}^\theta(t)\mathbf{X}(t)\rangle = |\theta(t)|\langle \mathbf{G}(t-, T) - \mathbf{1}, \mathbf{A}(t)\mathbf{X}(t)\rangle.$$

Consequently, the discounted bond price process under $\mathbb{P}^\theta$ in Eq. (A.8) becomes:

$$
\begin{aligned}
\widetilde{B}(t,T) = \widetilde{B}(0,T) + \int_0^t \widetilde{B}(t_1,T)\Big(& -\alpha^*(t_1,T,\mathbf{X}(t_1)) + \frac{1}{2}\sigma^*(t_1,T,\mathbf{X}(t_1))^2 \\
& +\theta(t_1)\sigma^*(t_1,T,\mathbf{X}(t_1)) - |\theta(t_1)|\langle \mathbf{G}(t_1-,T) - \mathbf{1}, \mathbf{A}(t_1)\mathbf{X}(t_1)\rangle \Big) dt_1 \\
& - \int_0^t \widetilde{B}(t_1,T)\sigma^*(t_1,T,\mathbf{X}(t_1)) dW^\theta(t_1) \\
& - \int_0^t \widetilde{B}(t_1-,T)\langle \mathbf{G}(t_1-,T) - \mathbf{1}, d\mathbf{M}^\theta(t_1)\rangle. \qquad \text{(A.9)}
\end{aligned}
$$

In what follows, we wish to find a martingale restriction, or condition on θ, such that $\mathbb{P}^\theta$ is an equivalent martingale measure using the no-arbitrage principle. Again, by the fundamental theorem for asset pricing in, for example, Harrison and Pliska (1981, 1983), there are no arbitrage opportunities if and only if there is an equivalent martingale measure under which the discounted bond price process $\{\widetilde{B}(t,T)|t \in \mathcal{T}\}$ is a martingale. This holds true if and only if the drift term in the discounted bond price process in Eq. (A.9) is identical to zero. That is,

$$
\begin{aligned}
& -\alpha^*(t,T,\mathbf{X}(t)) + \frac{1}{2}\sigma^*(t,T,\mathbf{X}(t))^2 + \theta(t)\sigma^*(t,T,\mathbf{X}(t)) \\
& -|\theta(t)|\langle \mathbf{G}(t-,T) - \mathbf{1}, \mathbf{A}(t)\mathbf{X}(t)\rangle = 0. \qquad \text{(A.10)}
\end{aligned}
$$

Differentiating Eq. (A.10) with respect to T gives:

$$
\begin{aligned}
\alpha(t,T,\mathbf{X}(t)) = \sigma(t,T,\mathbf{X}(t)) & \int_t^T \sigma(t,u,\mathbf{X}(t)) du + \theta(t)\sigma(t,T,\mathbf{X}(t)) \\
& -|\theta(t)| \left\langle \frac{\partial \mathbf{G}(t,T)}{\partial T}, \mathbf{A}(t)\mathbf{X}(t) \right\rangle, \qquad \text{(A.11)}
\end{aligned}
$$

where the partial derivative $\frac{\partial \mathbf{G}(t,T)}{\partial T}$ is given by Eq. (3.5). □

References

Brace, A., Gatarek, D., & Musiela, M. (1997). The market model of interest rate dynamics. *Mathematical Finance*, *7*(2), 127–154.

Chiarella, C., Maina, S. C., & Sklibosios, C. N. (2010). *Markovian defaultable HJM term structure models with unspanned stochastic volatility*. Research Paper 283, Quantitative Finance Research Centre, University of Technology Sydney, August 2010.

Dufour, F., & Elliott, R. J. (1999). Filtering with discrete state observations. *Applied Mathematics and Optimization*, *40*, 259–272.

Elhouar, M. (2008). Finite-dimensional realizations of regime-switching HJM models. *Applied Mathematical Finance*, *15*(4), 331–354.

Elliott, R. J. (1982). *Stochastic calculus and applications*. Berlin, Heidelberg, New York: Springer.

Elliott, R. J., Aggoun, L., & Moore, J. B. (1995). *Hidden Markov models: Estimation and control*. Berlin, Heidelberg, New York: Springer.
Elliott, R. J., & Kopp, E. (2005). *Mathematics of financial markets* (2nd ed.). Berlin, Heidelberg, New York: Springer.
Elliott, R. J., & Siu, T. K. (2016). Pricing regime-switching risk in an HJM interest rate environment. *Quantitative Finance*, *16*(12), 1791–1800.
Elliott, R. J., & Valchev, S. (2004). *Libor market model with regime-switching volatility*. Working Paper No. 228, National Centre of Competence in Research Financial Valuation and Risk Management.
Harrison, J. M., & Pliska, S. (1981). Martingales and stochastic integrals in the theory of continuous trading. *Stochastic Processes and Applications*, *11*, 215–260.
Harrison, J. M., & Pliska, S. (1983). A stochastic calculus model for continuous trading, complete markets. *Stochastic Processes and Applications*, *15*, 313–316.
Heath, D., Jarrow, R., & Morton, A. (1992). Bond pricing and the term structure of interest rate: a new methodology. *Econometrica*, *60*, 77–105.
Jamshidian, F. (1997). LIBOR and swap market models and measures. *Finance and Stochastics*, *1*, 293–330.
Miltersen, K., Sandmann, K., & Sondermann, D. (1997). Closed form solutions for term structure derivates with log-normal interest rates. *Journal of Finance*, *52*(1), 409–430.
Steinruecke, L., Zagst, R., & Swishchuk, A. (2014). The LIBOR market model: a Markov-switching jump-diffusion extension. In R. S. Mamon & R. J. Elliott (Eds.), *Hidden Markov models in finance: Further developments and applications* (Vol. 2, pp. 85–116). Berlin, Heidelberg, New York: Springer.
Steinruecke, L., Zagst, R., & Swishchuk, A. (2015). The Markov-switching jump diffusion LIBOR market model. *Quantitative Finance*, *15*(3), 455–476.
Valchev, S. (2004). Stochastic volatility Gaussian Heath-Jarrow-Morton models. *Applied Mathematical Finance*, *11*, 347–368.
Yin, G., & Zhang, H. (2007). Singularly perturbed Markov chains: Limit results and applications. *Annals of Applied Probability*, *17*(1), 207–229.

Explicit Computation of the Post-crisis Spot LIBOR in a Jump-Diffusion Framework

Luca Di Persio and Nicola Gugole

Abstract Starting from the worldwide financial crisis originated by the dramatic US economic events happened in 2007, many markets have seen a sudden growth of heterogeneous risk types, spanning from credit ones to liquidity ones. These abrupt changes in fundamentals, have produced the develop of significant spreads between the same interbank rate, e.g. the LIBOR rate, considered at different tenors. In the present chapter, we show how to explicitly compute the post-crisis spot LIBOR at different tenors, taking into account the possibility of jumps in the instantaneous spot rate trajectories, representing, in our setting, the so called OIS short-rate. Such an analysis is based on the intensity approach, where large and sudden movements can be modeled by adding marked point processes to the classical diffusion interest rate framework. Rigorous computations are also provided according with appropriate assumptions on the jumps intensity shape.

Keywords Post-crisis spot LIBOR · OIS short-rate · Affine jump-diffusions Multi-curve models · Marked point processes

1 Financial Terminology and Notation

Before entering into details about the mathematical setup, we would like to recall, for the sake of completeness, some financial terminology and notation.

- A *Zero Coupon Bond*, or ZCB in short, is a financial contract which pays, by convention, an amount of one currency unit to its holder at maturity date $T > 0$. Its price at time $0 \leq t \leq T$ is denoted by $P(t, T)$. By risk-neutral valuation arguments, see, e.g., Björk (2009), it is possible to write the t-price of the ZCB in the following

L. Di Persio (✉)
Department of Computer Science, Head of the HPA (High Performance Analytics) Research & Development unit, University of Verona, Verona, Italy
e-mail: luca.dipersio@univr.it

N. Gugole
Department of Computer Science, University of Verona, Verona, Italy
e-mail: nicola_gugole@outlook.it

M. Mili et al. (eds.), *New Methods in Fixed Income Modeling*, Contributions to Management Science, https://doi.org/10.1007/978-3-319-95285-7_4

form

$$P(t,T) = \mathbb{E}_t^{\mathcal{Q}}\left[\exp\left(-\int_t^T r_u\, du\right)\right], \tag{1}$$

where $\mathbf{r} \doteq \{r_t\}_{0\le t\le T}$ is the so called *short-rate process* and $\mathbb{E}_t^{\mathcal{Q}}$ denotes the t-conditional expectation under a *martingale measure* $\mathcal{Q}$.

- The LIBOR is a financial benchmark which stands for *London InterBank Offered Rate*, and reflects the short term funding costs of the major banks active in London. It comes in 7 maturities (from overnight to 12 months) and in 5 different currencies (US Dollar, Euro, British Pound Sterling, Japanese Yen, Swiss Franc). For each currency, a specific panel of representative banks is selected. Panel banks are required to submit a rate in answer to the question: *At what rate could you borrow funds, were you to do so by asking for and then accepting inter-bank offers in a reasonable market size just prior to 11 a.m.?*[1] The LIBOR daily quotes for the various currencies and maturities are computed as a *trimmed arithmetic mean* of all of the panel banks submitted rates. We will denote by $L(t; T, T+\delta)$ the *LIBOR* at time $t \in [0, T]$ for the period $[T, T+\delta]$. The quantity $\delta > 0$ identifies the *tenor* of the interest rate. In particular, by taking $t = T$, we obtain the *T-spot LIBOR* $L(T; T, T+\delta)$ at the same tenor, which is the quantity of interest in the present chapter. In the pre-crisis setting, see, e.g., Björk (2009), we could write

$$L(t; T, T+\delta) = \frac{1}{\delta}\left(\frac{P(t,T)}{P(t,T+\delta)} - 1\right), \ \forall\, \delta > 0. \tag{2}$$

In the next points, we will discover that Eq. (2) is not valid anymore since the crisis.

- A *Forward Rate Agreement* (FRA) is an Over The Counter derivative which allows the holder to lock in at any date $0 \le t \le T$ the interest rate between the inception date T and the maturity $T+\delta > T$ at a fixed value R. At maturity $T+\delta$, a payment based on R is made and another one based on a floating rate, generally the T-spot LIBOR $L(T; T, T+\delta)$, is received. In the pre-crisis setting, see again Björk (2009), it was possible to express the t-value of the fair fixed rate R of the FRA as

$$R(t; T, T+\delta) = L(t; T, T+\delta) = \frac{1}{\delta}\left(\frac{P(t,T)}{P(t,T+\delta)} - 1\right). \tag{3}$$

- The crisis implied the emergence of spreads between the LIBORs at different tenors, see Fig. 1. As a consequence of this fact we should introduce, in contrast with Eq. (2), a distinct discount curve for every possible tenor $\delta_i > 0$ by imposing

[1] Visit https://www.theice.com/index for more details.

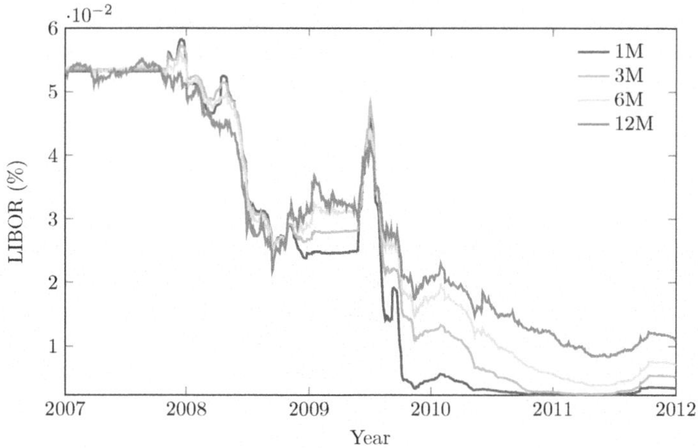

Fig. 1 Historical (daily) spot LIBOR measured at different tenors, in the period 2007–2011. The spread became particularly evident from the beginning of 2009. The data was downloaded from https://fred.stlouisfed.org

$$L(t;T,T+\delta_i) \doteq \frac{1}{\delta_i}\left(\frac{P^{\delta_i}(t,T)}{P^{\delta_i}(t,T+\delta_i)} - 1\right) \Rightarrow \frac{P^{\delta_i}(t,T)}{P^{\delta_i}(t,T+\delta_i)} = 1 + \delta_i L(t;T,T+\delta_i)\,, \tag{4}$$

where $P^{\delta_i}(t,T)$ is the t-price of a post-crisis bond with maturity T, affected by the same risk factors as for the LIBOR with tenor δ_i. We denote the δ_i-discount curve by $\{T \mapsto P^{\delta_i}(t,T),\ 0 \le t \le T\}$. An important point to underline is that only the quotient in the right hand of Eq. (4) is determined, i.e., there is no unique inverse relationship between the δ_i-LIBOR curve $\{T \mapsto L(t;T,T+\delta_i),\ 0 \le t \le T\}$ and the δ_i-discount curve, see, e.g., Ametrano and Bianchetti (2013) and Miglietta (2015). Clearly, the non uniqueness of the discount curve is an important issue and has many implications. For instance, we could treat each tenor market as separate from the others, imposing specific non-arbitrage conditions between them. However, in order to simplify the whole framework, and in line with a common approach in the post-crisis literature, we will construct a risk-free discount curve, called *Overnight Indexed Swap discount curve*, and we will consider it as common between all the tenors.

- An *Overnight Indexed Swap* (OIS) is a financial contract in which two counterparties exchange a stream of fixed rate payments for a stream of floating rate payments based on a notional amount, where the floating rate is obtained compounding in an appropriate way the reference overnight rate (the *Fed Funds rate* in the US), see, e.g., Grbac and Runggaldier (2015, Sect. 1.4.4) for more details,. The market swap rate of an OIS is simply called *OIS rate*. This is, usually, directly observable on the market. An OIS with a single payment date is in fact a *Forward Rate Agreement* (FRA), and we will refer to it using the shorthand OIS-FRA. Consider an OIS-FRA contract, having inception date T and maturity $T + \delta$. Following on from a pre-crisis fashion, we will assume that its fixed rate, i.e., the rate such that the value of the contract itself is null at time $t < T$, is given by

$$R^{OIS}(t; T, T + \delta) = \frac{1}{\delta}\left(\frac{P^{OIS}(t, T)}{P^{OIS}(t, T + \delta)} - 1\right), \ \forall \delta > 0,$$

where $P^{OIS}(t, T)$ stands for the price at time t of an *OIS bond* with maturity T. This bond has to be considered as hypothetical, since its price is, in the general case, the result of an averaging procedure applied on the OIS rates. As it is constructed starting from the reference overnight rate, which bears very little risk by its short life, it can be thought as a risk-free bond in the post-crisis setting. For this reason, we will consider the *OIS discount curve*, denoted by

$$\{T \mapsto P^{OIS}(t, T),\ 0 \le t \le T\},$$

as the common risk-free discount curve between all the tenors. By convention, we impose $P^{OIS}(T, T) = 1$.

- The *LIBOR-OIS spread* represents the difference between LIBORs and OIS rates related to the same date and tenor length. This difference can be measured at the spot level or at a forward one. As said in Thornton (2009), this spread is assumed to be a measure of the banks' health because it reflects what they believe is the risk of default associated with lending to other banks. The spot LIBOR-OIS spread related to various tenor lengths (1–3–6 months), remained approximately constant for long time, until August 9, 2007. In that date, it experienced a sharp rise, due to the well-known issues in the subprime mortgage market. In the following months, it fluctuated around a much higher level, until September 17, 2008, following the

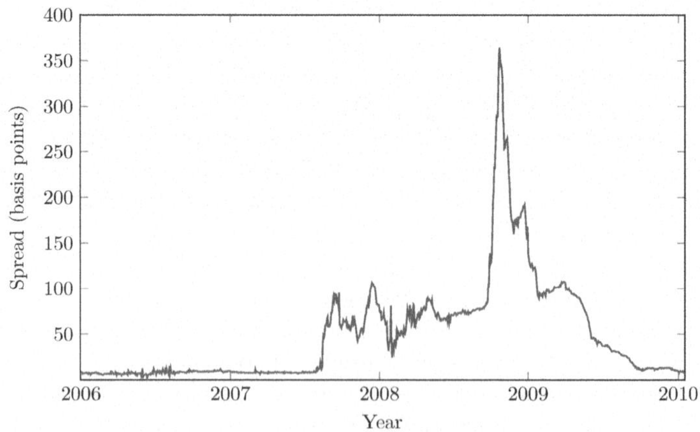

Fig. 2 Historical (daily) 3-month LIBOR-OIS spread. The data is from *Bloomberg*, whose access was kindly provided by the University of Verona

announcement of Lehman Brothers' bankruptcy. Figure 2 shows the evolution of the 3-month LIBOR-OIS spread along time. From a mathematical point of view, the LIBOR-OIS spread translates in the break of the classical relationship between the LIBOR and (post-crisis) risk-free bond prices, i.e.,

$$L(t; T, T+\delta) \neq \frac{1}{\delta}\left(\frac{P^{OIS}(t,T)}{P^{OIS}(t,T+\delta)} - 1\right) = R^{OIS}(t; T, T+\delta)\,,$$

from which, moving to the spot level,

$$L(T; T, T+\delta) \neq \frac{1}{\delta}\left(\frac{1}{P^{OIS}(T,T+\delta)} - 1\right) = R(T; T, T+\delta). \quad (5)$$

2 Introduction to the Problem

The recent financial crisis was characterized by a relevant growth of many interbank risks, such as the credit and liquidity risk. This fact gave rise to many spreads, even between the same interbank rate, e.g., the LIBOR, at different tenors, originating the so called *multi-curve* issue, clearly visible in Fig. 1. As seen in the previous section, a possible approach to manage this problem is to consider the OIS discount curve $\{T \mapsto P^{OIS}(t,T),\ 0 \leq t \leq T\}$ as the unique discount curve for all possible tenors. In line with Eq. (1), we assume

$$P^{OIS}(t,T) \doteq \mathbb{E}_t^{\mathbb{Q}}\left[\exp\left(-\int_t^T r_u^{OIS}\, du\right)\right], \quad 0 \leq t \leq T, \quad (6)$$

where the stochastic process $\{r_t^{OIS}\}_{t\geq 0}$, from now on simply $\{r_t\}_{t\geq 0}$, is the so called *OIS short-rate*. This quantity, even if not directly observable in the market, could be reasonably approximated by the reference overnight rate.

On the other hand, due to the LIBOR-OIS spread, the fixed income market experienced the break of the classical relationship between the spot LIBOR and the bond price, see Eq. (5). In order to face this issue, we will postulate a similar relationship, as done in Grbac and Runggaldier (2015),

$$L(T; T, T+\delta) \doteq \frac{1}{\delta}\left(\frac{1}{\overline{P}^{\delta}(T,T+\delta)} - 1\right), \quad (7)$$

where $\overline{P}^{\delta}(T, T+\delta)$ is the T-price of a fictitious risky bond supposed to be affected by the same risk factors as for the T-spot LIBOR with tenor δ. The value of these factors depends on the length of the tenor of interest; in general, one can assume that the longer is the tenor, the higher are the associated risks. In particular, we can incorporate them into $\overline{P}^{\delta}(T, T+\delta)$ by adding a tenor dependent *short-rate spread*

$\{s_t^\delta\}_{t\geq 0}$ to the OIS short-rate in formula (6), i.e., by defining

$$\overline{P}^\delta(t,T) \doteq \mathbb{E}_t^{\mathcal{Q}}\left[\exp\left(-\int_t^T \left(r_u + s_u^\delta\right)\,\mathrm{d}u\right)\right], \quad 0 \leq t \leq T. \tag{8}$$

An approach of this type is called *bottom-up*, since the OIS short-rate and the corresponding spread are the basic variables, used to model more complex objects, such as the OIS bonds, the risky bonds and the LIBORs, as we will see in Sect. 4.

The aim of this chapter, which retrieves elements from Grbac and Runggaldier (2015), Björk et al. (1997) and Runggaldier (2003), is to explicitly compute the post-crisis spot LIBOR related to some tenor $\delta > 0$, see Eq. (7), through the develop of a *multi-curve short-rate model*. This is an essential step in order to find the price of LIBOR-dependent fixed income derivatives. The choice to include both Brownian motions and marked point processes in the related dynamic has a clear empirical motivation: the underlying asset or interest rate trajectory of any financial instrument is not, in general, continuous but is characterized by the presence of jumps due to the incoming of new information about, e.g., companies' health, government decisions, natural phenomena, etc. Even by simply observing Figs. 1 and 2 we could get a first idea of how the LIBOR is empirically involved in this issue. We refer to Cont and Tankov (2004, Chap. 1) for a wide introduction on the topic. The inclusion of jumps represents an extension with respect to the diffusion framework developed in Grbac and Runggaldier (2015, Chap. 2), which is in turn based and/or extends previous works, see Filipović and Trolle (2013), Kijima et al. (2009), Kenyon (2010), Morino and Runggaldier (2014), and McNeil et al. (2015). However, before doing this we need some technical background concerning affine jump-diffusions, see Sect. 3, with the goal to generalize the diffusion framework of Grbac and Runggaldier (2015, Chap. 2).

3 Affine Jump-Diffusions

Let us start considering a filtered probability space $(\Omega, \mathcal{F}, \mathbf{F}, \mathcal{Q})$, where Ω is the space of events, $\mathbf{F} \doteq \{\mathcal{F}_t\}_{0\leq t\leq T}$ is a filtration made by sub-σ-algebras of $\mathcal{F}$, satisfying the condition $\mathcal{F}_T = \mathcal{F}$, and $\mathcal{Q}$ is a martingale measure.

In what follows, we give the notion of *marked point process*, see, e.g., Brémaud (1981) and Runggaldier (2003) for an exhaustive review on this topic.

Definition 3.1 Consider a measurable space $(E, \mathcal{E})$.[2] We define an *E-marked point process* as a double sequence $\boldsymbol{\mu} \doteq \{(T_n, Y_n)\}_{n\geq 1}$ where

[2]A measurable space is a couple $(E, \mathcal{E})$, where E is a non empty set and $\mathcal{E}$ is a σ-algebra of subsets of E.

- $\{T_n\}_{n\geq 1}$ is a univariate point process, namely a sequence of non-negative random variables $0 = T_0 < T_1 < T_2 < \dots$ with the assumption of non-explosion in finite time,
- $\{Y_n\}_{n\geq 1}$ is a sequence of E-valued random variables, called *marks*.

Fix $A \in \mathcal{E}$ and define the counting process $\{N(t, A)\}_{t\geq 0}$ by

$$N(t, A) \doteq \sum_{n\geq 1} \mathbb{1}_{\{T_n \leq t\}} \mathbb{1}_{\{Y_n \in A\}}. \tag{9}$$

Consider the filtration $\mathbf{F}^N \doteq \{\mathcal{F}_t^N\}_{t\geq 0}$ where $\mathcal{F}_t^N \doteq \sigma\{N(s, A), s \leq t, A \in \mathcal{E}\}$ for any $t \geq 0$. We define the *random counting measure associated to* $\mathbf{F}^N$ by

$$\mu\big((0, t], A\big) \doteq N(t, A), \quad t \geq 0, \ A \in \mathcal{E}.$$

We will often adopt the standard notation $N(t, A) = \int_0^t \int_A \mu(\mathrm{d}u, \mathrm{d}y)$, where $t \geq 0$, $A \in \mathcal{E}$.

In the following we will identify the double sequence $\boldsymbol{\mu}$ with the random counting measure $\mu(\mathrm{d}t, \mathrm{d}y)$, since they represent equivalent definitions for an E-marked point process. Assume now that for each $A \in \mathcal{E}$ the counting process $\{N(t, A)\}_{t\geq 0}$ in Eq. (9) has stationary independent increments. Then, according to Cont and Tankov (2004, Lemma 2.1), $\{N(t, A)\}_{t\geq 0}$ is a Poisson process and admits an intensity of type $\{\lambda(t, A)\}_{t\geq 0}$, representing the arrival rate on the set A over time. In mathematical terms, for each fixed $A \in \mathcal{E}$, $\delta t > 0$, the intensity $\{\lambda(t, A)\}_{t\geq 0}$ satisfies:

$$\mathcal{Q}\,(N(t + \delta t, A) - N(t, A) = 0) = 1 - \lambda(t, A)\delta t + o(\delta t),$$

$$\mathcal{Q}\,(N(t + \delta t, A) - N(t, A) = 1) = \lambda(t, A)\delta t + o(\delta t),$$

$$\mathcal{Q}\,(N(t + \delta t, A) - N(t, A) \geq 2) = o(\delta t).$$

This leads to a measure-valued intensity denoted by $\lambda(t, \mathrm{d}y)$ and we talk about *Poisson random counting measure* $\boldsymbol{\mu}$. From a practical point of view, the most common form of intensity process is

$$\lambda(t, \mathrm{d}y) = \lambda_t\, m(t, \mathrm{d}y),$$

where $\{\lambda_t\}_{t\geq 0}$ is a non-negative $\mathcal{F}_t$-predictable process and represents the intensity of the Poisson process $\{N(t, E)\}_{t\geq 0}$ which counts the jumps on the whole mark space, while $m(t, \mathrm{d}y)$ is a time-dependent probability measure on E, i.e., satisfies

$$\begin{cases} m(t, A) = \int_A m(t, \mathrm{d}y) \geq 0, \\ m(t, E) = \int_E m(t, \mathrm{d}y) = 1, \end{cases} \qquad A \in \mathcal{E}, \ 0 \leq t \leq T.$$

Assumption 3.2 Let us consider a $\mathcal{F}_t$-adapted process $\boldsymbol{\Psi} \doteq \{\Psi_t\}_{0\le t\le T}$. We assume that its dynamic under $\mathcal{Q}$ is given by

$$\begin{cases} d\Psi_t = \alpha(t, \Psi_t)\,dt + \beta(t, \Psi_t)\,dW_t + \int_E \gamma(t, \Psi_t, y)\mu(dt, dy), \\ \Psi|_{t=0} = \Psi_0 \in \mathbb{R}. \end{cases} \tag{10}$$

where the non-random functions $\alpha(t,\psi)$, $\beta(t,\psi)$, $\gamma(t,\psi,y)$ are sufficiently well behaved, $\mathbf{W} \doteq \{W_t\}_{0\le t\le T}$ is a $\mathcal{Q}$-Wiener process and $\boldsymbol{\mu}$ is a Poisson random counting measure related to some measurable mark space $(E, \mathcal{E})$, having a non-negative $\mathcal{F}_t$-predictable $\mathcal{Q}$-intensity $\lambda(t, dy)$, which is a deterministic measure for each value of t.

We make a little digression on the dynamic described by Eq. (10), especially on the relation with the classical notation for *jump-diffusion models*. First of all, starting from Definition 3.1, we can write

$$\int_0^t \int_E \gamma(u, \Psi_u, y)\mu(du, dy) = \sum_{n\ge1} \gamma(T_n, \Psi_{T_n}, Y_n)\mathbb{1}_{\{T_n\le t\}} = \sum_{n=1}^{N_t} \gamma(T_n, \Psi_{T_n}, Y_n),$$

where $N_t \doteq N(t, E)$ denotes, as said before, the Poisson process which counts the jumps on E and has intensity $\lambda(t, E) = \lambda_t$, since $m(t, E) = 1$. If we take the differential with respect to t in the previous equality we obtain

$$\int_E \gamma(t, \Psi_t, y)\mu(dt, dy) = \gamma(t, \Psi_t, Y_t)\,dN_t,$$

where $\{Y_t\}_{0\le t\le T}$ is a stochastic process obtained from the sequence $\{Y_n\}_{n\ge1}$ by a piecewise constant and left-continuous time interpolation, i.e.,

$$\gamma(t, \Psi_t, Y_t) \doteq \gamma(T_n, \Psi_{T_n}, Y_n), \quad t \in [T_n, T_{n+1}).$$

Now we can rewrite the dynamics for $\boldsymbol{\Psi}$, see Eq. (10), in an equivalent way as

$$\begin{cases} d\Psi_t = \alpha(t, \Psi_t)\,dt + \beta(t, \Psi_t)\,dW_t + \gamma(t, \Psi_t, Y_t)\,dN_t, \\ \Psi|_{t=0} = \Psi_0 \in \mathbb{R}. \end{cases}$$

The latter equation represents a generalization of the *Merton jump-diffusion model*, appeared for the first time in Merton (1976). If the sequence $\{Y_n\}_{n\ge1}$ is made by independent and identically distributed random variables with density function $f : E \to \mathbb{R}$ and $N_t \sim \text{Po}(\lambda_t t)$,[3] $\lambda_t > 0$, then we can express the intensity of $\boldsymbol{\mu}$ as

$$\lambda(t, dy) = \lambda_t f(y)\,dy\,. \tag{11}$$

[3]The shorthand Po indicates the Poisson distribution.

From now on we will assume this decomposition for the intensity. More details can be found for instance in Glasserman and Kou (2003) and Shreve (2004, Chap. 11), the latter proving such a special decomposition of the intensity for a compound Poisson process with a discrete mark space.

3.1 Exponentially Affine Term Structure

Definition 3.3 Let $T > 0$. A short-rate model is said to have an *exponentially affine term structure* if, for every $t \in [0, T]$, the ZCB price can be written in the form

$$P(t, T) = \exp\Big(A(t, T) - B(t, T)r_t\Big), \quad 0 \leq t < T, \; P(T, T) = 1,$$

for some deterministic functions $A(t, T)$ and $B(t, T)$.

It is evident that the property introduced in Definition 3.3 is very suitable in the context of interest rate models, since in general an analytical formula is simple to be implemented via software. The following theorem, which combines elements from Lamberton and Lapeyre (2007, Prop. 6.2.4), Grbac and Runggaldier (2015, Sect. 2.1.2) and Björk et al. (1997, Prop. 6.5), introduces a special setting in which the $\mathcal{F}_t$-adapted stochastic process $\boldsymbol{\Psi} \doteq \{\Psi_t\}_{0 \leq t \leq T}$, described in Assumption 3.2, leads to the exponentially affine term structure presented in Definition 3.3. In the next section we will call the same process in a more specific way, i.e., *exponentially affine factor*, since it will be used as a building block to construct a multi-factor short-rate process.

Theorem 3.4 *In the setting of Assumption* 3.2 *and Eq.* (11)*, suppose that*

$$\begin{cases} \alpha(t, \psi) \doteq a - b\psi, \quad a \geq 0, \; b > 0, \\ \beta(t, \psi) \doteq \sigma > 0, \\ \gamma(t, \psi, y) \doteq \gamma(t, y), \\ \lambda(t, \mathrm{d}y) \doteq \lambda_t f(y)\,\mathrm{d}y, \end{cases}$$

and that for every $t \in [0, T]$ there exist some well-defined deterministic functions $B(t, T)$ and $A(t, T)$ solving respectively

$$\begin{cases} \partial_t B(t, T) - bB(t, T) + \xi = 0, \\ B(T, T) = 0, \end{cases} \tag{12}$$

$$\begin{cases} \partial_t A(t, T) - aB(t, T) + \frac{1}{2}\sigma^2 B(t, T)^2 + \Theta\big(t, B(t, T)\big) = 0, \\ A(T, T) = 0, \end{cases} \tag{13}$$

where $\xi \in \mathbb{R}$ *and* [4]

$$\Theta(t,z) \doteq \lambda_t \int_E (e^{-\gamma(t,y)z} - 1) f(y)\,\mathrm{d}y, \quad z \in \mathbb{R}.$$

Then

$$\mathbb{E}_t^{\mathbb{Q}}\left[\exp\left(-\int_t^T \Psi_u\,du\right)\right] = \exp\Big(A(t,T) - B(t,T)\Psi_t\Big).$$

Proof See Lamberton and Lapeyre (2007, Prop. 6.2.4), Grbac and Runggaldier (2015, Sect. 2.1.2) and Björk et al. (1997, Prop. 6.5).

Remark 3.5 Notice that the choice of the diffusion coefficients α and β in the previous theorem, is consistent with the *Vasiček* model, which is well known in the context of exponentially affine models. For further details see Brigo and Mercurio (2007).

3.2 The Vasiček Model with Double Exponential Jumps

In the setting of Assumption 3.2, Eq. (11), and Theorem 3.4, suppose that $a \geq 0$, $b > 0$, and $\gamma(t,y) \doteq \gamma(y)$. Then we deal with the Vasiček model with jumps and constant coefficients (with respect to time), see also Remark 3.5, whose SDE is the following

$$\mathrm{d}\Psi_t = (a - b\Psi_t)\,\mathrm{d}t + \sigma\,\mathrm{d}W_t + \int_E \gamma(y)\mu(\mathrm{d}t,\mathrm{d}y). \tag{14}$$

An important property of this model is the so called *mean reversion*, i.e., the tendency to move (for sufficiently large times) around a *long term mean level*, more precisely the fixed constant a/b, which is well defined for $b > 0$ (notice that when $\Psi_t = a/b$ then the drift term in the related SDE is equal to 0).

Systems (12) and (13) become

$$\begin{cases} \partial_t B(t,T) - bB(t,T) + \xi = 0, \\ B(T,T) = 0, \end{cases}$$

and

$$\begin{cases} \partial_t A(t,T) - aB(t,T) + \frac{1}{2}\sigma^2 B(t,T)^2 + \lambda_t \int_E (e^{-\gamma(y)B(t,T)} - 1) f(y)\,\mathrm{d}y = 0, \\ A(T,T) = 0, \end{cases}$$

[4]The fact that $\Theta\big(t, B(t,T)\big) \in L^1\big([0,T];\mathbb{R}\big)$ is required in order to solve ODE (13) and find the function $A(t,T)$. In general it is not for free, and it will be achieved imposing restrictions on the parameters which characterize the intensity $\lambda(t,\mathrm{d}y)$.

so that after some computations we find that

$$B(t,T) = \frac{\xi}{b}(1 - e^{-b(T-t)}). \tag{15}$$

On the other hand we have to compute

$$\begin{aligned} A(t,T) &= -a\int_t^T B(u,T)\,\mathrm{d}u + \frac{\sigma^2}{2}\int_t^T B(u,T)^2\,\mathrm{d}u \\ &\quad + \int_t^T\int_E \lambda_u (e^{-\gamma(y)B(u,T)} - 1) f(y)\,\mathrm{d}y\,\mathrm{d}u \\ &\doteq A_1(t,T) + A_2(t,T) + A_3(t,T), \end{aligned} \tag{16}$$

where it is possible to find $A_1(t,T)$ and $A_2(t,T)$ explicitly, since they come directly from the classical diffusion framework, thus leading in general to a semi-analytical solution. In particular

$$\begin{aligned} A_1(t,T) &= a\frac{\xi}{b}\int_t^T e^{-b(T-u)}\,\mathrm{d}u - \frac{a\xi}{b}(T-t) \\ &= a\xi\frac{1-e^{-b(T-t)}}{b^2} - \frac{a\xi}{b}(T-t), \end{aligned}$$

while

$$A_2(t,T) = \sigma^2\xi^2\frac{1-e^{-2b(T-t)}}{4b^3} - \sigma^2\xi^2\frac{1-e^{-b(T-t)}}{b^3} + \frac{\sigma^2\xi^2}{2b^2}(T-t).$$

There exist some cases in which it is possible to compute the quantity $A_3(t,T)$ of Eq. (16) in an explicit way. This is the case for the *V-DEJ* (Vasiček with Double Exponential Jumps) model, as shown in the literature, see for instance (Das and Foresi 1996; Chacko and Das 2002). Consider Eq. (14) in which we choose $\gamma(y) \doteq \gamma y$, $\gamma \geq 0$, and we model the Poisson random counting measure $\boldsymbol{\mu}$ over the mark space $E \doteq \mathbb{R}$ by the measure-valued intensity

$$\begin{aligned} &\lambda(t,\mathrm{d}y) \doteq \lambda f(y)\mathrm{d}y\,, \\ &f(y) \doteq p\cdot\underbrace{\eta\exp\{-\eta y\}}_{f_U(y)}\mathbb{1}_{\{y\geq 0\}} + (1-p)\cdot\underbrace{\zeta\exp\{\zeta y\}}_{f_D(y)}\mathbb{1}_{\{y<0\}}\,, \end{aligned}$$

where $0 \leq p \leq 1$, $\lambda \geq 0$, and $\eta, \zeta > 0$.

Here p represents the probability of upward jumps, modeled by a density function $f_U(y)$ associated to a random variable $Y \sim \mathrm{Exp}(\eta)$, while $1-p$ is the probability of downward jumps, modeled by $f_D(y)$ which is the density of $-Y$, where $Y \sim \mathrm{Exp}(\zeta)$. If we set $p = 1$ then we obtain a model with only positive and exponentially distributed jumps and we will talk about *V-EJ+* (Vasiček with Exponential Jumps) model. The quantity $\lambda > 0$ represents the arrival rate for the Poisson process $N_t \doteq N(t, E)$. Now we can write

$$\int_{\mathbb{R}} \gamma(y)\, \mu(\mathrm{d}t, \mathrm{d}y) = \gamma\, Y_t\, \mathrm{d}N_t,$$

where the process $\{Y_t\}_{0 \le t \le T}$ is obtained from the sequence $\{(T_n, Y_n)\}_{n \ge 1}$ by a piecewise constant and left-continuous time interpolation, $\{Y_n\}_{n \ge 1}$ being an i.i.d. sequence with density $f(y)$. Then the model in Eq. (14) can be rewritten as

$$\mathrm{d}\Psi_t = (a - b\Psi_t)\, \mathrm{d}t + \sigma\, \mathrm{d}W_t + \gamma\, Y_t\, \mathrm{d}N_t\, .$$

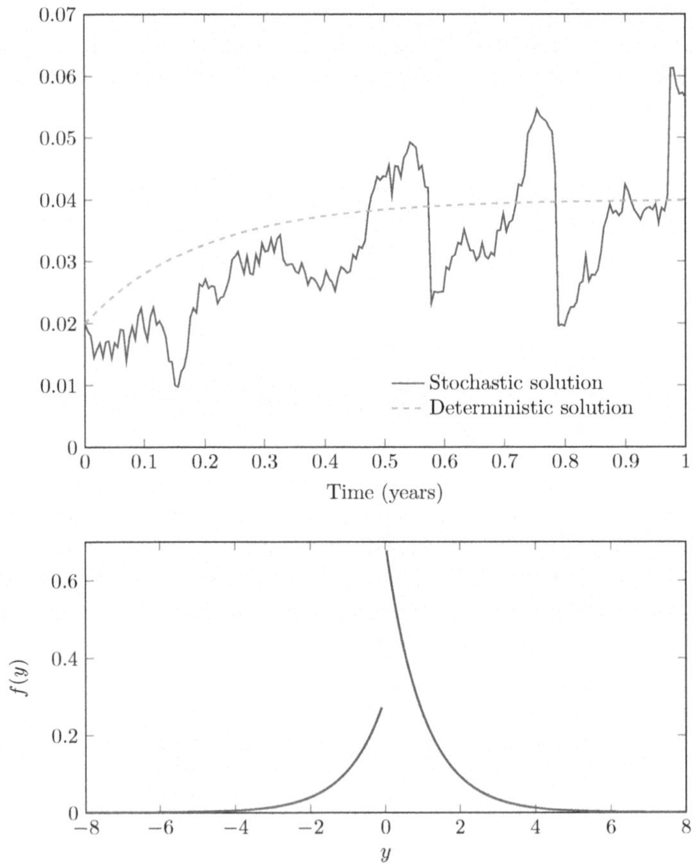

Fig. 3 Top: simulation of the V-DEJ model in comparison with the deterministic trend, with the following values for the parameters: $T = 1$ (horizon), $N = 400$ (number of time steps), $\mathrm{d}t = T/(N-1)$ (time step), $a = 0.2$, $b = 5$, $\sigma = 0.03$, $\lambda = 2$, $\gamma = 0.02$, $p = 0.7$, $\eta = 1$, $\zeta = 1$, $\Psi_0 = 0.02$. Bottom: plot of the intensity of the jumps. How we can see, upward jumps have greater probability to happen (fatter right tail)

Figure 3 shows a numerical simulation of the model, in comparison with the corresponding deterministic solution, i.e., the solution $\psi = \psi(t)$ of the differential equation

$$\frac{\mathrm{d}\psi(t)}{\mathrm{d}t} = a - b\,\psi(t)\,.$$

In particular, we have

$$\begin{aligned}
A_3(t,T) &= \int_t^T \Theta\big(u, B(u,T)\big)\,\mathrm{d}u = \lambda \int_t^T \int_{-+\infty}^{++\infty} (e^{-\gamma B(u,T)y} - 1) f(y)\,\mathrm{d}y\,\mathrm{d}u \\
&= (1-p)\lambda \int_t^T \int_{-+\infty}^{0} (e^{-\gamma B(u,T)y} - 1)\zeta e^{\zeta y}\,\mathrm{d}y\,\mathrm{d}u \\
&\quad + p\lambda \int_t^T \int_0^{++\infty} (e^{-\gamma B(u,T)y} - 1)\eta e^{-\eta y}\,\mathrm{d}y\,\mathrm{d}u \\
&= (1-p)\lambda\zeta \int_t^T \int_{-+\infty}^{0} e^{(\zeta-\gamma B(u,T))y}\,\mathrm{d}y\,\mathrm{d}u \\
&\quad + p\lambda\eta \int_t^T \int_0^{++\infty} e^{-(\eta+\gamma B(u,T))y}\,\mathrm{d}y\,\mathrm{d}u - \lambda(T-t) \\
&= (1-p)\lambda\zeta \int_t^T \frac{1}{\zeta - \gamma B(u,T)}\,\mathrm{d}u \\
&\quad + p\lambda\eta \int_t^T \frac{1}{\eta + \gamma B(u,T)}\,\mathrm{d}u - \lambda(T-t),
\end{aligned} \tag{17}$$

provided

$$\begin{aligned}
&\begin{cases} 0 < \zeta \neq \gamma B(u,T), \\ 0 < \eta \neq -\gamma B(u,T), \end{cases} \quad u \in [t,T] \\
&\iff \begin{cases} 0 < \zeta \notin \left[\gamma \min\limits_{u\in[t,T]} B(u,T), \gamma \max\limits_{u\in[t,T]} B(u,T)\right], \\ 0 < \eta \notin \left[-\gamma \max\limits_{u\in[t,T]} B(u,T), -\gamma \min\limits_{u\in[t,T]} B(u,T)\right], \end{cases}
\end{aligned} \tag{18}$$

in order to guarantee that $\Theta\big(u, B(u,T)\big)$ is well-defined and integrable in the interval $[t,T]$. Notice that

$$\frac{\mathrm{d}}{\mathrm{d}u} B(u,T) = -\xi \underbrace{e^{-b(T-u)}}_{>0},$$

which implies that

$$\min_{u\in[t,T]} B(u,T) = \begin{cases} B(t,T) = \dfrac{\xi}{b}(1-e^{-b(T-t)}), & \xi \le 0, \\ B(T,T) = 0, & \xi > 0, \end{cases}$$

$$\max_{u\in[t,T]} B(u,T) = \begin{cases} B(T,T) = 0, & \xi \le 0, \\ B(t,T) = \dfrac{\xi}{b}(1-e^{-b(T-t)}), & \xi > 0. \end{cases}$$

Hence (18) becomes

$$0 < \zeta \notin \begin{cases} [\gamma B(t,T), \gamma B(T,T)]\,, & \xi \le 0, \\ [\gamma B(T,T), \gamma B(t,T)]\,, & \xi > 0, \end{cases} \tag{19}$$

$$0 < \eta \notin \begin{cases} [-\gamma B(T,T), -\gamma B(t,T)]\,, & \xi \ge 0, \\ [-\gamma B(t,T), -\gamma B(T,T)]\,, & \xi > 0. \end{cases} \tag{20}$$

We compute the first integral in Eq. (17), in the case that $b\zeta \neq \gamma\xi$, as

$$\begin{aligned} \int_t^T \frac{1}{\zeta - \gamma B(u,T)}\,\mathrm{d}u &= \int_t^T \frac{b}{b\zeta - \gamma\xi + \gamma\xi e^{-b(T-u)}}\,\mathrm{d}u \\ &= -\frac{1}{b\zeta - \gamma\xi}\int_t^T \frac{-b(b\zeta - \gamma\xi)e^{b(T-u)}}{(b\zeta - \gamma\xi)e^{b(T-u)} + \gamma\xi}\,\mathrm{d}u \end{aligned}$$

$$\begin{aligned} &= -\frac{1}{b\zeta - \gamma\xi}\log\big((b\zeta - \gamma\xi)e^{b(T-u)} + \gamma\xi\big)\Big|_{u=t}^{u=T} \\ &= -\frac{1}{b\zeta - \gamma\xi}\log\left(\frac{b\zeta}{(b\zeta - \gamma\xi)e^{b(T-t)} + \gamma\xi}\right), \end{aligned}$$

while for $b\zeta = \gamma\xi$ (possible only when $\xi > 0$) we will write more simply

$$\int_t^T \frac{1}{\zeta - \gamma B(u,T)}\,\mathrm{d}u = -\frac{1}{\gamma\xi}\int_t^T -be^{b(T-u)}\,\mathrm{d}u = -\frac{1-e^{b(T-t)}}{\gamma\xi}.$$

For what concerns the second integral in Eq. (17) we can write, in the case $b\eta \neq -\gamma\xi$,

$$\begin{aligned}
\int_t^T \frac{1}{\eta+\gamma B(u,T)}\,\mathrm{d}u &= \int_t^T \frac{b}{b\eta+\gamma\xi-\gamma\xi e^{-b(T-u)}}\,du \\
&= -\frac{1}{b\eta+\gamma\xi}\int_t^T \frac{-b(b\eta+\gamma\xi)e^{b(T-u)}}{(b\eta+\gamma\xi)e^{b(T-u)}-\gamma\xi}\,du \\
&= -\frac{1}{b\eta+\gamma\xi}\log\big((b\eta+\gamma\xi)e^{b(T-u)}-\gamma\xi\big)\Big|_{u=t}^{u=T} \\
&= -\frac{1}{b\eta+\gamma\xi}\log\left(\frac{b\eta}{(b\eta+\gamma\xi)e^{b(T-t)}-\gamma\xi}\right).
\end{aligned}$$

while for $b\eta=-\gamma\xi$ (possible only when $\xi<0$) we will write more simply

$$\int_t^T \frac{1}{\eta+\gamma B(u,T)}\,du = -\frac{1}{\gamma\xi}\int_t^T be^{b(T-u)}\,\mathrm{d}u = -\frac{e^{b(T-t)}-1}{\gamma\xi}.$$

Summarizing, we get

$$A_3(t,T)=$$

$$\begin{cases}
-\lambda(T-t)-(1-p)\lambda\zeta\dfrac{1-e^{b(T-t)}}{\gamma\xi}-\dfrac{p\lambda\eta}{b\eta+\gamma\xi}\log\left(\dfrac{b\eta}{(b\eta+\gamma\xi)e^{b(T-t)}-\gamma\xi}\right), \\
\qquad b\zeta=\gamma\xi, \\
-\lambda(T-t)-\dfrac{(1-p)\lambda\zeta}{b\zeta-\gamma\xi}\log\left(\dfrac{b\zeta}{(b\zeta-\gamma\xi)e^{b(T-t)}+\gamma\xi}\right)-p\lambda\eta\dfrac{e^{b(T-t)}-1}{\gamma\xi}, \\
\qquad b\eta=-\gamma\xi, \\
-\lambda(T-t)-\dfrac{(1-p)\lambda\zeta}{b\zeta-\gamma\xi}\log\left(\dfrac{b\zeta}{(b\zeta-\gamma\xi)e^{b(T-t)}+\gamma\xi}\right) \\
-\dfrac{p\lambda\eta}{b\eta+\gamma\xi}\log\left(\dfrac{b\eta}{(b\eta+\gamma\xi)e^{b(T-t)}-\gamma\xi}\right), \\
\qquad b\zeta\neq\gamma\xi\wedge b\eta\neq-\gamma\xi.
\end{cases}$$

Notice that the quantity $A_3(t,T)$ satisfies the boundary condition $A_3(T,T)=0$.

4 A Model for the Post-crisis Spot LIBOR

After the necessary technical digression in Sect. 3, we are ready to develop the framework introduced in Sect. 2. The aim is to find an explicit formulation for the post-crisis spot LIBOR, in a jump-diffusion context.

From now on, the dynamics of all the involved quantities will be specified directly on a filtered probability space $(\Omega,\mathcal{F},\mathbf{F},\mathcal{Q})$, where Ω is the space of events, $\mathbf{F}\doteq\{\mathcal{F}_t\}_{0\le t\le T}$ is a filtration satisfying $\mathcal{F}_T=\mathcal{F}$ and $\mathcal{Q}$ is a martingale measure. In order to develop the model, and for the sake of simplicity, we will consider a single tenor $\delta>0$ and the associated *short-rate spread process* $\mathbf{s}^\delta\doteq\{s_t^\delta\}_{0\le t\le T}$, which we will add to the OIS short-rate $\mathbf{r} \doteq \{r_t\}_{0\le t\le T}$ to model the dynamics of a fictitious *post-crisis risky bond*, as seen in Eq. (8). The latter terminology reflects the fact that the spread $\mathbf{s}^\delta$ takes into account all the various risks affecting the interbank sector in the post-crisis setting, as the credit and the liquidity ones: intuitively, if the value of $\mathbf{s}^\delta$ increases over time, the corresponding value of the aforementioned risky bond will decrease because of the negative exponential. We consider also a correlation between $\mathbf{r}$ and $\mathbf{s}^\delta$, see, e.g., Grbac and Runggaldier (2015), modeling it by the following multi-factorial dynamics

$$\begin{cases} r_t \doteq \sum_{i=1}^{n} \Psi_t^i\,, \\ s_t^\delta \doteq \sum_{i=1}^{m} \rho^i \Psi_t^i + \sum_{i=n+1}^{l} \Psi_t^i\,, \end{cases} \tag{21}$$

where $m \le n < l$. We assume that, for all $1 \le i \le l$, the factor $\mathbf{\Psi}^i \doteq \{\Psi_t^i\}_{0\le t\le T}$, satisfies the following general jump-diffusion dynamics

$$\begin{cases} \mathrm{d}\Psi_t^i = \alpha^i(t, \Psi_t^i; \vartheta^i)\,\mathrm{d}t + \beta^i(t, \Psi_t^i; \vartheta^i)\,\mathrm{d}W_t^i + \int_{E^i} \gamma^i(t, y; \vartheta^i)\mu^i(\mathrm{d}t, \mathrm{d}y), \\ \Psi^i|_{t=0} \doteq \Psi_0^i \in \mathbb{R}, \end{cases}$$

where ϑ^i is the vector of all parameters concerning the i-th factor, $\mathbf{W}^i \doteq \{W_t^i\}_{0\le t\le T}$ is a $\mathcal{Q}$-Brownian motion, $\boldsymbol{\mu}^i$ is an E^i-marked point process having a predictable $\mathcal{Q}$-intensity $\lambda^i(t, \mathrm{d}y; \vartheta^i)$, and E^i is related to some real measurable space $(E^i, \mathcal{E}^i)$. Both $\mathbf{W}^i$ and $\boldsymbol{\mu}^i$ will be $\mathcal{F}_t$-adapted for every i. Moreover we suppose that all the involved processes are independent, so that $\mathbf{\Psi}^i$ is independent of $\mathbf{\Psi}^j$ for every $i \ne j$, and that the functions $\alpha^i(t, \psi; \vartheta^i)$, $\beta^i(t, \psi; \vartheta^i)$, $\gamma^i(t, \psi, y; \vartheta^i)$ are sufficiently well-behaved with respect to all the entries. We will drop the dependence on ϑ^i when not needed, in order simplify the notation. For all $1 \le i \le m$, $\rho^i \in \mathbb{R}$ represents a sort of correlation coefficient with the scope to link $\mathbf{s}^\delta$ to the i-th factor of $\mathbf{r}$. We define for $1 \le i \le l$,

$$\alpha^i(t, \psi) \doteq a^i - b^i\psi, \quad a^i \ge 0,\ b^i > 0, \tag{22}$$

$$\beta^i(t, \psi) \doteq \sigma^i > 0, \tag{23}$$

where all the coefficients are constant, in particular they do not depend on time, such a choice leading to a gain in analytical tractability. It is worth to mention that time-dependent coefficients would allow for a better fit with respect to the observed term structure of interest rates, but at the cost of a rather problematic calibration procedure.

The choices for $\alpha^i(t, \psi)$ and $\beta^i(t, \psi)$ in Eqs. (22) and (23), are consistent with the well-known *Vasiček* model, which is commonly used when talking about interest rate modeling, because of some of its properties, as mean-reversion. For further details see Brigo and Mercurio (2007).

Moving to the jump part, we suppose that $\boldsymbol{\mu}^i$ is a *Poisson random measure* defined by

$$\mu^i\big((0,t],A\big) \doteq N^i(t,A), \quad 0 \le t \le T,\ A \in \mathcal{E}^i,$$

where $\{N^i(t,A)\}_{0\le t\le T}$ is a Poisson process having intensity

$$\lambda^i(t,A) \doteq \lambda^i_t \int_A f^i(y)\,\mathrm{d}y, \tag{24}$$

where $\{\lambda^i_t\}_{t\ge 0}$ is a non-negative $\mathcal{F}_t$-predictable process and represents the intensity of the Poisson process $\{N^i(t,E)\}_{0\le t\le T}$, while $f^i : E^i \to \mathbb{R}$ is some probability measure which models the distribution of the jumps. In general, it is not possible to obtain such a decomposition for every jump process, but this is the case for a compound Poisson process, see Glasserman and Kou (2003) and Shreve (2004), the latter proving such a special property of the intensity for a compound Poisson process with a discrete mark space.

Remark 4.1 In order to treat more than a single tenor, let us say δ^1 and $\delta^2 > \delta^1$, we should rewrite system (21) as follows:

$$\begin{cases} r_t \doteq \sum_{i=1}^{n} \Psi^i_t\,, \\ s^{\delta^1}_t \doteq \sum_{i=1}^{m^1} \rho^{1,i}\,\Psi^i_t + \sum_{i=n+1}^{l^1} \Psi^i_t\,, \\ s^{\delta^2}_t \doteq s^{\delta^1}_t + \sum_{i=1}^{m^2} \rho^{2,i}\,\Psi^i_t + \sum_{i=l^1+1}^{l^2} \Psi^i_t\,, \end{cases}$$

where $m^1, m^2 \le n < l^1 < l^2$ and $s^{\delta^2}_t - s^{\delta^1}_t$, which is the difference between the spreads for tenors δ^2 and δ^1, should be modeled as a positive quantity, reflecting the fact that, in general, a longer tenor brings with it more risks than a shorter one.

Now, recalling Theorem 3.4 we know that if for every $t \in [0,T)$ there exist well-defined deterministic functions $B^i(t,T;\xi^i)$ and $A^i(t,T;\xi^i)$ solving respectively[5]

$$\begin{cases} \partial_t B^i(t,T) - b^i B^i(t,T) + \xi^i = 0, \\ B^i(T,T) = 0, \end{cases} \tag{25}$$

and

$$\begin{cases} \partial_t A^i(t,T) - a^i B^i(t,T) + \dfrac{1}{2}(\sigma^i)^2\big(B^i(t,T)\big)^2 + \Theta^i\big(t, B^i(t,T)\big) = 0, \\ A^i(T,T) = 0, \end{cases} \tag{26}$$

$$\Theta^i(t,z) \doteq \lambda^i_t \int_{E^i} (e^{-\gamma^i(t,y)z} - 1) f^i(y)\mathrm{d}y\,,$$

then we have

[5] We will often omit the dependence on ξ^i.

$$\mathbb{E}_t^{\mathbb{Q}}\Big[\exp\Big(-\int_t^T \xi^i \Psi_u^i du\Big)\Big] = \exp\Big(A^i(t,T) - B^i(t,T)\Psi_t^i\Big). \tag{27}$$

Assumption 4.2 From now on we will suppose that every factor $\mathbf{\Psi}^i$, $i = 1, \dots, l$, leads to an exponentially affine term structure as in Eq. (27).

Given Assumption 4.2, we have that systems (25) and (26) hold for $1 \le i \le n$ with $\xi^i = 1$, and so Eq. (6) for the price of an OIS bond can be rewritten, exploiting the independence of the factors and defining

$$A(t,T) \doteq \sum_{i=1}^{n} A^i(t,T;\xi^i),$$

as follows

$$\begin{aligned} P^{OIS}(t,T) &= \mathbb{E}_t^{\mathbb{Q}}\Big[\exp\Big(-\sum_{i=1}^{n}\int_t^T \Psi_u^i\, du\Big)\Big] \\ &= \prod_{i=1}^{n} \mathbb{E}_t^{\mathbb{Q}}\Big[\exp\Big(-\int_t^T \Psi_u^i\, du\Big)\Big] \\ &= \exp\Big(A(t,T) - \sum_{i=1}^{n} B^i(t,T)\Psi_t^i\Big). \end{aligned}$$

At this point, we can model the post-crisis spot LIBOR. We recall that, differently from the pre-crisis framework, the classical relationship between the spot LIBOR for the period $[T, T+\delta]$ and the bond price $P^{OIS}(T, T+\delta)$ generally does not hold, see Eq. (5), i.e.,

$$L(T;T,T+\delta) \neq \frac{1}{\delta}\Big(\frac{1}{P^{OIS}(T,T+\delta)} - 1\Big),$$

because of the so called LIBOR-OIS spread, which incorporates various interbank risks such as the credit risk. In order to face such a problem we adopt the same strategy used in Grbac and Runggaldier (2015).

Assumption 4.3 We postulate the relationship

$$L(T;T,T+\delta) \doteq \frac{1}{\delta}\Big(\frac{1}{\overline{P}^{\delta}(T,T+\delta)} - 1\Big), \tag{28}$$

where $\overline{P}^{\delta}(t,T)$ is a fictitious risky bond supposed to be affected by the same risk factors as for the T-spot LIBOR with tenor δ.

As anticipated by Eq. (8), we assume

$$\overline{P}^{\delta}(t, T) \doteq \mathbb{E}_t^{\mathbb{Q}}\left[\exp\left(-\int_t^T (r_u + s_u^{\delta})\, du\right)\right], \tag{29}$$

where the process $\mathbf{s}^{\delta}$ is usually known in the literature as *hazard rate*, or *default intensity*. We use the same relationship in Eq. (29) to model the price of post-crisis risky bonds, however as anticipated before, in our context $\mathbf{s}^{\delta}$ denotes, more generally, a short-rate spread. Starting from Eq. (29) and inserting the multifactorial form of $\mathbf{r}$ and $\mathbf{s}^{\delta}$ from system (21), we obtain

$$\begin{aligned} \overline{P}^{\delta}(t, T) &= \mathbb{E}_t^{\mathbb{Q}}\left[\exp\left(-\sum_{i=1}^{m}\int_t^T (1+\rho^i)\Psi_u^i\, du - \sum_{i=m+1}^{l}\int_t^T \Psi_u^i\, du\right)\right], \\ &= \exp\left(\overline{A}(t, T) - \sum_{i=1}^{l}\overline{B}^i(t, T)\Psi_t^i\right), \end{aligned} \tag{30}$$

exploiting the independence of the factors, where

$$\overline{A}(t, T) \doteq \sum_{i=1}^{l}\overline{A}^i(t, T; \bar{\xi}^i),$$

with

$$\bar{\xi}^i = \begin{cases} 1+\rho^i, & 1 \le i \le m, \\ 1, & m+1 \le i \le l. \end{cases}$$

Then, inserting Eq. (30) in Eq. (28), it is possible to obtain the desired representation for the post-crisis spot LIBOR,

$$L(T; T, T+\delta) = \frac{1}{\delta}\left[\exp\left(-\overline{A}(T, T+\delta) + \sum_{i=1}^{l}\overline{B}^i(T, T+\delta)\Psi_T^i\right) - 1\right]. \tag{31}$$

In the next subsection we will see a case in which Eq. (31) can be computed in an explicit way, making some specific assumptions on the jump component.

4.1 V-DEJ/EJ+ Model

In this subsection we propose a *multi-curve short-rate model* with jumps which leads to an analytical formula for the OIS bond prices and subsequently for the post-crisis spot LIBOR.

The system of factors, simplifying the general structure introduced in Eq. (21) for the sake of clarity, has the following form

$$\begin{cases} r_t \doteq \Psi_t^1 \,, \\ s_t \doteq \rho \Psi_t^1 + \Psi_t^2 \,, \end{cases}$$

where $\boldsymbol{\Psi}^i$, $i = 1, 2$, satisfy the dynamics

$$\begin{aligned} \mathrm{d}\Psi_t^i &= (a^i - b^i \Psi_t^i)\,\mathrm{d}t + \sigma^i \,\mathrm{d}W_t^i + \gamma^i \int_{E^i} y\, \mu^i(\mathrm{d}t, \mathrm{d}y) \\ &= (a^i - b^i \Psi_t^i)\,\mathrm{d}t + \sigma^i \,\mathrm{d}W_t^i + \gamma^i \, Y_t^i \,\mathrm{d}N_t^i \,, \end{aligned}$$

where $a^i \geq 0, b^i > 0, \gamma^i \geq 0, E^1 = \mathbb{R}, E^2 = \mathbb{R}^+, N_t^i \sim \mathrm{Po}(\lambda^i t)$ and $\{Y_t^i\}_{0 \leq t \leq T}$ is a stochastic process obtained from the sequence $\{(T_n^i, Y_n^i)\}_{n \geq 1}$ (associated to the Poisson random measure $\boldsymbol{\mu}^i$) by a piecewise constant and left-continuous time interpolation, where $\{Y_n^i\}_{n \geq 1}$ is i.i.d. and has density

$$f^i(y) \doteq \begin{cases} p^1 \cdot \eta^1 \exp\{-\eta^1 y\} \mathbb{1}_{\{y \geq 0\}} + (1 - p^1) \cdot \zeta^1 \exp\{\zeta^1 y\} \mathbb{1}_{\{y < 0\}}, & i = 1, \\ \eta^2 \exp\{-\eta^2 y\} \mathbb{1}_{\{y \geq 0\}}, & i = 2, \end{cases}$$

with $\zeta^1, \eta^1, \eta^2 > 0$, in particular $0 < p^1 < 1$ and $1 - p^1$ denoting respectively the probability of upward and downward jumps for $\boldsymbol{\Psi}^1$. We allow only positive jumps for $\boldsymbol{\Psi}^2$, a choice which helps to diminish its probability to reach negative values. Recall that the short-rate spread process $\mathbf{s}^\delta \doteq \{s_t\}_{0 \leq t \leq T}$ should be a non-negative quantity. We can imagine $p^2 = 1$. We deal with the measure valued intensity $\lambda^i(\mathrm{d}y) = \lambda^i f^i(y)\,\mathrm{d}y$ for $\boldsymbol{\mu}^i$.

In other words $\boldsymbol{\Psi}^1$ is a V-DEJ model and $\boldsymbol{\Psi}^2$ a V-EJ+ model, according to the terminology used in Sect. 3.2. Then we write the price of the OIS bond as

$$P^{OIS}(t, T) = \mathbb{E}_t^{\mathbb{Q}}\left[\exp\left(-\int_t^T \Psi_u^1 \,du\right)\right] = \exp\left(A^1(t, T) - B^1(t, T)\Psi_t^1\right), \quad (32)$$

$$B^1(t, T) \doteq \frac{1 - e^{-b^1(T-t)}}{b^1},$$

$$A^1(t, T) \doteq A_1^1(t, T) + A_2^1(t, T) + A_3^1(t, T),$$

where $A^1(t, T)$ has the same shape of Eq. (16), the superscript 1 relates with the first factor. According to Sect. 3.2, the model leads to a completely explicit framework, under some suitable conditions on the parameters. In particular, in order to compute the quantity $A_3^1(t, T)$ in $P^{OIS}(t, T)$, we have to impose

$$\begin{cases} \zeta^1 > \gamma^1 B^1(t, T), \\ \eta^1 > 0, \end{cases}$$

a condition which can be obtained from Eqs. (19) and (20), noticing that $\xi^1 = 1$ and $B^1(T,T) = 0$.
The other quantity of interest is the price of the risky bond, given by

$$\begin{aligned} \overline{P}^{\delta}(t,T) &= \mathbb{E}_t^{\mathcal{Q}}\left[\exp\left(-\int_t^T (1+\rho)\Psi_u^1\,du - \int_t^T \Psi_u^2\,du\right)\right], \\ &= \exp\left(\overline{A}(t,T) - \sum_{i=1,2} \overline{B}^i(t,T)\Psi_t^i\right), \end{aligned} \tag{33}$$

where

$$\overline{B}^1(t,T) = (1+\rho)\frac{1-e^{-b^1(T-t)}}{b^1}, \quad \overline{B}^2(t,T) = \frac{1-e^{-b^2(T-t)}}{b^2} \geq 0,$$

while

$$\overline{A}(t,T) \doteq \overline{A}^1(t,T) + \overline{A}^2(t,T) = \overline{A}^1(t,T;1+\rho) + \overline{A}^2(t,T;1).$$

$$\overline{A}^1(t,T) = \overline{A}^1_1(t,T) + \overline{A}^1_2(t,T) + \overline{A}^1_3(t,T),$$

$$\overline{A}^2(t,T) = \overline{A}^2_1(t,T) + \overline{A}^2_2(t,T) + \overline{A}^2_3(t,T),$$

similarly to Eq. (16). Notice that ζ^1, η^1 need to satisfy conditions as in (19) and (20), with adapted coefficients, more precisely,

$$\zeta^1 > \begin{cases} \gamma^1\overline{B}^1(t,T), & \rho > -1, \\ 0, & \rho \leq -1, \end{cases}$$

$$\eta^1 > \begin{cases} 0, & \rho > -1, \\ -\gamma^1\overline{B}^1(t,T), & \rho \leq -1. \end{cases}$$

We have no conditions on η^2, indeed $\overline{B}^2(t,T) \geq 0$. In particular, to compute the T-spot LIBOR $L(T;T,T+\delta)$ in Eq. (31) we will need

$$\zeta^1 > \begin{cases} \gamma^1\overline{B}^1(T,T+\delta) = \gamma^1(1+\rho)\dfrac{1-e^{-b^1\delta}}{b^1}, & \rho > -1, \\ 0, & \rho \leq -1, \end{cases}$$

$$\eta^1 > \begin{cases} 0, & \rho > -1, \\ -\gamma^1\overline{B}^1(T,T+\delta) = -\gamma^1(1+\rho)\dfrac{1-e^{-b^1\delta}}{b^1}, & \rho \leq -1. \end{cases}$$

These constraints depend only on δ and not on T.

5 Conclusions

We have seen that it is possible to explicitly compute the post-crisis spot LIBOR at tenor δ in a jump-diffusion framework, obtaining a first extension to the work contained in Grbac and Runggaldier (2015, Chap. 2). However, in general, this is not for free, and some assumptions on the shape of the jumps intensity, as well as conditions on the related parameters, are required, see Sect. 4.1. Recent works, see Bormetti et al. (2017), have also suggested to treat spot Libor rates as market primitives rather than being defined by no-arbitrage relationships, also pointing out limitations of multiple curve models with deterministic basis, particularly when dealing with sensitive products, as in the case of basis swaps, see also Fanelli (2016). We aim at further develop the present work, particularly following the line adopted in Grbac and Runggaldier (2015, Chap. 3), to compute even more complex quantities, such as the forward LIBOR $L(t; T, T + \delta)$ with tenor δ. The latter can be thought as the market expectation of the spot LIBOR under a suitable forward measure. It is worth to mention that its computation is an essential step to price many interest rate derivatives as Forward Rate Agreements, Floating Rate Bonds, Interest Rate Swaps, etc., also in view of different applications, as in the case of energy markets, see, e.g., Hikspoors and Jaimungal (2007). We remark that, differently from the pre-crisis setting, we have to take care about the simultaneous presence of various interest rate curves, known as the multi-curve issue, which implies that the pricing procedure must be adapted accordingly.

Acknowledgements We would like to thank Prof. Wolfgang J. Runggaldier (University of Padova) for the useful suggestions and comments about the present work.

References

Ametrano, F., & Bianchetti, M. (2013). Everything you always wanted to know about multiple interest rate curve bootstrapping but were afraid to ask.

Björk, T. (2009). *Arbitrage theory in continuous time*. Oxford: OUP.

Björk, T., Kabanov, Y., & Runggaldier, W. J. (1997). Bond market structure in the presence of marked point processes. *Mathematical Finance*, *7*(2), 211–239.

Bormetti, G., Brigo, D., & Francischello, M., Pallavicini. A. (2017). Impact of multiple curve dynamics in credit valuation adjustments under collateralization. *Quantitative Finance*, 1–14.

Brémaud, P. (1981). *Point processes and queues*. Springer: Halsted Press.

Brigo, D., & Mercurio, F. (2007). *Interest rate models—Theory and practice: With smile, inflation and credit*. Springer.

Chacko, G., & Das, S. (2002). Pricing interest rate derivatives: A general approach. *The Review of Financial Studies*, *15*(1), 195–241.

Cont, R., & Tankov, P. (2004). *Financial modelling with jump processes*. Chapman & Hall.

Das, S. R., & Foresi, S. (1996). Exact solutions for bond and option prices with systematic jump risk. *Review of Derivatives Research*, *1*(1), 7–24.

Fanelli, V. (2016). A defaultable HJM modelling of the libor rate for pricing basis swaps after the credit crunch. *European Journal of Operational Research*, *249*(1), 238–244.

Filipović, D., & Trolle, A. B. (2013). The term structure of interbank risk. *Journal of Financial Economics*, *109*(3), 707–733.

Glasserman, P., & Kou, S. G. (2003). The term structure of simple forward rates with jump risk. *Mathematical Finance*, *13*(3), 383–410.

Grbac, Z., & Runggaldier, W. J. (2015). *Interest rate modeling: Post-crisis challenges and approaches*. SpringerBriefs in quantitative finance: Springer.

Hikspoors, S., & Jaimungal, S. (2007). Energy spot price models and spread options pricing. *International Journal of Theoretical and Applied Finance*, *10*(7), 1111–1135.

Kenyon, C. (2010). Short-rate pricing after the liquidity and credit shocks: Including the basis.

Kijima, M., Tanaka, K., & Wong, T. (2009). A multi-quality model of interest rates. *Quantitative Finance*, *9*(2), 133–145.

Lamberton, D., & Lapeyre, B. (2007). *Introduction to stochastic calculus applied to finance*. CRC Press.

McNeil, A. J., Frey, R., & Embrechts, P. (2015). *Quantitative risk management: Concepts, techniques and tools*. Princeton University Press.

Merton, R. C. (1976). Option pricing when underlying stock returns are discontinuous. *Journal of financial economics*, *3*(1–2), 125–144.

Miglietta. G. (2015). *Topics in Interest Rate Modeling* (Ph.D. thesis). Università di Padova.

Morino, L., & Runggaldier, W. J. (2014). On multicurve models for the term structure. In *Nonlinear economic dynamics and financial modelling* (pp. 275–290). Springer.

Runggaldier, W. J. (2003). Jump-diffusion models. In *Handbook of heavy tailed distributions in finance*, (Vol. 1, pp. 169–209).

Shreve, S. (2004). *Stochastic calculus for finance II: Continuous-time models*. Number v. 11 in Springer Finance Textbooks. Springer.

Thornton, D. (2009). What the LIBOR-OIS spread says. *Economic Synopses*.

An Overview of Post-crisis Term Structure Models

Marcus R. W. Martin

Abstract This chapter is intended to provide an overview of state-of-the-art term structure models used in pricing and risk managing interest rate dependent financial products as well as forecasting interest rates in economic scenario generators for market and counterparty credit risk management purposes. After introducing a general overview of the post-crisis markets environment we will provide insight into post-crisis modelling of term structures via short rate models and Libor Market models for multiple curves and show how these models are applied in economic scenario generators used for risk management and pricing purposes alike.

Keywords Post-crisis term structure models · Short rate models for multiple curve pricing · LIBOR market models for multiple curve pricing · Economic scenario generators · Model risk

1 Introduction

The beginnings of term structure modelling date back to the late 1970s with the famous Vasicek equilibrium model as important initial starting point. This development cumulated in the well-known Heath-Jarrow-Morton framework, Markov Functional Models and Libor/Swap Market Models of the late 1990s. Clearly, particular attention has to be given to these "classical" or "pre-crisis" term structure models in a low or even negative interest rate environment we observe by the time of writing this book: More precisely, the global financial crisis (GFC) of 2007 and 2008 as well as the Eurozone sovereign debt crisis in 2011 and 2012 irreversibly changed not only the mechanics of how markets work but also the theoretical framework under which these markets have to be described and modelled.

M. R. W. Martin (✉)
Department of Mathematics, Natural, and Computer Sciences (MND), Technische Hochschule Mittelhessen - University of Applied Sciences,
Wilhelm-Leuschner-Straße 13, 61169 Friedberg, Germany
e-mail: marcus.martin@mnd.thm.de

M. Mili et al. (eds.), *New Methods in Fixed Income Modeling*, Contributions to Management Science, https://doi.org/10.1007/978-3-319-95285-7_5

Historically, the immense impact of the fundamental and dramatic changes in the fixed income markets on risk management was first observed at the beginning of the GFC. At that time, a horrendous number of Value at Risk outliers was reported by several large banks using regulatory approved internal market risk models. In particular, a whole bunch of outliers (also called VaR exceptions) were due to realized basis risks in interest rate driven financial products which could be interpreted as manifestations of counterparty credit, liquidity and funding risks. E.g., UBS had to report 29 VaR outliers for 2007 (on a 99% confidence level) while Bear Stearns experienced 27 VaR exceptions (on a 95% confidence level) during the last year as an independent entity (cf. Table 1 in Jorion 2009, p. 931).

The different pricing paradigms already show significant differences in the present values and risk sensitivities even for simple products which necessitates a change of the fundamental concept: Instead of one single term structure curve multiple curves have to be introduced reflecting the different levels of counterparty credit, liquidity and funding risks observed in the fixed income markets.

Hence, this chapter is organized as follows: In the following section we will discuss the post-crisis approaches for pricing plain vanilla interest products (like forwards or swaps) taking into account the post-crisis multiple curve approach. Since these plain vanilla (linear) derivatives have an immediate impact on the way nonlinear derivatives have to be priced and hedged (risk managed), we provide an overview of different post-crisis term structure modelling approaches in the following section. Furthermore, the consequences on Economic Scenario Generators were discussed before we conclude this chapter with an outlook.

2 Post-crisis Interest Markets: Single- Versus Multi-curve Universe

When considering the annualized 3-month- and 6-month-deposit rates by 12th of November, 2008, given by $L_{3m} = 4.286\%$ and $L_{6m} = 4.286\%$, classical pre-crisis no-arbitrage arguments imply that the identity

$$\left(1 + \frac{1}{4} L_{3m}\right)^{-1} \cdot \left(1 + \frac{1}{4} F(0, 3m, 6m)\right)^{-1} = \left(1 + \frac{1}{2} L_{6m}\right)^{-1}$$

(using 30/360 day count convention for simplicity) has to hold for the forward rate $F(0, 3m, 6m)$. While simple calculation yields $F(0, 3m, 6m) = 4.357\%$, the FRA rate quoted on 12th of November was at 2.85%, i.e. more than 150 basis points below the calculated forward rate (cf. Mercurio 2009; Morini 2013): Why was this arbitrage opportunity not exploited?

Following Grbac and Runggaldier (2015), the recent financial crisis has led to paradigm shifting events in interest rate markets because substantial spreads have

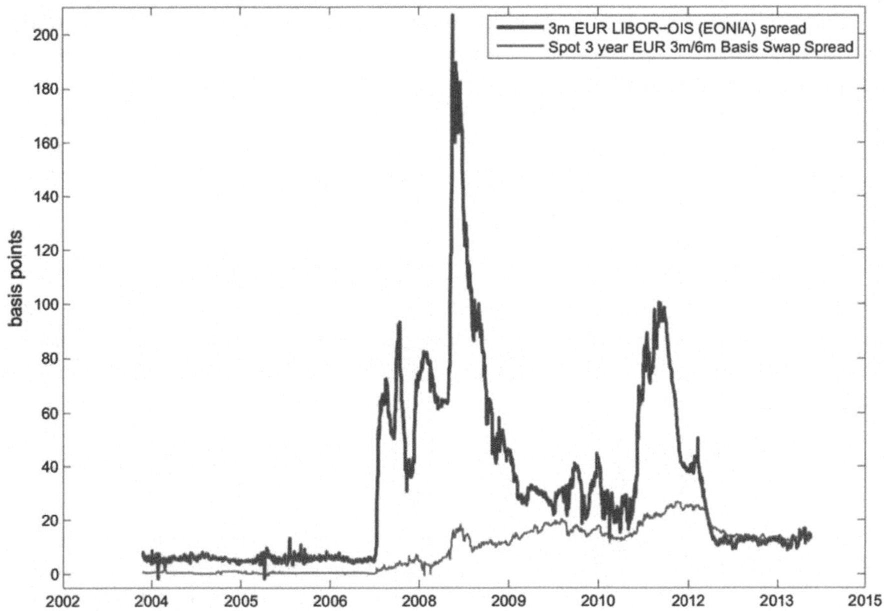

Fig. 1 Spread development (taken from Grbac and Runggaldier 2015, Fig. 1.4)

appeared between rates that used to be closely matched before as was the case in the simple example before but see also Fig. 1 for an illustration.

Since these spreads persisted since the crisis, these arbitrage opportunities have not been exploited, implying that the spreads are due to various risk levels. To explain and take into account these risk levels multiple curves were introduced (cf. Morini and Bianchetti (2013)).

In post-crisis markets most OTC derivatives are traded *secured* or *collateralized*, i.e. the contract is periodically marked-to-market where the party whose position has lost value has to post collateral to the other party (cf. Grbac and Runggaldier 2015). The posted collateral remains property of the collateral payer and is, hence, remunerated for the time it is used as collateral. I.e., in case no default occurs during the lifetime of the contract, the collateral payer receives the collateral back together with accumulated interest based on the collateral rate. The International Swaps and Derivatives Association (ISDA) provides Standardized Credit Support Annexes (S-CSA) which prescribe the frequency of marking-to-market and posting collateral, the collateral rate, eligible currencies and securities which are allowed to serve as collateral, and specifications of close-out cash flows. As alternative, central clearing is provided via central counterparties (CCP) as SwapClear, ICE Trust US and ICE Clear Europe (but has still to be aware of liquidity risk due to collateral postings and variation margin calls in adverse market conditions).

In the following, we assume that counterparty and liquidity risk of the interbank market still directly influence the reference rates in these contracts and create the multiple curve phenomenon.

Fixing a finite time horizon $T^* > 0$ for all market activities and products, a discrete tenor structure with tenor x is defined as $\mathcal{T}^x = \{T_k^x : k \in \{1, \ldots, M_x\}\}$, where

$$0 \leq T_1^x < T_2^x < \cdots < T_{M_x-1}^x < T_{M_x}^x \leq T^*$$

and $\delta_k^x := T_k^x - T_{k-1}^x$ denotes the year fraction (where we will not discuss the day count and business day conventions here for sake of clarity of exposition). According to the simple introductory example above under the multiple curve framework one assumes that

$$\frac{P^x(0, T_{k-1}^x)}{P^x(0, T_k^x)} = 1 + \delta_k^x \cdot F^x(0; T_{k-1}^x, T_k^x), \quad k \in \{1, \ldots, M_x\},$$

holds for each tenor $x \in \{1, 3, 6, 9, 12\}$ [months], where $P^x(t, T_k^x)$ denotes the present value (at $t \in [0; T^*]$) of the time-T_k^x zero bond for the tenor x. In this framework the collection of all zero bonds under all tenors x and maturities T_k^x denote the building blocks of an arbitrage-free and complete market on a filtered probability space $(\Omega, \mathcal{F}, (\mathcal{F}_t)_{t\in[0;T^*]}, \mathbb{Q})$ with risk-neutral equivalent martingale measure $\mathbb{Q}$ where the filtration $(\mathcal{F}_t)_{t\in[0;T^*]}$ satisfies the usual conditions.

Bootstrapping the discount factors or zero bond prices $P^x(0, T)$ from tradable instruments has to be done according to the tenor structure $\mathcal{T}^x$ (representing a certain risk level as mentioned above). According to Ametrano and Bianchetti (2013), the post-crisis standard market practice for the construction of multiple interest rate yield curves can be summarized in the following procedure:

1. Interbank credit/liquidity issues do matter for pricing, Libor rates are risky, Basis Swap spreads are no longer negligible.
2. The collateral does matter for pricing, *OIS discounting* is adopted, i.e. the discounting factors $P^{OIS}(0, T)$ are based on overnight indexed swap rates.
3. Decide the appropriate funding rates of the derivatives to be priced, then select the corresponding market instruments and build one single discounting curve using the classical single-curve bootstrapping technique.
4. Select *multiple separated sets of vanilla interest rate instruments* traded in real time on the market with increasing maturities $\mathcal{T}^x$, each set homogeneous in the underlying rate (typically with tenors $x \in \{1, 3, 6, 9, 12\}$ [months]).
5. Build *multiple separated FRA curves* using the selected instruments plus their bootstrapping rules and the unique discounting curve.
6. Compute the relevant FRA rates and the corresponding cash flows from the FRA curve with the appropriate tenor.
7. Compute the relevant discount factors from the discounting curve with the appropriate funding characteristics.

8. Work out prices by summing the discounted cash flows;
9. Compute the delta sensitivity and hedge the resulting delta risk using the suggested amounts (hedge ratios) of the *corresponding set* of vanillas.

Ametrano and Bianchetti (2013) provide a detailed and in-depth overview of the procedure of calibration and term structure construction,

$$[0; T^*] \ni T \mapsto P^x(0, T) \in [0; 1],$$

for the instruments traded in the European interest rate markets. The present value of future cash flows $c_k^x := c^x(t, F^x(t, T_{k-1}^x, T_k^x), T_k^x)$ depending on a tenor-specific forward rate $F^x(t, T_{k-1}^x, T_k^x)$ fixed in-arrears at T_{k-1}^x and paid at T_k^x is given by

$$PV_t = \sum_{k=1}^{m} P^{OIS}(t, T_k^x) \cdot c^x(t, T_k^x) \tag{1}$$

(which means that we assume the collateral rate as the OIS or S-CSA rate to discount the future cash flows). Hence, FRA and swaps can be priced using (1) where, e.g.,

$$c^x(t, T_k^x) = \delta_k^x \cdot (F^x(t; T_{k-1}^x, T_k^x) - S)$$

is used for payer swaps and

$$c^x(t, T_k^x) = \delta_k^x \cdot (S - F^x(t; T_{k-1}^x, T_k^x))$$

is used for receiver swaps with fair (forward) swap rate S. Note that (1) implies that the price of a plain vanilla (linear) interest rate derivative depends on multiple curves – one for discounting (OIS or S-CSA curve) and at least one for the cash flows that depend on reference or benchmark rates, i.e. tenor-specific forward rates. This fact necessitates the introduction of corresponding risk factors in (internal) market risk models for adequate risk management and hedging portfolios of these products.

Swaptions, i.e. options on swaps, are the core instruments for managing and hedging interest rate risks in the banking and trading books of financial institutions and in insurance portfolios. Consequently, it is of crucial importance that pricing and risk measurement of swaptions is as precise as possible according to best market practises. While the usual methods for pricing European swaptions can be extended to negative rate multiple curve environments (cf., e.g., Mercurio 2009 or Grbac and Runggaldier 2015) the pricing and calculating of sensitivities of non-standard swaptions (as, e.g., Bermudan Swaptions) need to be performed applying an adequate term structure model. In Sects. 3.1 and 3.2 below we will present some models capturing the different risk levels, i.e. multiple curve models.

3 Post-crisis Term Structure Models

According to the motivation provided in the previous section, we deal with two well-known classes for modelling the term structure of interest rates, namely short-rate models and Libor market models. The purpose of this section is to provide an overview of how to extend these pre-crisis model classes to capture the multiple curves in a consistent arbitrage-free framework.

3.1 Short-Rate Models for Multiple Curves

As with the pioneering articles of Mercurio (2009), Bianchetti (2010), Morini (2013), Morini and Bianchetti (2013), Ametrano and Bianchetti (2013), Pallavicini and Tarenghi (2010) and others on multiple curves (cf. Henrard 2014; Grbac and Runggaldier 2015), we proceed as follows: The stochastic evolution of the multiple curves representing the term structures

$$\left[0; T^*\right] \ni T \mapsto P^x(t, T) \in [0; 1], \quad t \in \mathbb{R}^+, x \in \{OIS, 1, 2, 3, 6, 9, 12\},$$

where each (hypothetical) tenor-specific zero bond is given under the equivalent martingale measure $\mathbb{Q}$ according to

$$P^x(t, T) = \mathbb{E}^{\mathbb{Q}}\left(\exp\left(-\int_{u=t}^{T} r_u^x du\right)|\mathcal{F}_t\right)$$

can be described using multi-factor short-rate models in a similar way as, e.g., in the classical pre-crisis textbooks (cf. Brigo and Mercurio 2006; Filipovic 2009; Andersen and Piterbarg 2010).

Grbac and Runggaldier (2015) describe the dynamics of the term structure using **exponentially affine factor models**. We sketch this approach which includes the most prominent multiple curve short rate models.

Using the intensity approach to credit risk (cf. Bielecki and Rutkowski 2002) a defaultable (non-observable or fictitious) zero bond is given by

$$P^x(t, T) = 1_{\{\tau > t\}} \cdot \mathbb{E}^{\mathbb{Q}}\left(\exp\left(-\int_{u=t}^{T} (r_u + s_u) du\right)|\mathcal{F}_t\right)$$

where $(s_t)_{t\in\mathbb{R}_0^+}$ represents the hazard rate (default intensity): This hazard rate is assumed to incorporate not only the credit risk but also liquidity risk as well as all other risk types observed in the interbank sector that affect the corresponding Libor rate with tenor $x \in \{1, 2, 3, 6, 9, 12\}$. Grbac and Runggaldier (2015) propose to model the short-rate $(r_t)_{t\in\mathbb{R}_0^+}$ and the "short-rate spread" $(s_t)_{t\in\mathbb{R}_0^+}$ driven by one common market factor Ψ_t^1 and one idiosyncratic factor Ψ_t^j, $j \in \{2, 3, 4\}$, for the short-rate and each of the spreads via

$$\begin{cases} r_t = \Psi_t^1 + \Psi_t^2 + \phi_t \\ s_t = \kappa^s \Psi_t^1 + \Psi_t^3 \\ \varrho_t = \kappa^\varrho \Psi_t^1 + \Psi_t^4 \end{cases}$$

where ϕ_t is a deterministic shift extension function (cf. Brigo and Mercurio 2006) used for reproducing today's term structure $[0; T^*] \ni T \mapsto P^x(t, T) \in [0; 1]$. By specifying the stochastic dynamics of the factors Ψ_t^k, $k \in \{1, 2, 3, 4\}$, the model is fully specified. In order to obtain simple "adjustment factors" between the pre- and post-crisis prices of plain vanilla linear derivatives, the common market factor is assumed to follow classical Vasicek or Hull-White stochastic dynamics

$$d\Psi_t^1 = (a^1 - b^1 \Psi_t^1)dt + \sigma^1 dW_t^1$$

while the idiosyncratic factors may follow Cox-Ingersoll-Ross-type (CIR) dynamics

$$d\Psi_t^j = \left(a^j - b^j \Psi_t^j\right)dt + \sigma^j \sqrt{c^j \Psi_t^j + d^j} dW_t^j$$

depending on the parameters $\{a^j, b^j, c^j, d^j\}$ with independent Wiener processes $\left(W_t^j\right)_{t \in \mathbb{R}_0^+}$. The time-$t$ price of the OIS zero bond is obtained as usual in affine term structure models as

$$P(t, T) = \exp\left(A(t, T) - B^1(t, T)\Psi_t^1 - B^2(t, T)\Psi_t^2\right)$$

and the time-t price of the fictitious zero bond by

$$P^x(t, T) = P(t, T) \cdot \exp\left(\tilde{A}(t, T) - \kappa^s B^1(t, T)\Psi_t^1 - \overline{B}^3(t, T)\Psi_t^3\right)$$

with analytically given deterministic functions $A, \widetilde{A}, B^1, B^2, \overline{B}^3$ (see Grbac and Runggaldier 2015 for details). In a first step, the Libor rate can be explicitly expressed by

$$L(T; T, T + \delta) = \frac{1}{\delta}\left(\frac{\exp\left(-\widetilde{A}(T, T + \delta) + \kappa^s B^1(T, T + \delta)\Psi_t^1 + \overline{B}^3(T, T + \delta)\Psi_t^3\right)}{P(T, T + \delta)} - 1\right)$$

The prices of plain vanilla linear interest rate derivatives can now be derived in an analytic way based on these building blocks (where analytical "adjustment factors" can be derived to transform pre-crisis prices for FRA and swaps into post-crisis ones). These analytic formulae (and adjustment factors) for linear derivatives have the advantage to simplify the calibration of model parameters (even if the building blocks consist in part of non-observable zero bond prices). Using the deterministic shift extension function ϕ_t, a perfect fit to the initial term structure can be achieved. This is of critical importance for practical purposes to ensure that the present value of future cash flows is identical to that produced by the term structure model chosen which furthermore guarantees that a model consistent hedge position to replicate the derivative can be set up.

As in pre-crisis multi-factor short-rate models the above post-crisis short-rate models can be used to price and hedge caps, floors, and swaptions (even deriving analytical or semi-analytical formulae for prices). Furthermore, more complex products can be priced based on these post-crisis term structure models via Monte Carlo simulation or trinomial trees, as e.g. Bermudan swaptions which can be used to hedge multi-callable termination rights in different investment strategies or insurance portfolios.

Note that **Gaussian exponentially quadratic models** can be specified in a similar way by

$$\begin{cases} r_t = \Psi_t^1 + \left(\Psi_t^2\right)^2 + \phi_t \\ s_t = \kappa^s \Psi_t^1 + \left(\Psi_t^3\right)^2 \\ \varrho_t = \kappa^\varrho \Psi_t^1 + \left(\Psi_t^4\right)^2 \end{cases}$$

where the dynamics of the idiosyncratic factors are assumed to be Gaussian (i.e. $c^j = 0$): Several prominent post-crisis short-rate models can be interpreted as special cases of exponentially affine factor models or Gaussian exponentially quadratic models (cf. Grbac and Runggaldier 2015). Another short-rate model of the term structure of interest rates for multiple tenors is given by Alfeus et al. (2017) who explicitly take into account the risk incurred when borrowing at a shorter tenor versus lending at a longer tenor ("roll-over risk"). The paper includes a detailed calibration study to the bid- and ask-prices of vanilla interest rate derivatives.

As a final remark, one should notice that negative Libor rates can either be achieved within the above set-up via using Vasicek- or Hull-White-based driving dynamics leading to normally distributed rates. The disadvantage of this idea is that—at least in principle—the negative rates might take arbitrarily large absolute values. Even if this is the case only on a rather small number of scenarios (i.e. with an almost neglecting probability) one might prefer to choose so-called shifted lognormal models or shifted CIR models which bear the disadvantage of prescribing the level at which the (formerly model-implied) rates are shifted to become negative (which gives rise to some model risk).

3.2 *Libor Market Models for Multiple Curves*

The forward Libor rate can be interpreted as the sum of

- the forward OIS rate plus
- the corresponding tenor-specific basis/spread

where the forward OIS rate is given by

$$F_k^x(t) := F\left(t; T_{k-1}^x, T_k^x\right) = \frac{1}{\delta_k^x} \cdot \left(\frac{P\left(t, T_{k-1}^x\right)}{P\left(t, T_k^x\right)} - 1\right)$$

and the additive spread is defined by

$$S_k^x(t) := L_k^x(t) - F_k^x(t)$$

with the FRA rates given by

$$L_k^x(t) := FRA\left(t; T_{k-1}^x, T_k^x\right) = \mathbb{E}^{\mathbb{Q}^{T_k^x}}\left(L\left(t; T_{k-1}^x, T_k^x\right) | \mathcal{F}_t\right)$$

under the T_k^x-terminal measure $\mathbb{Q}^{T_k^x}$. Note that using additive spreads ensures that the tenor-specific spread $S_k^x(t)$ is a $\mathbb{Q}^{T_k^x}$-martingale, too, and that according to the definition of $S_k^x(t)$ we only have to specify the joint stochastic dynamics (including volatility and correlation structure) of two of the quantities $S_k^x(t)$, $L_k^x(t)$, and $F_k^x(t)$ to prescribe the full model dynamics.

Following Mercurio and Xie (2012), the forward OIS rates (for tenors x_i) are usually modeled according to the following shifted-type dynamics

$$dF_k^{x_i}(t) = \sigma_k^{x_i}(t) V_t^F \left[\Theta_k^{x_i} + \frac{1}{\delta_k^{x_i}} + F_k^{x_i}(t)\right] dW_t^{F,x_i,k}$$

where $\sigma_k^{x_i}$ are deterministic functions, $\Theta_k^{x_i}$ is a deterministic shift (for modeling negative Libor rates), and $\left(W_t^{F,x_i,k}\right)_{t\in\mathbb{R}_0^+}$ are Wiener processes under the $T_k^{x_i}$-terminal measure $\mathbb{Q}^{T_k^{x_i}}$. The process $\left(V_t^F\right)_{t\in\mathbb{R}_0^+}$ is a common factor process with $V_0^F = 1$ and independent of all driving Wiener processes $\left(W_t^{F,x_i,k}\right)_{t\in\mathbb{R}_0^+}$. The $F_k^{x_i}$ have to satisfy no-arbitrage conditions

$$\sigma_k^{x_i}(t) = \sum_{j=i_{k-1}+1}^{i_k} \sigma_j^{x_1}(t)$$

namely the deterministic volatility function $\sigma_k^{x_i}$ of forward OIS rate $F_k^{x_i}$ has to equal the sum of volatility functions $\sigma_j^{x_1}$ of the rates $F_j^{x_1}$ for $j \in \{i_{k-1}+1, i_{k-1}+2, \dots, i_k\}$ where j corresponds to the indices of the tenor dates

$$T_{k-1}^{x_i} = T_{i_{k-1}+1}^{x_1} < T_{i_{k-1}+2}^{x_1} < \cdots < T_{i_k}^{x_1} = T_k^{x_i}$$

between the tenor dates $T_{k-1}^{x_i}$ and $T_k^{x_i}$.

Libor-OIS spreads can be modeled via

$$S_k^{x_i}(t) = \psi_k^{x_i}\left(F_k^{x_i}(t), X_k^{x_i}(t)\right)$$

where $X_k^{x_i}(t)$ are factor processes and the functions $\psi_k^{x_i}$ have to be chosen such that the spreads $S_k^{x_i}(t)$ are martingales under $\mathbb{Q}^{T_k^{x_i}}$. Using affine functions $\psi_k^{x_i}$ the parameters of these can be explained with the correlations between Libor rates and basis spreads and, furthermore, caps and swaptions can be priced in (semi-) analytical form.

Alternatively, the Libor-OIS spread can be modeled by

$$dS_k^{x_i}(t) = S_k^{x_i}(0) \cdot M^{x_i}(t)$$

where $M^{x_i}(t)$ follows a stochastic alpha beta rho-type process (SABR-type process)

$$\begin{cases} dM^{x_i}(t) = (M^{x_i}(t))^{\beta_i} v_t^{x_i} dZ^{x_i}(t) \\ dv_t^{x_i} = \varepsilon^{x_i} v_t^{x_i} dB^{x_i}(t) \end{cases}$$

with $\beta_i \in (0, +1]$, $\varepsilon^{x_i} > 0$ and where $(Z^{x_i}(t))_{t \in \mathbb{R}_0^+}$ and $(B^{x_i}(t))_{t \in \mathbb{R}_0^+}$ are correlated Wiener processes independent of $\left(W_t^{F,x_i,k}\right)_{t \in \mathbb{R}_0^+}$ where

$$dZ^{x_i}(t)dB^{x_i}(t) = \varrho^{x_i} dt$$

with $\varrho^{x_i} \in [-1, +1]$. According to Mercurio (2010) this model allows for convenient caplet and, in some particular cases, swaption prices which is convenient for simultaneous option pricing across different tenors, as well as for calibration.

For the classes of **affine multiple curve Libor models** and **multiplicative spread models** we refer to Sects. 4.2 and 4.3 of Grbac and Runggaldier (2015).

3.3 *Economic Scenario Generators for Term Structures*

Banks, (life) insurance companies and asset management firms alike are forced to implement economic scenario generators (ESG) for different purposes: Banks usually apply these for computing economic capital for counterparty credit risk and several valuation adjustments (XVA) for pricing and risk management. Insurance companies and asset manager use ESG to focus on risk management and stress testing purposes (cf. Cesari et al. 2009; Gregory 2015 for an overview).

The specification of post-crisis multiple curve term structure models used for generating future economic scenarios is essential for long-term portfolio management since swaps and swaptions are the most important hedging instruments used. Hence, the prices and sensitivities of these instruments have to be matched perfectly in order to avoid mis-management.

There are some natural requirements a post-crisis term structure model has to fulfil in order to produce adequate future interest rate scenarios in an ESG:

- The chosen term structure model has to allow for negative Libor rates and negative bond yields, as these benchmark rates became negative in the near-past (as of 2017): Classical (pre-crisis) short-rate or Libor market models that produce positive rates (as, e.g., Black-Derman-Toy, Black-Karasinski, or CIR, cf. Brigo and Mercurio 2006), a deterministic shift can be used to solve that problem.
- In general, the model has to produce mid- to long-term scenarios of a low or even negative interest rate environment (sometimes called "Japan scenario") with an adequate frequency: Notice that these scenarios were observed in the past (i.e., under the empirical measure) and should, therefore, be observed under the empirical or equivalent risk neutral (pricing) measure with a positive probability as well.
- Usually, a principal component analysis (PCA) on the movements of the different term structures is performed to clarify the number of factors necessary to explain the variance of the different yield and Libor curves. By the same argument as used above, the model-generated yield curves should display similar dynamics as observed in the past (under the empirical as well as under the equivalent pricing measure) to ensure that interest rate risks in a portfolio are adequately captured.
- Ideally, the initial term structures of discounting and benchmarking curves as well as post-crisis prices of plain vanilla interest derivatives (as swaps and swaptions) are matched perfectly: This is sometimes called "market consistent valuation".
- Furthermore, insurance firms usually perform a so-called *martingale test* on their ESG (cf. Varnell 2011): Since the main mathematical concept used for the "market consistent valuation" is the use of martingales under risk neutral pricing measure it is best practice to show that the average of the discounted cash flows from any asset (or combination of assets) would equal the starting value of the asset (or combination of assets). This test should also be applied for ESG used for calculating XVA.
- In as far as termination or cancellation rights might influence the overall positions also an adequate volatility structure and modelling has to be taken into account in ESG.

The post-crisis short-rate and Libor market models presented above are the most prominent term structure models used in pricing and ESG (for HJM-type post-crisis models see Grbac and Runggaldier 2015).

4 Conclusion and Outlook

Post-crisis term structure modeling is by far more complex than before: Multiple curve models have to be implemented to fit today's prices and the initial term structure as well as its' future dynamics. Applications include XVA and exposure calculations for pricing derivatives or risk-management of insurance and pensions plans which are influenced by the particular features observed in the post-crisis interest rate markets.

The increased complexity in term structure modeling gives rise to a not negligible amount of model risk that has to be taken care of in risk management processes (cf. Morini 2011).

Acknowledgements The author would like to express his sincere gratitude to Dr. Mehdi Mili and Dr. Javier Fernández Navas for providing helpful remarks.

References

Alfeus, M., Grasselli, M., & Schlögl, E. (2017). *A consistent stochastic model of the term structure of interest rates for multiple tenors. Quantitative finance research*. Research Paper No. 384, University of Technology Sidney. Retrieved January 21, 2015 from https://www.uts.edu.au/sites/default/files/QFR-rp384.pdf.

Ametrano, F. M., & Bianchetti, M. (2013). *Everything you always wanted to know about multiple interest rate curve bootstrapping but were afraid to ask*. Preprint, SSRN. October 01, 2017 from https://papers.ssrn.com/sol3/papers.cfm?abstract_id=2219548.

Andersen, L. G. B., & Piterbarg, V. (2010). *Interest rate modeling*. (Vol. I–III). London, New York: Atlantic Financial Press.

Bianchetti, M. (2010). Two curves, one price. *Risk,* 74–80.

Bielecki, T. R., & Rutkowski, M. (2002). *Credit risk: Modeling, valuation and hedging*. Berlin: Springer.

Brigo, D., & Mercurio, F. (2006). *Interest rate models—Theory and practice* (2nd ed.). Springer: Berlin.

Cesari, G., Aquilina, J., Charpillon, N., Filipovic, Z., Lee, G., & Manda, I. (2009). *Modelling, pricing, and hedging counterparty credit exposure*. Berlin: Springer.

Filipovic, D. (2009). *Term structure models*. Berlin: Springer.

Grbac, Z., & Runggaldier, W. J. (2015). *Interest rate modeling: post-crisis challenges and approaches.*, Springer Briefs in Quantitative Finance Heidelberg, New York: Springer.

Gregory, J. (2015). *The xVA challenge: Counterparty credit risk, funding, collateral, and capital* (3rd ed.). New York: Wiley.

Henrard, M. (2014). *Interest rate modeling in the multi-curve framework: Foundations, evolution and implementation*. London: Palgrave Macmillan.

Jorion, P. (2009). Risk management lessons from the credit crisis. *European Financial Management, 15*(5), 923–933.

Mercurio, F. (2009). *Interest rates and the credit crunch: New formulas and market models*. New York: Bloomberg Portfolio Research Paper.

Mercurio, F. (2010). *Libor market models with stochastic basis*. Preprint, SSRN. Retrieved October 01, 2017 from https://papers.ssrn.com/sol3/papers.cfm?abstract_id=1563685.

Mercurio, F., & Xie, Z. (2012). The basis goes stochastic. *Risk, 25*(12), 78–83.

Morini, M. (2011). *Understanding and managing model risk: A practical guide for quants, traders and validators*. Chichester: Wiley.

Morini, M. (2013). Solving the puzzle in the interest rate market. In M. Morini, M. Bianchetti (Eds.) *Interest rate modelling after the financial crisis* (pp. 61–105). London: RISK Books.
Morini, M., & Bianchetti, M. (Eds.). (2013). *Interest modeling after the financial crisis*. London: RISK Books.
Pallavicini, A., & Tarenghi, M. (2010). *Interest-rate modelling with multiple yield curves*. Preprint, SSRN. Retrieved January 21, 2018 from https://papers.ssrn.com/sol3/papers.cfm?abstract_id=1629688.
Varnell, E. M. (2011). Economic scenario generators and solvency II. *British Acturial Journal, 16*(1), 121–159.

A Comparison of Estimation Techniques for the Covariance Matrix in a Fixed-Income Framework

Marco Neffelli and Marina Resta

Abstract We compare various methodologies to estimate the covariance matrix in a fixed-income portfolio. Adopting a statistical approach for the robust estimation of the covariance matrix, we compared the Shrinkage (SH), the Nonlinear Shrinkage (NSH), the Minimum Covariance Determinant (MCD) and the Minimum Regularised Covariance Determinant (MRCD) estimators against the sample covariance matrix, here employed as a benchmark. The comparison was run in an application aimed at individuating the principal components of the US term structure curve. The contribution of the work mainly resides in the fact that we give a freshly new application of the MRCD and the NS robust covariance estimators within the fixed-income framework. Results confirm that, likewise financial portfolios, also fixed-income portfolios can benefit of using robust statistical methodologies for the estimation of the covariance matrix.

Keywords Covariance matrix estimators · Fixed-income framework
PCA analysis · Shrinkage · Minimum covariance determinant

1 Introduction

The estimation of the variance-covariance matrix (covariance matrix) -CVm- plays an important role in many financial applications, including risk management, option pricing and portfolio selection. In fact, the CVm conveys all the information related to the co-movements among a bunch of financial securities, and hence allows the investors to allocate resources following the principle of diversification. Such universal "consensus" around the role of the CVm holds also when focusing on fixed-

M. Neffelli (✉) · M. Resta
Department of Economics and Business Studies, University of Genova,
Via Vivaldi 5, 16126 Genoa, Italy
e-mail: marco.neffelli@edu.unige.it

M. Resta
e-mail: marina.resta@economia.unige.it

M. Mili et al. (eds.), *New Methods in Fixed Income Modeling*, Contributions to Management Science, https://doi.org/10.1007/978-3-319-95285-7_6

income instruments. Even in this case, in fact, the covariance matrix plays a crucial role for at least three reasons. First, in the term structure analysis, the CVm helps to estimate co-movements across different maturities. Second, the CVm can assess the main factors driving fixed-income securities prices for risk metrics calculation and stress testing (Litterman and Scheinkman 1991). Third, inside the portfolio allocation framework, the analysis of the CVm lets to determine how fixed-income instruments react in combination to other financial products, and helps the investor in assessing the risk profile of fixed-income securities (Martellini et al. 2003).

The covariance matrix is generally calculated by an inference procedure: being the dataset in use just a mere sample of the real population, which it is not known, the covariance matrix estimates the true relationships among population components. However, far from being an easy task, inferring the CVm is affected by a trade-off between the estimation error and the model error: model-free approaches, in fact, guarantee the unbiasedness of the estimator suffering only from estimation error, while parametric approaches impose some structure for the CVm, lowering the estimation error but enlarging the model error (Bengtsson and Holst 2002). The more commonly applied estimation technique relies on the so-called Sample Covariance matrix (SCVm): the work of (Markowitz 1952) is a key example of the use of SCVm for portfolio selection. The most appealing feature of this technique is that it does not require specifying any structure for the CVm, that is: it implies a model-free approach (Anderson 1963). Moreover, when the data sample is Normally distributed, the SCVm is an unbiased estimator for the population moments (Bengtsson and Holst 2002). However, estimating the SCVm has also pernicious drawbacks, especially in portfolio optimization (Ledoit and Wolf 2004b). To make an example, when the sample size is smaller than the number of considered variables, the sample covariance matrix is ill-conditioned by construction (Ledoit and Wolf 2004a). This issue has been variously addressed, by proposing alternative methodologies to estimate the covariance matrix. In the case of portfolio selection, the attempts of enhancing the covariance matrix include the contribution of (Jorion 1986) who used a Stein-type estimator. Moreover, (Michaud 1989) demonstrated that Bayes-Stein estimators can positively impact on portfolio selection procedure in presence of outliers in the assets time series; (Black and Litterman 1992) tackled the Michaud's issue by proposing a global equilibrium extension of the Markowitz's model; (Jagannathan and Ma 2003) demonstrated that constraints on portfolio optimization based on the SCVm generates the same positive outcome as applying shrinkage on the SCVm. On the other hand, problems arise also when, on the opposite, the sample size is too wide. The main point is that when the sample dimension is NxT, where N is the number of assets and T is the number of observations for each asset, the CVm is of dimensions N. However, as the maximum rank of the sample covariance matrix should not exceed T-1 (Bengtsson and Holst 2002), when N is greater than T (and hence, a fortiori greater than T-1) the CVm cannot be inverted (Schäfer and Strimmer 2005). This can represent a serious issue in financial applications, especially within the fixed-income framework, where large bond portfolios are the rule more than the exception, then dealing with high-dimensional portfolios. Finally, as demonstrated in (Ledoit and Wolf 2004b) the SCVm is particularly sensitive to outliers in the dataset: sample eigenvalues are systematically spiked upward or downward, depending on the observations are either too much large or small, respectively. As pointed out by (Fan et al. 2005), we therefore cannot trust on the sample covariance matrix, since it is hazardous to estimate it without imposing any structure.

Over the years these problems have been variously faced in the financial literature. Pantaleo et al. (2011) presented a review of the solutions that deal with this model-free curse, and classified the resulting estimators into three research strands, relying on: (i) spectral properties of the CVm, (ii) the hierarchical clustering approach, and (iii) statistical models. The spectral properties of the covariance matrix inspire a research strand in connection to factor modelling which is based, in turn, on the Arbitrage Pricing Theory (Ross 1976) and the Capital Asset Price Model (Sharpe 1964; Lintner 1965). In this class we can also include the estimators based on Random Matrix Theory -RMT- (Mehta 1990), which operate a mathematical transformation directly on each sample eigenvalue by adding or subtracting value whether the eigenvalue is above or below a selected threshold. Main contributions include the work of (Bengtsson and Holst 2002), who used the RMT to identify the leading factors for his factor model for portfolio applications; moreover, (Pafka et al. 2004) used an exponential moving average model together to the RMT to improve the portfolio optimization methodology, while (Frahm and Jaekel 2005) applied the concept of RMT to minimize the risk of a portfolio based on the S&P500. Finally, (Wolf 2004), highlighted that the resampled efficiency (Michaud and Michaud 1998) is a very similar tool to RMT, since the portfolio optimization input are calculated via a Monte Carlo resampling. The second research strand is linked to the hierarchical clustering approach (Anderberg 1973). This assumes that data can be clustered in groups according to a convenient similarity measure. By changing this similarity measure, data can be grouped in several different ways, thus enhancing the results of the portfolio optimization procedure. The approach has been widely discussed by Tola et al. (2008), in comparison to the RMT for optimizing a portfolio composed by highly capitalized stocks from NYSE, as well as in (Pantaleo et al. 2011) who used the hierarchical clustering in a comparative study among nine covariance estimators. Finally, the class of statistical estimators is perhaps the most representative one: it includes the Shrinkage (SH) technique (Ledoit and Wolf 2003, 2004a, b), as well as other robust estimators like the Minimum Covariance Determinant -MCD- due to (Rousseeuw 1984). The MCD estimator is commonly employed to identify robust estimates for the parameters of multivariate distributions and it has applications in many scientific areas. To cite some examples, the MCD is of common use in multivariate data analysis, and it is frequently used as input for other procedures likewise multivariate linear regression (Rousseeuw et al. 2004; Agulló et al. 2008), discriminant analysis (Hawkins and McLachlan 1997) and factorial analysis (Pison et al. 2003). More recently, the MCD has been extended to high-dimensional problems by (Boudt et al. 2017), with the Minimum Regularised Covariance Determinant (MRCD). The MRCD is suitable for outlier detection, observations ranking and clustering analysis (Boudt et al. 2017). However, while the MCD has been mainly used outside finance, the shrinkage (SH) has played a crucial role in fostering the Markowitz portfolio optimization framework. In particular, the SH procedure (Ledoit and Wolf 2003,

2004b) revisits the concept of Stein estimators (Stein 1956), proposing a series of alternatives for the target matrix, including: the covariance matrix as derived from the Single Index model (Sharpe 1964), likewise in (Jorion 1986), and the constant correlation matrix, i.e. a matrix where the pairwise correlation is treated as a constant, as in (Ledoit and Wolf 2004b) who used a five factors model for comparison purposes against the shrinkage methodology for the covariance among equities in portfolios of different size. Furthermore, (Schäfer and Strimmer 2005) extended the approach of (Ledoit and Wolf 2003) to small size samples, and considered additional types of target matrix for the shrinkage. Finally, (Ledoit and Wolf 2012) introduced the Nonlinear Shrinkage (NSH) technique and opened a new way for enhancing the shrinkage covariance estimator.

This chapter benefits from the previous rows review to focus on the following research question: which statistical estimator works at best to estimate the covariance matrix among interest rates? In detail, the chapter is structured as follows. Section 2 gives some brief remarks and basic analytics for the methodologies we employed to calculate the CVm, and namely: Sample, Shrinkage, Nonlinear Shrinkage, MCD and MRCD estimators. A case study is presented in Sect. 3: following the seminal work of (Litterman and Scheinkman 1991), in light of the robust PCA introduced in (Hubert et al. 2005), we analyse the US term structure curve through a robust PCA based on Sampling, SH, NS, MCD and MRCD approaches. To the best of our knowledge, the extension of MCD and MRCD estimators to this branch of finance is quite new, as well as the case study under discussion. Section 4 concludes.

2 Methodology

2.1 Notational Conventions

From now on, we denote by Σ the true covariance matrix, by S_{SCV} the sample covariance matrix and by: S_{Shrink}, $S_{NLShrink}$, S_{MCD} and S_{MRCD} the covariance matrices obtained with the Shrinkage, Nonlinear Shrinkage, MCD and MRCD methodologies, respectively. We also denote by MSE $(\cdot)$ the Mean Squared Error of behind introduced matrices: to make an example, the MSE for the true covariance matrix is:

$$MSE(\Sigma) = Var(\Sigma) + Bias(\Sigma)^2$$

where $Bias(\Sigma)$ represents the bias induced by the estimation error and $Var(\Sigma)$ is the variance of the CVm.

The CVm, by definition, is a squared, symmetric, positive (semi-)definite matrix, with the variances on the main diagonal and the covariances elsewhere. Thus, it should be invertible and well-conditioned (Fisher and Sun 2011).

2.2 Sample Estimator

Let us denote by R a NxT matrix, where N is the number of assets and T the number of observations. Following (Bengtsson and Holst 2002), the sample covariance matrix S is given by:

$$S_{SCV} = \frac{1}{T-1} R\left(I - \frac{1}{T}\mathbf{1}\mathbf{1}'\right)R', \tag{1}$$

where I is the identity matrix of dimensions N, $\mathbf{1}$ is the $Nx1$ unitary vector, and the punctuation mark indicates the transposition operator. The maximum rank of S is $T-1$.

As said in Sect. 1, the sample estimator is unbiased and easy to estimate but contains a huge amount of estimation errors, because residing on a model-free approach, it does not require any structure for R, and therefore it is very sensitive to outliers in the dataset.

2.3 Shrinkage and Nonlinear Shrinkage Estimators

The rationale behind the introduction of the shrinkage procedure within the portfolio framework relates to the low accuracy of traditional estimators when describing the very underlying features of stocks, since their efficiency decreases more and more, as the sample size decreases (Bengtsson and Holst 2002). Using the Shrinkage estimator makes possible to limit the potential error on estimates by reducing the Mean Squared Error (MSE) of the SCVm.

Furthermore, by recalling Sect. 2.1, even if the SCVm is an unbiased estimator, it does not minimize the MSE (Chen et al. 2010), since MSE(S) is formed only by the variance, and $Bias(S) = 0$. However, (Stein 1956) demonstrated how shrinkage estimators can reduce the MSE, and (Ledoit and Wolf 2003, 2004b) improved this assertion as they highlighted that reducing the MSE can be reached by imposing some structure to the sample covariance matrix via a proper target matrix, F. In other words, instead of using a model-free approach, their CVm estimator is based on a convex linear combination between the sample covariance matrix S and a target matrix F:

$$S_{Shrink} = \delta F + (1-\delta)S_{SCV}, \tag{2}$$

where $\delta \in [0,1]$ is the shrinkage intensity. For $\delta = 1$, S_{Shrink} equals the target matrix, while for $\delta = 0$, we have: $S_{Shrink} = S_{SCV}$, i.e. the SCVm.

In order to calculate the optimal shrinkage intensity (Ledoit and Wolf 2004b) derived an optimal value δ^* based on minimizing the expected value of the loss function given by the Frobenius norm of the quadratic distance between the true and the shrinkage covariance matrix:

$$L(\delta) = \left|\left|[\delta F + (1-\delta)S_{SCV} - \Sigma]^2\right|\right|, \tag{3}$$

When N is fixed and T tends to infinity, the optimal value δ^* asymptotically behaves like a constant (Ledoit and Wolf 2004b).

The estimate of the covariance matrix obtained with the shrinkage technique is always positive definite and well-conditioned, making it a good candidate for computational implementations (Ledoit and Wolf 2004b). Moreover, shrinking the SCVm towards a more structured matrix makes the covariance matrix less sensitive to estimation errors; however, it can be extremely biased if the assumptions of the underlying model diverge from those of the true covariance. This trade-off is carefully described in Jagannathan and Ma (2003), asserting that the SH estimator looks like a compromise between the bias of the target matrix and the variability of the traditional SCVm.

As seen in Sect. 1, we can find various characterizations of the target matrix F. In this work, we followed the approach of (Ledoit and Wolf 2004b), assuming the target matrix being equal to the constant correlation matrix. This choice can be easily motivated in the following way: let us consider a set of N perspective interest rates with different maturity and same tenor, daily observed along a time horizon of length T. Since variations among interest rates are quite slow, we can assume that the relationships among different maturities do not change daily.

Then, the covariance and correlation between the assets i and j are given by:

$$\begin{aligned} Cov(i,j) = s_{i,j} &= \tfrac{1}{T-1}\sum_{n=1}^{N}\left(i_n - \bar{i}\right)\left(j_n - \bar{j}\right), \\ Cor(i,j) = \rho_{ij} &= \tfrac{s_{ij}}{s_i s_j}, \end{aligned} \tag{4}$$

where $\bar{i}$, $\bar{j}$ are the mean value for assets i and j, respectively. The average of the sample correlation is then given by:

$$\bar{\rho} = \frac{2}{(T-1)T}\sum_{i=1}^{T-1}\sum_{j=i+1}^{T}\rho_{i,j} \tag{5}$$

Putting together (4) and (5), by assuming the target F being the constant correlation matrix, we have:

$$f_{ii} = s_{ii} = s_i^2 \;\; f_{ij} = \rho_{ij} s_i s_j$$

where f_{ii} is the variance of every asset, lying on the main diagonal of F, and f_{ij} is the average of correlations between each couple of assets, elsewhere.

This procedure generates a CVm with a more imposed structure and a lower estimation error. Combining F with S in the shrinkage procedure gives then an improved estimator of the true covariance matrix.

However, the linear Shrinkage is just a first order approximation to the nonlinear problem of calculating sample eigenvalues (Ledoit and Wolf 2012), so that each sample eigenvalue is shifted towards the grand mean of all sample eigenvalues with the same intensity. On the contrary, the rationale behind Nonlinear Shrinkage (NSH) is that different sample eigenvalues should be differently moved. (Ledoit and Péché 2011), expanded the shrinkage concept to the nonlinear case, yet. However, they improved the linear estimator by constructing a target matrix based on the distribution of the sample eigenvalues, only: in this way, the target matrix is independent from the structure of the true covariance matrix, but generates an *oracle* estimator, which is reliable only in a very limited number of cases. Conversely, the *bona-fide* estimator in (Ledoit and Wolf 2012) is basically an oracle estimator, but consistently estimated: Monte Carlo stress tests highlighted that this estimator is at least as good as the 2004 Shrinkage procedure of (Ledoit and Wolf 2004b) or even better, making it an improved candidate for portfolio optimization.

From the analytic viewpoint, the NSH procedure aims at calculating:

$$S_{NLShrink} = YDY'$$

where Y is the matrix formed by the eigenvectors $y_{i,N}$ of the sample matrix S, and D is a diagonal matrix whose elements capture the link between the eigenvectors and the true covariance matrix. The procedure consists in two steps that are below summarised.

Step 1. Find the matrix A that is closest to the true covariance matrix Σ according to the Frobenius norm.

Step 2. Solve the minimization problem:

$$\min_{D} ||YDY' - \mathrm{A}||,$$

to find $D^* \equiv Diag\left(d_1^*, \ldots d_N^*\right)$, with $d_i^* \equiv \boldsymbol{y}_i' \Sigma \boldsymbol{y}_i,\ i = 1, \ldots, N.$

2.4 Minimum Covariance Determinant and Minimum Regularised Covariance Determinant

The Minimum Covariance Determinant (MCD) is a robust covariance estimator that allows detecting outliers in multivariate data by calculating the Mahalanobis distances between every observation and the central value of the data (Rousseeuw and Yohai 1984). More precisely, the aim of the estimation process is to find a subset of the data with the lowest value of the determinant and containing h observations, with: $\frac{T+N+1}{2} < h < T$, where N and T are as defined in previous rows. The choice of h

depends only on the covariance between the data, thus the MCD covariance matrix is given by the SCVm of the subset of the original data that minimizes the dispersion of the observations.

The procedure to obtain S_{MCD} can be summarised into five steps.

Let us denote by R the $N \times T$ matrix with N components $\mathbf{r}_i \in \mathbb{R}^T$, $i = 1, \ldots, N$. Each vector $\mathbf{r}_i$ $(i = 1, \ldots, N)$ can be viewed as the random variable whose realizations are the T observations of the ith asset.

Step 1. Identify an initial subset H_O of h observations, with $\frac{T+N+1}{2} < h < T$:

$$H_0 = argmin(\det(cov(\boldsymbol{r}_i | i \in H_0)), \tag{6}$$

From H_0 we can evaluate the sample mean vector $\boldsymbol{\mu}_0$ and the sample covariance matrix S_0 than can be then employed as indicators of location and scatter.

Step 2. Compute the Mahalanobis distances between each component of R and $\boldsymbol{\mu}_0$:

$$d_i = Dist_{S_0}(\boldsymbol{r}_i, \boldsymbol{\mu}_0) = \sqrt{(\boldsymbol{r}_i - \boldsymbol{\mu}_0)' S_0^{-1} (\boldsymbol{r}_i - \boldsymbol{\mu}_0)},$$

where S_0^{-1} is the inverse of S_0, and $d_i \in \mathbb{R}$.

Step 3. The assets whose distance is behind the acceptance region are assigned a weight equal to zero and henceforth excluded, while those whose distance falls inside the acceptance region are kept and receive a weight equal to one. Weights are assigned with the following:

$$w_i = \begin{cases} 0, d_i > \sqrt{\chi^2_{N,0.975}}, \\ 1, d_i \leq \sqrt{\chi^2_{N,0.975}}. \end{cases} \tag{7}$$

where $\sqrt{\chi^2_{N,0.975}}$ represents the cut-off value for detecting outliers, and $\chi^2_{N,0.975}$ is the 0.975 quantile of the χ^2_N distribution. Thus, a new subset H_1 is derived with center $\boldsymbol{\mu}_1$ and scatter S_0.

Step 4. Using the weights derived by (7), the estimators of position and scatter $\boldsymbol{\mu}_{MCD}$ and S_{MCD} are computed:

$$\boldsymbol{\mu}_{MCD} = \frac{\sum_{i=1}^n w_i \boldsymbol{r}_i}{\sum_{i=1}^n w_i},$$

$$S_{MCD} = \frac{c_1 \left(\sum_{i=1}^n w_i (\boldsymbol{r}_i - \boldsymbol{\mu}_{MCD})(\boldsymbol{r}_i - \boldsymbol{\mu}_{MCD})' \right)}{\sum_{i=1}^n w_i}.$$

Here c_1 is a constant that ensures the asymptotic consistency towards a normal distribution (Croux and Haesbroeck 1999), so that the robust Mahalanobis distances for each asset become:

$$Rd_i = Dist_{S_{MCD}}(\boldsymbol{r}_i, \boldsymbol{\mu}_{MCD}) = \sqrt{(\boldsymbol{r}_i - \boldsymbol{\mu}_{MCD})' S_{MCD}^{-1}(\boldsymbol{r}_i - \boldsymbol{\mu}_{MCD})}.$$

Step 5. The final step consists then in marking as outliers and excluding the observations outside the acceptance region, i.e. those for which we have:

$$w_{Rd_i} = \{\boldsymbol{r}_i | Rd_i \geq \sqrt{\chi^2_{N,0.975}}\}.$$

However, applying the MCD estimator has remained difficult for many years, because too computationally expensive. Branch and bounds algorithms (Candela 1996), heuristics (Woodruff and Rocke 1994) and relaxation techniques of the exact solution (Schyns 2008) have been suggested. The FAST-MCD algorithm developed by (Rousseeuw and Van Driessen 1999) tackled down the problem, making the MCD computation more efficient. FAST-MCD is a deterministic procedure which produces a good approximation of the MCD for both small and large datasets. The main idea consists in considering a small random subset of dimension $N+1$, instead of looking for the subset of h. The procedure replicates the steps already highlighted in the MCD algorithm, working on the new reduced dimensions subset. The resulting subset $H_{FAST\text{-}MCD}$ is then locally improved thanks to sequential concentration steps (C-steps) in which the mechanism of computation of the Mahalanobis distances and outlier elimination is repeated until any further reduction of the determinant of the covariance matrix becomes unfeasible.

In 2017, (Boudt et al.) proposed an alternative estimator called Minimum Regularized Covariance Determinant estimator (MRCD). The MRCD is still based on subsets of the covariance matrix, this time selected through a convex combination between a target matrix and the SCVm. The main steps leading to the MRCD are below illustrated.

Step 1. The dataset is standardised according to the:

$$\boldsymbol{u}_i = P_R^{-1}(\boldsymbol{r}_i - \boldsymbol{v}_R), \quad (i = 1, \ldots, N)$$

where P_R^{-1} is the NxN diagonal matrix containing the scatter estimations, $\boldsymbol{r}_i$ are the components of R, and $\boldsymbol{v}_R$ is the Nx1 median vector.

Step 2. Compute:

$$K(H) = \kappa F + (1 - \kappa)c_\alpha S_U(H)$$

where κ is the shrinkage weight or regularization parameter, F the target matrix, $S_U(H)$ the sample covariance matrix calculated on $U = \{\boldsymbol{u}_i\}$ in the subset H, and c_α a consistency factor as defined in (Croux and Haesbroeck 1999), with $\alpha = (T - h)/T$ being the so-called trimming percentage.

Step 3. Find the subset of the original dataset minimising the determinant of the regularised matrix $K(H)$.

$$H_{MRCD} = \underset{H}{\operatorname{argmin}}\left(det(K(H))^{1/N}\right)$$

Step 4. Compute the MRCD covariance matrix:

$$K_{MRCD} = P_r Q \Lambda^{1/2}\left[\kappa \mathrm{I} + (1-\kappa) S^*(H_{MRCD})\right]\Lambda^{1/2} Q' P_r$$

where S^* is an ad hoc diagonal matrix[1] that rescales the diagonal elements of the final covariance matrix K_{MRCD}, κ is the regularization parameter, Q is the orthogonal matrix of eigenvectors from the target matrix and $\Lambda^{1/2}$ is the square root eigenvalues matrix.

3 Case Study

It is generally acknowledged that bond prices are sensitive not only to parallel shifts in the yield curve, but also to non-parallel shifts; in particular, (Litterman and Scheinkman 1991) found out that bond prices are mainly sensitive to three factors, which can explain almost the 99% of the total variance: level, steepness, and curvature.

Evaluating the exposure to these three factors is the leading feature of Litterman and Scheinkman's (LS, thereinafter) approach to hedging, and it is generally ruled out by applying the Principal Component Analysis (PCA)[2] on the covariance matrix of changes in interest rates, usually estimated via the sample estimator. With PCA, the LS approach can be characterised as follows: the first Principal Component (PC) should equally affect all the maturities in the term structure, and should be regarded as the response in shifts of the term structure. The second PC response should look like an upward sloping curve: it should affect closer maturity with the same intensity, but with different sign. This should be regarded as the response in changes of the slope of term structure. The third PC should affect in the same way the extremes of

[1]As explained in Boudt et al. (2017), S^* is a transformation of the sample covariance matrix $S_U(H)$, obtained as: $S^* = \Lambda^{1/2} Q' S_U(H) Q \Lambda^{1/2}$.

[2]Principal Component Analysis (Hotelling 1933) is a dimension reduction technique that works on a covariance (or correlation) matrix identifying the volatility factors that drive the time series under investigation. The PCA relies on the spectral decomposition of the covariance matrix Σ:

$$\Sigma = \mathrm{G}\Omega\mathrm{G}'$$

where G is the square matrix of the eigenvalues of Σ, and Ω is a diagonal matrix filled with the eigenvalues of the covariance matrix. The principal components are given by the normalized eigenvectors, ranked in descendant order according to the size of related eigenvalues. This because the total variance is equal to the sum of all the eigenvalues, so that the size of a single eigenvalues is the percentage of total variance explained. As a limited number of eigenvalues is usually enough to explain at least the 99% of total variation, the reduction of the covariance matrix can be performed by retaining only the eigenvalues that explain a certain threshold of the variance, eliminating the others.

Table 1 US term structure composition for the examined instruments

Instrument	Maturity	No. of instruments
USD LIBOR	Overnight, 1 week, 1, 2, 3, 6, 12-month	7
USD swap	2, 3, 4, 5, 6, 7, 8, 9, 10, 11, 12, 13, 14,15, 20, 25 and 30-year	17

the term structure, with a change of sign in the middle maturities. This should be regarded as the response in changes of the curvature of the term structure.

3.1 Data Description

Robust methodologies in estimating the covariance matrix might improve PCA results, hence fostering the reliability of LS hedging approach. To such aim, we checked the LS assertion considering the US term structure with 3-month tenor for the instruments highlighted in Table 1, with maturity varying from overnight to 30 years.

The dataset, retrieved from Bloomberg database, consists of 24 daily time series in the period: 02/01/2014–08/09/2017, for an overall number of 962 observations per series. In detail, we selected the US LIBOR interest rate with 7 different maturities, ranging from overnight to 12 months, to describe the short-term interest rates; 17 US swap contracts with maturity from 2 to 30 years to represent medium and long-term interest rates, hence fully characterizing the US term structure. The behaviour of the 24 time series is depicted in Fig. 1.

3.2 Results

In order to find relationships among the term structure components, we estimated the covariance matrix of daily changes in the yields of the 24 curves under examination, testing five different covariance estimators: Sample, Shrinkage, Nonlinear Shrinkage, MCD and MRCD.[3]

The estimates highlighted how the quality of the sample covariance matrix tends to deteriorate increasing the number of interest rate curves from 1 to 24, as testified by looking at the condition number, i.e. the ratio between the largest and the smallest eigenvalue, as depicted in Fig. 2.

[3]The sample and MCD methodologies are estimated using MATLAB embedded functions. The Shrinkage code is from Ledoit and Wolf (2004b), while the Nonlinear Shrinkage is estimated with the nlshrink R package, https://rdrr.io/cran/nlshrink/man/nlshrink.html, and the MRCD is from the KU Leuven webpage, https://wis.kuleuven.be/stat/robust/Programs/MRCD.

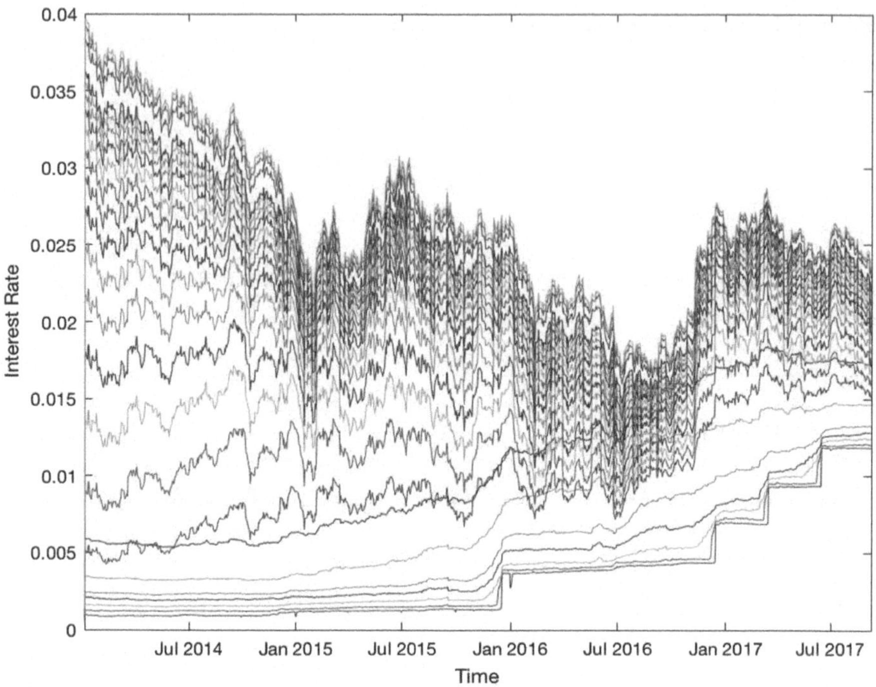

Fig. 1 Daily spot interest rates for the period 02/01/2014–08/09/2017. From bottom to top, the plot depicts the behaviour of: USD LIBOR with maturity overnight, 1 week, 1, 2, 3, 6, 12 months, and USD Swap with maturity 2, 3, 4, 5, 6, 7, 8, 9, 10, 11, 12, 13, 14,15, 20, 25 and 30-year

Looking at the plots in Fig. 2, we can highlight that the sample covariance method appears being very sensitive to the matrix size, when the number of examined interest rates is greater than eight; while the Shrinkage, MCD and MRCD adapt well in estimating the covariance for high-dimensional arrays. The Nonlinear Shrinkage tends to deteriorate the performance behind the same "magic" threshold as the sample approach, but it stabilizes after the 15th interest rate to converge to zero. As a preliminary conclusion, we can then argue that estimating the covariance matrix with a statistical estimator likewise SH, MCD and MRCD should lead to more robust results than in the case of both the sample covariance estimator and NS.

We then run the PCA on the five covariance matrices, obtained with the abovementioned estimators, and we monitored the percentage of explained variance by different and uncorrelated risk factors. According to the LS approach, this should lead to find 3 factors explaining the 99% of the overall variance; for this reason, we set the value three as a threshold for our analysis. Results are displayed in Table 2. Values in Panel A represent the cumulative variance in percentage for earlier five factors, sorted in ascendant order by percentage of explained variance, while Panel B shows the number of factors explaining 99% of total variance. We employed the following shortcuts: Sample (SCV), Shrinkage (SH), Nonlinear Shrinkage (NSH),

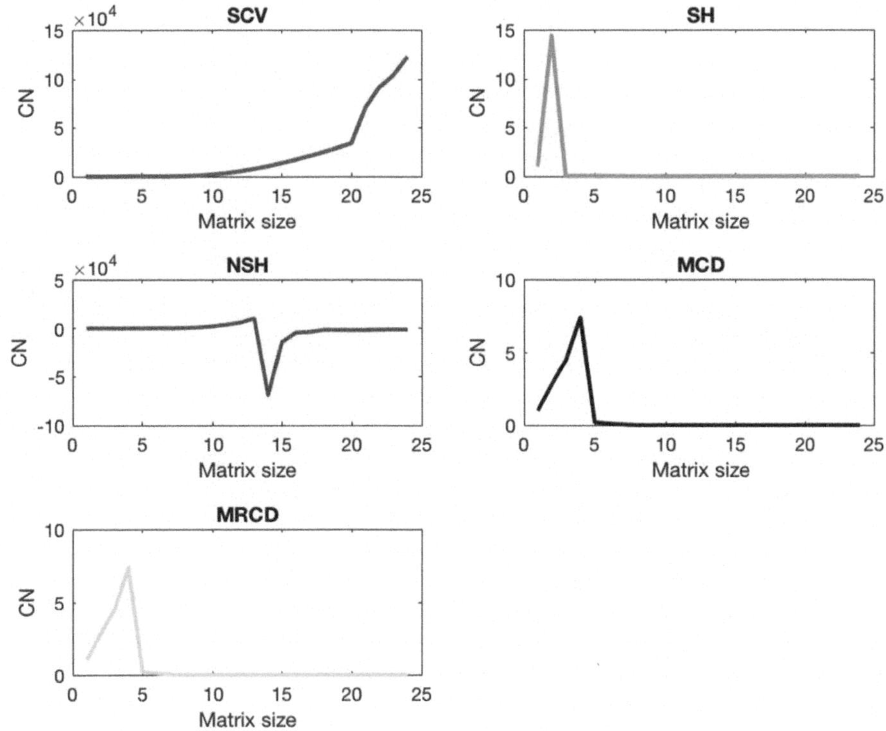

Fig. 2 Condition Number (CN) varying the size of the covariance matrix. From left to right: CN behaviour for the Sample covariance matrix (SCV—top-left), Shrinkage (SH—top-right), Nonlinear Shrinkage (NSH—middle-left), MCD (middle-right) and for MRCD (bottom-left). On the x-axis we reported the matrix size, while the y-axis reports the value of the condition number

Table 3.2 Earlier five factors explaining the overall variance, for various estimation methodologies of the covariance matrix

Factor	SCV	SH	NSH	MCD	MRCD
Panel A					
F1	97.50	75.41	97.18	94.84	81.49
F2	99.28	78.03	98.95	99.07	95.88
F3	99.52	80.57	99.22	99.37	97.09
F4	99.68	82.96	99.37	99.60	97.97
F5	99.76	85.20	99.48	99.72	98.54
Panel B					
Factors	2	15	3	2	7

Minimum Covariance Determinant (MCD) and Minimum Regularized Covariance Determinant (MRCD).

Looking at the results, the sample and MCD estimators are the fastest to reach the 99% threshold of explained variance, with just 2 factors. Nonlinear Shrinkage is also very fast, with 3 factors, confirming the LS view. On the other hand, MRCD and

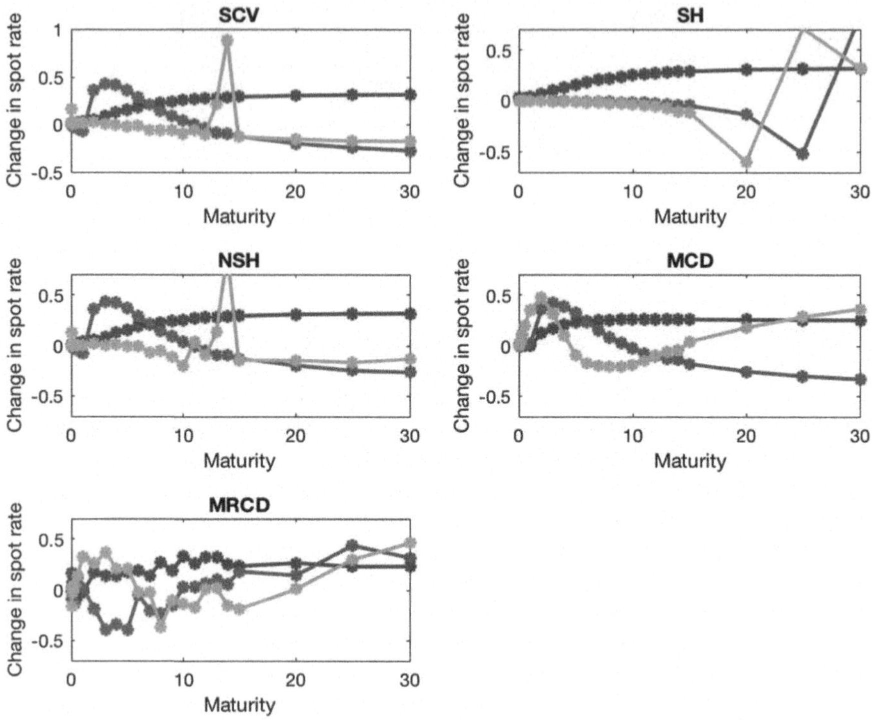

Fig. 3 Impact on the US term structure of earlier three Principal Components (PC), varying the estimation procedure of the covariance matrix. Curves represent the sensitivity of changes (returns) in interest rate against increasing maturity. The sensitivity to the first PC is blue, the one to the second PC is orange and the last to the third PC is yellow

Shrinkage are very slow, taking 7 and 15 factors to explain the threshold, respectively. The relation between the change in interest rates and different maturities should be captured by the number of the above listed factors for each methodology. However, there is no economic or financial motivation for explaining the total variation of the US term structure with more than three factors. Thus, in order to have a common ground of comparison with the LS model, in Fig. 3 we plot the sensitivity of changes in interest rate to increasing maturity, as explained by earlier three PCs for all the covariance methodologies in use. The slowest methodologies in explaining the total variation of the US term structure (SH and MRCD) clearly fail in delivering good insights about the risk factors afflicting the US term structure. On the other hand, the sample, NSH and MCD present results in line with the LS model.

A risk manager should find these results useful, because all those three methodologies (sample, Nonlinear Shrinkage and MCD) characterise in the same way the response to earlier two risk factors: the first PC (in blue), in fact, expresses the sensitivity to parallel shifts, while the second (orange) the change in the slope; moreover, the magnitude is the same across all the maturities but the sign changes. However, both sample and Nonlinear Shrinkage fail to deliver a good representation of the third component, which is well represented only by the MCD methodology. The US term structure, in fact, when the estimation is performed with either sample or Nonlinear Shrinkage estimator, shows a strange peak between 10 and 12-year maturity: this seems to be originated from some outliers in the term structure. The MCD methodology, on the other hand, confirms the LS assertion also for the third component: it has same sign for both closer and long-term maturities, while in the medium term the sign is different. This should be regarded as the response in changes of the curvature of the term structure.

In conclusion, we found that three out of five covariance methodologies, namely sample, Nonlinear Shrinkage and MCD, are able to represent the US term structure response with respect to the three factors introduced by the LS model. In particular, the MCD is the methodology that achieves more aligned results to the conclusions stated by the LS model. Looking in conjunction at the results of Table 2 and Fig. 3 we can then argue that the MCD arise as the best estimation method among those examined, as it maintains well-conditioned the CVm, and offers results well-fitting to the LS assertion.

4 Conclusion

In this work we compared various statistical methodologies aimed at providing a robust estimate of the covariance matrix, namely: Shrinkage, Nonlinear Shrinkage, Minimum Covariance Determinant (MCD) and Minimum Covariance Regularized Determinant (MRCD) estimators. Results were evaluated using the sample covariance matrix estimator as benchmark. These techniques were tested within the fixed-income framework. To this extent, we analysed the performances of the examined estimators in evaluating the US term structure with the PCA technique, according to what stated by (Litterman and Scheinkman 1991). Strong evidences highlight the benefit of switching from the sample covariance matrix to robust variants. First, by looking at the condition numbers, we highlighted that while the sample covariance estimator is very sensitive to the matrix size, on the other hand, the Shrinkage, MCD and MRCD well adapt in estimating the covariance for high-dimensional arrays, which in our example are composed by 24 interest rates. Second, the PCA approach seems to give more precise results when applied on the robust estimates: MCD and Nonlinear Shrinkage explain the 99% of total variance in a slower way than the sample methodology, thus highlighting that the second and the third factor represent a not negligible percentage of the overall variation. In particular, the MCD is the only methodology that well characterised the third factor too: this is crucial especially for risk management applications, as it expresses the response in changes of the curvature of the term structure.

To conclude, we can state that statistical covariance estimators can help in modelling the factors that drive the term structure curve, as they make possible to highlight relationships among different maturities in a fashion which is less likely to be biased by outlier observations. In detail, the PCA approach is enhanced, and this in turn should signify a better hedging power for financial products relying on fixed-income instruments. Future works are planned to assess the impact of such robust estimators on a wider range of fixed-income instruments applications including risk and portfolio management.

References

Agulló, J., Croux, C., & Van Aelst, S. (2008). The multivariate least-trimmed squares estimator. *Journal of Multivariate Analysis, 99*(3), 311–338.

Anderberg, M. R. (1973). Cluster analysis for applications. Monographs and textbooks on probability and mathematical statistics.

Anderson, T. W. (1963). Asymptotic theory for principal component analysis. *The Annals of Mathematical Statistics, 34*(1), 122–148.

Bengtsson, C., & Holst, J. (2002). *On portfolio selection: Improved covariance matrix estimation for Swedish asset returns*. Univ.

Black, F., & Litterman, R. (1992). Global portfolio optimization. *Financial Analysts Journal, 48*(5), 28–43.

Boudt, K., Rousseeuw, P., Vanduffel, S., & Verdonck, T. (2017). The minimum regularized covariance determinant estimator.

Candela, J. A. (1996). Exact iterative computation of the multivariate minimum volume ellipsoid estimator with a branch and bound algorithm. In *COMPSTAT* (pp. 175–180). Springer.

Chen, Y., Wiesel, A., Eldar, Y. C., & Hero, A. O. (2010). Shrinkage algorithms for MMSE covariance estimation. *IEEE Transactions on Signal Processing, 58*(10), 5016–5029.

Croux, C., & Haesbroeck, G. (1999). Influence function and efficiency of the minimum covariance determinant scatter matrix estimator. *Journal of Multivariate Analysis, 71*(2), 161–190.

Fan, J., Peng, H., Huang, T., & Ren, Y. (2005). Theory and methods-rejoinder-semilinear high-dimensional model for normalization of microarray data: A theoretical analysis and partial consistency. *Journal of the American Statistical Association, 100*(471), 808–813.

Fisher, T. J., & Sun, X. (2011). Improved Stein-type shrinkage estimators for the high-dimensional multivariate normal covariance matrix. *Computational Statistics & Data Analysis, 55*(5), 1909–1918.

Frahm, G., & Jaekel, U. (2005). Random matrix theory and robust covariance matrix estimation for financial data. *arXiv preprint physics/0503007*.

Hawkins, D. M., & McLachlan, G. J. (1997). High-breakdown linear discriminant analysis. *Journal of the American Statistical Association, 92*(437), 136–143.

Hotelling, H. (1933). Analysis of a complex of statistical variables into principal components. *Journal of Educational Psychology, 24*(6), 417.

Hubert, M., Rousseeuw, P. J., & Vanden Branden, K. (2005). ROBPCA: A new approach to robust principal component analysis. *Technometrics, 47*(1), 64–79.

Ledoit, O., & Péché, S. (2011). Eigenvectors of some large sample covariance matrix ensembles. *Probability Theory and Related Fields, 151*(1), 233–264.

Ledoit, O., & Wolf, M. (2003). Improved estimation of the covariance matrix of stock returns with an application to portfolio selection. *Journal of Empirical Finance, 10*(5), 603–621.

Ledoit, O., & Wolf, M. (2004a). A well-conditioned estimator for large-dimensional covariance matrices. *Journal of Multivariate Analysis, 88*(2), 365–411.

Ledoit, O., & Wolf, M. (2004b). Honey, I shrunk the sample covariance matrix. *The Journal of Portfolio Management, 30*(4), 110–119.

Ledoit, O., & Wolf, M. (2012). Nonlinear shrinkage estimation of large-dimensional covariance matrices. *The Annals of Statistics, 40*(2), 1024–1060.

Lintner, J. (1965). The valuation of risk assets and the selection of risky investments in stock portfolios and capital budgets. *The Review of Economics and Statistics*, 13–37.
Litterman, R. B., & Scheinkman, J. (1991). Common factors affecting bond returns. *The Journal of Fixed Income, 1*(1), 54–61.
Markowitz, H. (1952). Portfolio selection. *The Journal of Finance, 7*(1), 77–91.
Martellini, L., Priaulet, P., & Priaulet, S. (2003). *Fixed-income securities: Valuation, risk management and portfolio strategies* (Vol. 237). New York: Wiley.
Mehta, M. L. (1990). '*Random matrices.*' Academic Press, 2nd edn.
Michaud, R. O. (1989). The Markowitz optimization enigma: Is 'optimized' optimal? *Financial Analysts Journal, 45*(1), 31–42.
Michaud, R. O., & Michaud, R. O. (1998). *Efficient asset management: A practical guide to stock portfolio optimization and asset allocation*. Oxford University Press.
Jagannathan, R., & Ma, T. (2003). Risk reduction in large portfolios: Why imposing the wrong constraints helps. *The Journal of Finance, 58*(4), 1651–1683.
Jorion, P. (1986). Bayes-Stein estimation for portfolio analysis. *Journal of Financial and Quantitative Analysis, 21*(3), 279–292.
Pafka, S., Potters, M., & Kondor, I. (2004). Exponential weighting and random-matrix-theory-based filtering of financial covariance matrices for portfolio optimization. *arXiv preprint cond-mat/0402573*.
Pantaleo, E., Tumminello, M., Lillo, F., & Mantegna, R. N. (2011). When do improved covariance matrix estimators enhance portfolio optimization? An empirical comparative study of nine estimators. *Quantitative Finance, 11*(7), 1067–1080.
Pison, G., Rousseeuw, P. J., Filzmoser, P., & Croux, C. (2003). Robust factor analysis. *Journal of Multivariate Analysis, 84*(1), 145–172.
Ross, S. A. (1976). The arbitrage theory of capital asset pricing. *Journal of Economic Theory, 13*(3), 341–360.
Rousseeuw, P. J. (1984). Least median of squares regression. *Journal of the American Statistical Association*, *79*(388), 871–880.
Rousseeuw, P., & Yohai, V. (1984). Robust regression by means of S-estimators. In *Robust and nonlinear time series analysis* (pp. 256–272). Springer, New York.
Rousseeuw, P. J., & Driessen, K. V. (1999). A fast algorithm for the minimum covariance determinant estimator. *Technometrics, 41*(3), 212–223.
Rousseeuw, P. J., Van Aelst, S., Van Driessen, K., & Gulló, J. A. (2004). Robust multivariate regression. *Technometrics, 46*(3), 293–305.
Schäfer, J., & Strimmer, K. (2005). A shrinkage approach to large-scale covariance matrix estimation and implications for functional genomics. *Statistical Applications in Genetics and Molecular Biology*, *4*(1).
Schyns, M. (2008). Robust portfolio selection. *JSM Proceedings, Statistical Computing Section*.
Sharpe, W. F. (1964). Capital asset prices: A theory of market equilibrium under conditions of risk. *The Journal of Finance, 19*(3), 425–442.
Stein, C. (1956, January). Inadmissibility of the usual estimator for the mean of a multivariate normal distribution. In *Proceedings of the Third Berkeley Symposium on Mathematical Statistics and Probability* (Vol. 1, No. 399, pp. 197–206).
Tola, V., Lillo, F., Gallegati, M., & Mantegna, R. N. (2008). Cluster analysis for portfolio optimization. *Journal of Economic Dynamics and Control, 32*(1), 235–258.
Wolf, M. (2004). Resampling vs. shrinkage for benchmarked managers.
Woodruff, D. L., & Rocke, D. M. (1994). Computable robust estimation of multivariate location and shape in high dimension using compound estimators. *Journal of the American Statistical Association, 89*(427), 888–896.

The Term Structure Under Non-linearity Assumptions: New Methods in Time Series

José Carlos Vides, Jesús Iglesias and Antonio A. Golpe

Abstract In this chapter, we summarized an empirical review of the EHTS aiming to establish the adequate procedures for its measurement by using time series. On one hand, the chapter discusses the main findings in the literature in the USA and the EMU and, on the other hand, analyses the linearity restrictions associated with the traditional approaches used in time series applications on term structure. The use of FCVAR represents a novel procedure to solve the linearity restrictions. Finally, this application allows the economic policies that derive from its results to be more appropriate for the objectives of the design of monetary policies.

Keywords Expectation hypothesis of term structure · Monetary policy · Term spread · Fractional cointegration

JEL Classification C32 · E43 · N11 · N12

1 Introduction

The relationship between interest rates or bond yields and different maturities has been widely studied in the finance literature from a theoretical and empirical point of view. The so-called term structure of interest rates has always been of fundamental

J. C. Vides
International University of Andalusia (Spain). (SERG), Seville, Spain
e-mail: joseca.vides@gmail.com

J. Iglesias (✉)
Department of Financial Economics and Operations Management,
University of Seville (Spain), (SERG), Seville, Spain
e-mail: jiglesias10@us.es

A. A. Golpe
Department of Economics, University of Huelva (Spain).
Spanish Entrepreneurship Research Group (SERG), Huelva, Spain
e-mail: antonio.golpe@dehie.uhu.es

M. Mili et al. (eds.), *New Methods in Fixed Income Modeling*, Contributions to Management Science, https://doi.org/10.1007/978-3-319-95285-7_7

importance to financial economists, investors and practitioners (see Campbell and Shiller (1991) for an influential approach or Gürkaynak and Wright (2012) for a survey).[1] In particular, understanding the term structure of interest rates is essential for the assessment of the effects of monetary and macroeconomic policies (Mankiw and Summers 1984) in the context of monetary policy as an indicator of market expectations (Rudebusch 1995); the term structure of interest rates contains useful information regarding future real activity and inflation and has prediction power (Estrella and Mishkin 1997).

According to this framework, the term structure should move in line with the predictions of the expectations hypothesis of the term structure (EHTS) so that returns respond to international market forces. Given the connection of the implications of monetary policy decisions to the future of financial markets, the literature has emphasized the understanding of how this relationship works in the long term. Bernanke and Blinder (1992) supported that this relationship among short and long-term interest rates implies that their spread contains significant information on future changes in the short-term rates and has an important role in the potential effectiveness of monetary policy. Similarly, Holmes et al. (2015) sustained that if a monetary policy is effective, changes in short-term policy interest rates should have an impact on long-term ones.

Empirical works have examined this hypothesis in different regions, focusing on the analysis of the relationship in the long term and, consequently, on the study of the linearity of this relationship by using cointegration analysis tools. However, a controversial framework is derived for this empirical review due to recent arguments that question the usefulness of linear cointegration because it provides less power and fails to detect a long-run relationship among short and long-term interest rates (Araç and Yalta 2015). Perhaps the most determinant contribution about the treatment of these series in the long-run is the one made by Hassler and Nautz (2008), which evidenced the presence of the fractional I(d) process in the long-run relationship between interest rates, creating a novel path in the treatment of the fractional time series.

Concerning the United States of America (USA), the EHTS is frequently accepted as a forecasting tool (Poole et al. 2002), and its implications in the monetary policy are also incorporated in the design of the fiscal policy (see Weber and Wolters 2012, 2013). Nevertheless, Mili et al. (2012) show nonlinearities in the relationship between interest rates in the USA. On the other hand, concerning the Euro Zone or the European Monetary Union (EMU), in the seminal paper of Hassler and Nautz (2008), the long-run relationship among European rates is explained by a fractional perspective, giving interpretations as a measure of the central bank's ability to control the overnight rate. Cossetti and Guidi (2009) denote that the actions of the European Central Bank (ECB) in monetary policy do not have substantial impacts on the yield curve; Nautz and Scheithauer (2011) also indicate that the monetary policy design determines the strength of the relationship between the overnight rate and the cen-

[1]The study of term structure has been done for a long time, going back to Macaulay (1938).

tral bank's policy rate. Finally, Tamakoshi and Hamori (2014) reject the presence of linearity in the Eonia—3-month Euribor rate relationship.

In this regard, a new approach in the study of the relationship between short- and long-term interest rates has arisen in the existing literature, considering that the standard unit root and cointegration test might be too restrictive (*I(1)/I(0)* dichotomy). Indeed, the rejection of the assumption that both short- and long-term interest rates follow the dichotomy *I(1)/I(0)* displaying the long-memory process (I(*d*)-type) in the case of the cointegration of both interest rates. The spread could also be measured as $I(d - b)$. To the best of our knowledge, the rigidity of the traditional approach, the linear cointegration, is broken to allow the series to be cointegrated, and the spread does not necessarily need to be stationary-*I(0)*. Overall, this new approach consists of the fractionally cointegrated vector autoregressive (FCVAR) model (Johansen 2008a, b) and Johansen and Nielsen (2012), which was further developed by Nielsen and Popiel (2016).

The rest of the chapter aims to establish an empirical framework that is useful to analyse the EHTS under improved tools at the time that this approach is implemented in a monetary policy portfolio. Therefore, the next section presents a review and a definition of the EHTS, and Sect. 3 shows the empirical evidence by region. Later, in Sect. 4, we develop the FCVAR model. Additionally, the monetary policy and controllability of interest rates are discussed in Sect. 5, and the conclusions are discussed in Sect. 6.

2 The Term Structure and the Expectations Hypothesis of the Term Structure

The term structure of interest rates analyses the relationship between the time remaining until the expiration of the various obligations or bonds and their returns during that period, provided that they all have the same degree of risk, liquidity and tax (Schaefer 1981). It is also called yield curve. The most well-known term structure is formed by financial assets issued by the state because of (a) care solvency risk and (b) problems for caregiver country titles when the market for such assets for the liquidity is problematic.

The term structure of interest rates has multiple applications, which can be divided into four large groups. First the financial economy allows the evaluation of multiple financial assets and the design of the investment or hedging strategies (Bansal and Shaliastovich 2012). Second economic theory allows for the study of issues such as the formation of expectations and the relationship between short- and long-term interest rates and the transmission of the monetary policy to the relevant macroeconomic variables (Mankiw and Summers 1984). Third, the Treasury contributes to analysing the constraints of funding. Finally, the term structure is an indicator for monetary policy that is useful for analysing, along with other tools, the conditions in which this theory acts, the prospects of achieving the target set, the perception of the tone of politics in the monetary policy and the degree of confidence in maintaining it in the future (Cassola and Morana 2008).

As it has been mentioned previously, one of the applications of the term structure is in the formation of expectations. In this sense, one of the most influential theories of term structure emerged as a way to explain the possible relationship between short- and long-term interest rates. This hypothesis that we are introducing is the EHTS, which establishes that an average of the current and expected short-term rates determines long-term rates with an inter-temporal term premium (Bekaert et al. 1997) and has economic implications in macroeconomics or finance and in the shape of the yield curve (see Shiller 1990 for a survey).

This hypothesis was initially defined by Lutz (1940), although it was also confirmed by different authors recently. He starts with the hypothesis that investors have homogeneous but not identical expectations and that the interest rates can be predicted with certainty. Thus, the basic hypotheses formulated by Lutz are as follows:

(a) The markets are efficient; the new information is rapidly reflected in the share prices.
(b) The investors maximize their expected profit by using short- and long-term securities.
(c) There are no transaction costs, and there is freedom of capital movements.
(d) Both the payment of the coupons and the return of the principal are known with certainty.

This hypothesis also explains the behaviour of the yield curve, since an upward sloping yield curve implies that future short-term rates are expected to rise. Conversely, with a downward sloping yield curve, the future short-term rates are expected to fall, i.e., the slope of the yield curve is an important source of information on the real economy evolution. In consequence, Estrella and Hardouvelis (1991) found that a positive curve slope is associated with future increases in real economic activity using macroeconomic variables and by providing a significant predictive power. One implication of the EHTS stated by Fama (1984) and Fama and Bliss (1987) is that the forward rate is an unbiased predictor of future short-term rates. Another implication of this hypothesis is that the spread between the long-term interest rate and the short-term interest rate—the term spread—is an unbiased predictor of future short-run changes in long-term rates (Mankiw 1986; Campbell and Shiller 1991; Campbell 1995). The potential effectiveness of the monetary policy is revealed by this relationship, which consists of the control of short-term policy rates by central banks (Bernanke and Blinder 1992) that will be explained in the next sections.

The fundamental equation of the EHTS of an $n > 1$ period bond R_t (i.e., long-term interest rate) is equal to an average of the current and expected r_t (i.e., short-term interest rate) set of a $n \leq 1$ period plus a constant term. The relationship can be expressed in the following form:

$$R_t = \frac{1}{n}\sum_{k=0}^{n-1} E_t[r_{t+k}] + \Phi_t^*, \qquad (1)$$

where Φ_t^* is a possible stationary term and E_t is the expectation operator at time t for the evolution of short-term interest rates driving long-term interest rates.

3 Evidence by Region

The term structure and the EHTS have been analysed in different contexts and economic regions, although the USA and the EMU are the main regions studied. In this section, we summarize the body of empirical papers that have arisen in the literature supporting (or not) the EHTS, distinguishing between both regions.

3.1 The USA

Concerning the studies on the EHTS in the USA, these works have been reviewed, and it is possible to find studies showing certain controversy in relation to the confirmation of the EHTS. Several studies find evidence in support of the EHTS (e.g., Campbell and Shiller 1987; Hamilton 1988; Hall et al. 1992). The evolution of this analysis has aimed to explore changes in the analysed periods. Engsted and Tanggaard (1994) and Enders and Granger (1998) show asymmetries in the movements towards the long-run equilibrium relationship. Additionally, Hansen (2003), Hansen and Seo (2002), Seo (2003), Junker et al. (2006), Clarida et al. (2006) and Mili et al. (2012) support the EHTS using cointegrating techniques, showing evidence supporting the nonlinearity in the term structure of interest rates. In this context, Sarno and Thornton (2003) used non-linear error-correction equations, finding that the adjustment of the overnight rate to the Treasury bill rate is asymmetric. There is also evidence in support of the EHTS in the relation between short- and long-term rates among the European and the USA rates (Lanne 2003; Brüggemann and Lütkepohl 2005) and in the combining of yield factors and macroeconomic variables to relate with the EHTS, which is evidence in favour of certain regimes (Diebold et al. 2006). Weber and Wolters (2012, 2013) applied the vector error correction model (VECM) to US term structure in order to contribute an economic explanation of the deviations from the EHTS. Recently, Kishor and Marfatia (2013) showed that the future rate is cointegrated with the 3-month rate. Holmes et al. (2015) also examined the term structure of interest rates using a pairwise stationary approach supporting that the EHTS holds in the long-run, i.e., the short-run policy changes affect the long-term rates.

However, some reasons to reject the EHTS validation for the USA have also emerged in the recent literature. According to Bekaert and Hodrick (2001) (see for a survey), there are three potential reasons for the rejection of the EHTS:

> First, the EHTS is based on the assumption of rational expectations and unlimited arbitrage. It may be that irrational investors make systematic forecast errors, and the ability of rational investors to profit from this situation is limited by their risk aversion. Second, the presence of time-varying risk premiums means that standard tests of the EHTS omit the variables

> capturing the risk premium. Whether these variables are related with interest rates, the estimated coefficients would be pulled away from those implied by the EHTS. Third, the tests themselves may lead to false rejections because of their poor properties in finite samples. (p. 1358)

In this respect, against the fulfilment of the EHTS in the USA, Engle et al. (1987) demonstrated the failure of the EHTS. In a wide sample examining the term spread of G7 countries, Hardouvelis (1994), who used a VAR model that attempted to forecast changes in long-term interest rates, showed that EHTS is supported in all countries except the USA, while Bekaert and Hodrick (2001) also showed evidence against the EHTS using different methodologies. Nevertheless, Thornton (2005) tested the EHTS in federal fund rates in order to determine whether the market's expectation is less able to forecast the federal fund rates. Sarno et al. (2007) also find mixed results in a bivariate analysis; therefore, using maturities from one month to 10 years and a powerful test (Lagrange Multiplier test), the EHTS is rejected. Conversely, Guidolin and Thornton (2010) concluded that future short-term rates have deep implications for policy makers, suggesting that whether or not EHTS is true, the inability to predict the future short-term rate would imply that both long-term and short-term rates are equal, suggesting that this relation would be inconsistent, hence, the conventional theory of the term structure of interest rates is threatened. Finally, Bulkley et al. (2011, 2015) determined the failure of the EHTS using bond yields on US Treasury securities. Overall, this subsection has summarized the empirical puzzle that is derived from the review of this literature in the USA.

3.2 European Monetary Union

In the European Monetary Union, the empirical framework is similar to that of the USA, accentuated by the numerous applications dedicated to each of the different countries that make up this region. This topic has been studied in different ways, highlighting the papers that relate long-term interest rates, i.e., sovereign bonds or interest rates, with longer maturities and short-term interest rates; some authors believe that the Euro OverNight Index Average (Eonia) rate is crucial for the signalling and transmission of the ECB monetary policy, using the Eonia as an indicator of the behaviour of the interest rates (Benito et al. 2007). For almost all central banks, the interbank money market for overnight lending is the key channel through which the monetary policy is executed. In this sense, overnight rates are the operational target of monetary policy that anchors the term structure of interest rates (Nautz and Offermanns 2007).

Initially, the study of this topic in the Eurozone was limited because studies were focused on specific countries rather than the whole region. Therefore, papers based in some European countries are noteworthy. Hurn et al. (1995) obtain results in favour of the EHTS from interest rates in the UK interbank market. Dahlquist and Jonsson (1995) were unable to reject the EHTS based on interest rates from Sweden. In the case of German studies, Hardy (1998), Hammersland and Vikøren (1997) and

Hafer et al. (1997), demonstrated that German term structure occupies a dominant position in the future EMU. Gerlach and Smets (1997) find empirical support for the EHTS for Belgium, France, Germany, Italy, and Spain. For the Eurozone, we have to go back to Gerlach and Smets (1997), who did not reject the EHTS in a sample of 17 Euro-countries. However, this issue acquired relevance once the European Monetary Union was born. Thus, Ayuso and Repullo (2003) show that non-symmetric adjustment of the Eonia would also be induced by an asymmetric loss of function of the central bank. This latter issue will be important for the development of future studies concerning the term structure associated with the monetary policy in the Eurozone. Meanwhile, Nautz and Offermanns (2007) confirmed the expectations with an asymmetric response for the Euro area. Importantly, we find a seminal paper (Hassler and Nautz 2008) that explained the long-run relationship between European rates from a fractional perspective and gives some advice about the controllability of interest rates by Central Banks, changing the traditional assumption of the term structure treatment and using the cointegration techniques. For its part, Cossetti and Guidi (2009) denote that the actions of the ECB in monetary policy do not have substantial impacts on the yield curve because the presence of cointegration was rejected for maturities longer than six years, which means that for shorter rates, the presence of expectations would not be rejected. Regarding the pressure on the Eonia, Linzert and Schmidt (2011) show that the rate expectations are not relevant in a scenario with a reduction of liquidity. Otherwise, Nautz and Scheithauer (2011) indicate that the strength of the relation between the overnight rate and the central bank's policy rate is determined by monetary policy design. Similarly, Belke et al. (2013) used a linkage between short-term interbank interest rates, i.e., the Eonia and the 3-month Euribor rate, to study the persistence of the spread due to the importance of the market expectations of the European monetary policy attitudes in the near future. Likewise, Tamakoshi and Hamori (2014) rejected the presence of linearity in the Eonia—3-month Euribor rate relationship using a threshold cointegration. They also determined that the Eonia plays a crucial role in signalling the target of the monetary policy. Araç and Yalta (2015) consider whether the recent financial and debt crises may have affected the relation between short-term and long-term interest rates, indicating that the EH holds in Greece, Ireland and Portugal. Meanwhile, for the other countries in the sample, there is no cointegration between short- and long-term interest rates.

Finally, in order to simplify and clarify this section, a summary is provided in Table 1 in order to show the authors, year of publication, the concerning country or region, the technique used and whether the EHTS is supported (or not).

Table 1 Summary of EHTS evidence by regions

Authors, year	Region/Country	Technique	EHTS
Araç and Yalta (2015)	Eurozone	Nonlinear cointegration	Only in Greece, Ireland and Portugal
Ayuso and Repullo (2003)	European Monetary Union	Generalized method of moments	Yes
Baillie and Bollerslev (1994)	Canada, West Germany, Japan, United Kingdom, France, Italy and Switzerland	Fractional cointegration	Yes
Bekaert and Hodrick (2001)	The USA, Germany and United Kingdom	VAR and Lagrange multiplier	No
Bekaert et al. (1997)	(Bootstrap approach)	VAR—GARCH	No
Belke et al. (2013)	European Monetary Union	VAR and Lagrange multiplier	Yes
Benito et al. (2007)	European Monetary Union	ARCH-Poisson-Gaussian process	
Brüggeman and Lütkepohl (2005)	The USA and the Eurozone	VECM	Yes
Bulkley et al. (2011)	The USA	VAR	No
Bulkley et al. (2015)	The USA	The law of small numbers	No
Busch and Nautz (2010)	European Monetary Union	Fractional integration	Yes
Camarero et al. (2008)	European Monetary Union	Pooled and panel cointegration	Yes
Campbell (1995)	The USA	Regressions of long-run changes	Yes
Campbell and Shiller (1987)	The USA	CVAR	Yes
Campbell and Shiller (1991)	The USA	VAR	Yes
Clarida et al. (2006)	The USA, Germany and Japan	Nonlinear VECM	Yes

(continued)

Table 1 (continued)

Authors, year	Region/Country	Technique	EHTS
Cömert (2012)	The USA	Simple OLS and generalized method of moments	Yes
Cossetti and Guidi (2009)	European Monetary Union	EGARCH and cointegration	Yes
Dahlquist and Jonsson (1995)	Sweden	Cointegration and ECM	Yes
Diebold et al. (2006)	The USA	Non-structural VAR	Yes
Enders and Granger (1998)	The USA	Momentum Threshold Autoregressive (M-TAR) and ECM	Yes
Engle et al. (1987)	The USA	ARCH	No
Engsted and Tanggaard (1994)	The USA	VECM	Yes
Estrella and Hardouvelis (1991)	The USA	OLS	Yes
Estrella and Mishkin (1997)	The USA and Germany	VAR	Yes
Evans and Marshall (1998)	The USA	VAR and impulse—response functions	Yes
Fama (1984)	The USA	OLS	Yes
Fama and Bliss (1987)	The USA	OLS	Yes
Gerlach and Smets (1997)	Europe and The USA	Cross—sectional regressions	Yes
Guidolin and Thornton (2010)	The USA	Diebold and Li Model and OLS	No
Hafer et al. (1997)	Belgium, France, Germany and Netherlands	VAR and permanent-transitory decomposition	Yes
Hall et al. (1992)	The USA	Cointegration	Yes
Hamilton (1988)	The USA	Markov processes and OLS	Yes

(continued)

Table 1 (continued)

Authors, year	Region/Country	Technique	EHTS
Hansen and Seo (2002)	The USA	Threshold VECM	Yes
Hansen (2003)	The USA	VAR	Yes
Hardouvelis (1994)	G7 countries	VAR	Yes (except The USA)
Hardy (1998)	Germany	OLS	Yes
Hassler and Nautz (2008)	European Monetary Union	Fractional integration	Yes
Holmes et al. (2015)	The USA	Pair-wise cointegration	Yes
Hurn et al. (1995)	United Kingdom	VAR	Yes
Junker et al. (2006)	The USA	Copula functions	Yes
Kishor and Marfatia (2013)	The USA	Dynamic OLS and VECM	Yes
Lanne (2003)	The USA	Markov switching model	Yes
Mankiw (1986)	The USA, Canada, United Kingdom and Germany	GLS	Yes
Mankiw and Summers (1984)	The USA	WLS	No
Mili et al. (2012)	The USA	Parametric nonlinear inference approach	Yes
Nautz and Offermanns (2007)	European Monetary Union	Nonlinear cointegration	Yes
Nautz and Scheithauer (2011)	European Monetary Union, The USA, United Kingdom and Switzerland	Fractional integration	Yes
Poole et al. (2002)	The USA	OLS and Poole/Rasche and Kuttner methodology	No
Sarno and Thornton (2003)	The USA	Non-linear asymmetric vector equilibrium correction model	Yes
Sarno et al. (2007)	The USA	Lagrange multiplier	No

(continued)

Table 1 (continued)

Authors, year	Region/Country	Technique	EHTS
Seo (2003)	The USA	Threshold cointegration	Yes
Strohsal and Weber (2014)	The USA	GARCH and cointegration	Yes
Tamakoshi and Hamori (2014)	European Monetary Union	Threshold cointegration	Yes
Thornton (2005)	The USA		No
Weber and Wolters (2012)	The USA	VECM	Yes
Weber and Wolters (2013)	The USA	VECM	Yes

4 Time Series Applications of the Term Structure: The FCVAR

Regarding the methodology used in the previous section, the majority of the literature has shown that it is possible to establish a relationship between short- and long-term rates using, mainly, cointegration techniques. Engle and Granger (1987) developed this concept. Initially, there were studies focused on whether interest rates can be characterized as an $I(0)$ or $I(1)$ series. For instance, Cox et al. (1985) concluded that the short-term nominal interest rate is a stationary and mean-reverting $I(0)$ process, whereas Campbell and Shiller (1987) assumed a unit root. To solve this restriction, many authors used threshold cointegration, such as Hansen and Seo (2002) and Seo (2003), who showed evidence supporting the nonlinear mean-reversion in the term structure of interest rates. Nevertheless, as it has been measured traditionally, we believe that the standard unit root and cointegration test might be too restrictive ($I(1)/I(0)$ dichotomy); the choice of such models is hotly debated, since it is unclear whether $I(0)$ or $I(1)$ processes are more appropriate (Caporale and Gil-Alana 2016). In this sense, different authors have indicated that term structure could display long-memory processes. In this regard, Hassler and Nautz (2008), Cassola and Morana (2008), Busch and Nautz (2010), Caporale and Gil-Alana (2016) and Nautz and Scheithauer (2011) determine that an $I(d)$ process could provide additional flexibility to the relationship behaviour, with d values different from 0 or 1.

According to this idea, a novel methodology emerges in order to avoid the problems with the axioms of traditional cointegration associated with rigidity, rejecting the assumption that both short- and long-term interest rates follow the dichotomy $I(1)/I(0)$, and the spread follows a stationary process ($I(0)$), in line with Pérez-Quirós and Rodríguez-Mendizábal (2006) or Nautz and Offermanns (2007). The model is a generalization of Johansen's (1995) cointegrated vector autoregressive (CVAR) model to allow for fractional processes of order d that co-integrate to order $d-b$. This model has the advantage of being used for stationary and non-stationary time series and is presented by Johansen (2008a, b) and further developed by Johansen and Nielsen (2012) and Nielsen and Popiel (2016).

To introduce the FCVAR model, we begin with the well-known, non-fractional, CVAR model with $Y_t = 1, \ldots, T$ a p-dimensional $I(1)$ time series. Therefore, the CVAR model is:

$$\Delta Y_t = \alpha\beta' Y_{t-1} + \sum_{i=1}^{k} \Gamma_i \Delta Y_{t-i} + \varepsilon_t = \alpha\beta' L Y_t + \sum_{i=1}^{k} \Gamma_i \Delta L^i Y_t + \varepsilon_t \quad (2)$$

The fractional difference operator introducing persistence in the model is Δ, and the fractional lag operator is L = (1 − Δ). Replacing lag operators with their fractional counterparts Δ^b and $L_b = \left(1 - \Delta^b\right)$, we obtain

$$\Delta^b Y_t = \alpha\beta' L_b Y_t + \sum_{i=1}^{k} \Gamma_i \Delta^b L_b^i Y_t + \varepsilon_t, \quad (3)$$

and we apply this equation to $Y_t = \Delta^{d-b} X_t$, such that

$$\Delta^d X_t = \alpha\beta' L_b \Delta^{d-b} X_t + \sum_{i=1}^{k} \Gamma_i \Delta^d L_b^i X_t + \varepsilon_t. \quad (4)$$

As always, ε_t is the p-dimensional independent and is identically distributed with the mean zero and covariance matrix Ω. The parameters α and β are $p \times r$ matrices, where $0 \leq r \leq p, d \geq b > 0$. In matrix β, the columns are the cointegrating relationships, and $\beta' X_t$ assumes the existence of a common stochastic trend, which is integrated with order d. the short-term parts from the long-run equilibrium are integrated in order—$d - b$, but if—$d - b$<1/2, then it is asymptotically a zero-mean stationary process. The coefficients in α correspond to the speed of adjustment of the equilibrium. Therefore, $\alpha\beta'$ is the adjustment long-run, ρ' is the restricted constant term, and Γ_i represents the short-run behaviour of the. We reach the final model

$$\Delta^d X_t = L_d \alpha\left(\beta' X_t + \rho'\right) + \sum_{i=1}^{k} \Gamma_i \Delta^d L_d^i X_t + \varepsilon_t. \quad (5)$$

When the VAR model is in the case of $d = b = 1$ (CVAR), X_t is integrated with order d, and b is the strength of the cointegrating relationships (as the value of b is higher, the persistence is lower in the cointegrating relationships). The error correction term is integrated with order $(d - b)$, which is $I(0)$ in this case. However, in the fractional cointegration, these axioms are relaxed because $(d - b) = 0$, i.e., the error correction term shows a short-run stationary behaviour or $(d - b) > 0$, i.e., there is a long memory process, and the error correction term will revert in the long run.

Johansen and Nielsen (2012) show that the maximum likelihood estimators $(d, \alpha, \Gamma_i, \ldots, \Gamma_k)$ are asymptotically normal, and the maximum likelihood estimator of (β, ρ) is asymptotically mixed normal.

To determine the number of stationary cointegrating relations following the hypotheses in the rank test based on a series of LR tests, in the FCVAR model, we test the hypothesis $H_0 : rank(\Pi) = r$ against the alternative $H_1 : rank(\Pi) = p$. As $L(d, b, r)$, the profile likelihood function is given a rank *r*, where (α, β, Γ) has been reduced by rank regression (see Johansen and Nielsen 2012). In the case of a model with a constant, we test $H_0 : rank(\Pi, \mu) = r$ against the alternative $H_1 : rank(\Pi, \mu) = p$, and the profile likelihood function given rank *r* is $L(d, r)$, where the parameters $(\alpha, \beta, \rho, \Gamma)$ have been focused. Note that matrix α and β are normalized separately in the same way for the CVAR model because the degrees of freedom are non-standard.

Maximizing the profile likelihood distribution under both hypothesis, the LR test statistics are now $LR_t(q)$. The asymptotic distribution of $LR_t(q)$ depends on the parameter *b* and on $q = n - r$. MacKinnon and Nielsen (2014), based on their numerical distribution functions, provide asymptotic critical values of the LR rank test. In the case of "weak cointegration", i.e., $0 < b < 1/2$, $LR_t(q)$ has a standard asymptotic distribution, $LR_t(\mathrm{q}) \xrightarrow{D} \chi^2(q^2)$ To summarize, by estimating the FCVAR model, we extract richer information from what was mentioned in the previous sections. Importantly, by separately parameterizing the long-run and the short-run dynamics of the series, the model is able to accommodate empirically realistic $I(\mathrm{d})$ long-memory and fractional cointegration while maintaining that the returns are $I(0)$ (Bollerslev et al. 2013).

As we said, this methodology allows testing of the long-run relationship between interest rates with different maturities, the measurement of the spread[2] and the implications for monetary policy in a joint estimation. For this reason, Table 2 proposes a possible strategy of empirical research, allowing for the study of long-run relationships of interest rates and testing the spread persistence in order to achieve monetary policy conclusions.

5 Monetary Policy and Controllability of Interest Rates

Studies concerning the term structure of interest rates have tried to evaluate their impact and how they are affected by the monetary policy of Central Banks. The term structure has long been established as reflecting economic agents' anticipations of future events and as an indicator of monetary policy, as seen in the volume of academic articles written over the past century dealing with term structure, which is testimony to both the practical importance of the topic as well as its intrinsic academic

[2]When the relationship between interest rates with different maturities is supported, it has to be restricted by a cointegrating vector of $(1, -1)$, then, the difference between those interest rates are interpreted as the spread.

Table 2 Strategy of empirical research

	Procedure	Hypotheses
Step 1	Fractional cointegration?	H_1^d: Is the fractional cointegration more appropriate than traditional cointegration?
Step 2	Estimation of β	H_1^β: Cointegrating vector is $(1, -1)$
Step 3	Estimation of adjustment coefficients (α_R, α_r)	$H_1^\beta \cap H_1^{\alpha_i}$: The interest rates are weakly exogenous under the restriction of the cointegrating vector $(1, -1)$
Step 4	Degree of spread persistence, i.e. order of integration $(d-b)$	H_1^{d-b}: Is the spread a long memory process?

appeal (see Vetzal 1994 for a survey). In consequence, changes in the economy could affect the EHTS; therefore, if a variation in short-term policy impacts the long term, monetary policy is effective (Holmes et al. 2015).

As mentioned in the previous sections, the potential effectiveness of monetary policy is revealed by this relationship, which consists of the control of short-term policy rates by central banks. In principle, since authorities can control the path of short-term interest rates, they should also be able to influence the wide movements in long-term interest rates sufficiently for policy objectives, provided traditional term structure relations hold up reasonably well (Christiansen and Pigott 1997). This does not require that the classical term structure theory holds exactly but only that expectations about future short-term interest rates have a major influence on long-term rates, as is suggested by traditional studies of the term structure (Shiller 1990). Although the connection between monetary policy and long-term interest rates appears to be weaker and less reliable, monetary policy can readily influence short-term interest rates (Roley and Sellon 1995; Camarero et al. 2008). In this sense, monetary policy shocks primarily affect short-term interest rates with a diminishing effect on longer-term rates, which can be explained by the EHTS (Evans and Marshall 1998). Hence, under the expectations hypothesis, changes in the term structure can be used to infer changes in investors' expectations concerning the path of monetary policy. If, in addition, the central bank's rule relating monetary policy to macroeconomic conditions is known by those investors, then we could also read off changes in their expectations of the state of the economy (Gürkaynak and Wright 2012) (see for a survey).

More importantly, several studies have treated the spread between long- and short-term interests. These studies have focused on different regions in the world; for the EMU, the impact of monetary policy shocks on bond yields declines with the maturity of the bonds, and this impact is significantly lower when the shock stems from a monetary policy meeting of the ECB (Perez-Quiros and Sicilia 2002). Regarding policy implications, Hassler and Nautz (2008), Cassola and Morana (2008) and Nautz and Scheithauer (2011) reveal that the Eonia spread is *I(0)* before but fractionally integrated with long memory when the order of fractional integration *d* has increased to approximately 0.25. Since $d < 0.5$, the Eonia is still under the ECB's control. Additionally, the increased persistence of the Eonia spread suggests that the degree of controllability of the Eonia spread may have declined. Meanwhile, Busch and Nautz (2010) estimated a long memory process and found that the persistence of deviations in long-term money market rates from the European central bank's policy rate has decreased, implying that monetary policy has become more effective in controlling interest rates. Caporale and Gil-Alana (2016) suggest that the ECB, and member state central banks, have controlled money market rates in a strict way, particularly at the short-end of their term structure. In the case of the USA, the work of Strohsal and Weber (2014) and Holmes et al. (2015) supports the EHTS; however, the degree of integration of the spread would be different from I(*0*). Previously, Cömert (2012) related overnight interest rates and long-term rates in the US and offered evidence that the Fed has been gradually losing its control over long-term interest rates.

Another tool to control the interest rates is the study of the spread. In this regard, Bernanke and Blinder (1992) showed that this relationship among short- and long-term interest rates implies that their spread contains significant information on future changes in short-term rates and has an important role in the potential effectiveness of monetary policy, which consists of the control of short-term policy rates by central banks. The economy is affected by the monetary impulses through long-term interest rate movements. Therefore, if spreads are highly persistent, the lasting impact of shocks may impede the transparency of policy signals and, thus, the central bank's impact on longer-term rates. In this respect, Nautz and Offermanns (2007) found evidence that the reaction of the Eonia rate to the spread is non-symmetric but, interestingly, the ECB did not lose control over the Eonia rate. From a fractional integration perspective, the spread could exhibit long memory ($d > 0$), but non-stationarity ($d \geq 0.5$) can be rejected in most cases. An explicit test for a change in the order of fractional integration is provided by Sibbertsen and Kruse (2009). Additionally, Baillie and Bollerslev (1994), Tkacz (2001)[3] and Cassola and Morana (2008), among others, suggest that the spread could follow a fractional order of integration, which could be an indicator of the power that the authorities have over the interest rates.

In summary, recent developments based on fractional integration and cointegration allow the obtained results to be useful tools for economic policy design. Therefore, the application of the FCVAR is crucial because this methodology permits, in a joint estimation, the study of the long-run and the short-run dynamics of the series, the measurement of the spread, and the implications for monetary policy, depending on the value of the order of integration of the spread.

[3]Following Tkacz (2001), when $(d - b) = 0$, the spread follows a stationary process and the shock duration is short-lived, i.e. this means that a shock would show a slow return towards the long-run equilibrium. If $0 < (d - b) < 0.5$, there is a stationary process, and the shock duration is long-lived, and finally, if $0.5 < (d - b) < 1$, the spread follows a non-stationary process, although it is mean-reverting and the shock duration is long-lived.

6 Conclusions

Corresponding to the EHTS, long-term rates could explain changes in future short-term rates. Understanding the term structure of interest rates has always been viewed as crucial to assess the impact of monetary policy and its transmission mechanism. Indeed, if a monetary policy is effective, changes in short-term policy interest rates should impact long-term ones. Despite this hypothesis being widely known, major contributions arose in the end of the last century and the beginning of the 21st century. Notably, the studies carried out by Campbell and Shiller (1987) and Fama and Bliss (1987) contributed to establishing the main implications of the EHTS. More recently, several papers have examined the existence of the EHTS by using time series methodologies, using different perspectives, i.e., selecting different maturities for interest rates and/or different countries, and providing conclusions for investors and policy makers.

Initially, the research concerning the EHTS was focused on the study of the interest rates under the lens of the existence (or not) of unit roots, i.e., the series would be I(0)/I(1). Nonetheless, authors such as Mili et al. (2012) and Hassler and Nautz (2008), for instance, showed the presence of non-linarites and a fractional I(d) process in the long-run relationship between interest rates, respectively. According to these results, a novel technique in the treatment of the fractional time series emerges, i.e., the FCVAR applied to the long-run relationship between short- and long-term interest rates. Under the FCVAR assumptions, it could be considered that the standard unit root and cointegration test might be too restrictive ($I(1)/I(0)$ dichotomy). Indeed, the rejection of the assumption that both short- and long-term interest rates follow the dichotomy $I(1)/I(0)$ displaying the long-memory process (I(d)-type) is similar to the case of the cointegration of both interest rates. Additionally, the spread could be measured as I($d - b$). Therefore, the rigidity of the traditional approach is broken in favour of allowing the series to be cointegrated, and the spread does not necessarily need to be $I(0)$.

Finally, by testing the term structure of interest rates, it is possible to study the behaviour of the long-run relationship between interest rates and how the term structure would change in the time after a shock. According to this idea, involving the concept of long memory and fractional integration and cointegration, the joint estimation of the long-run relationship and the study of the persistence of the spread are possible. In this respect, the long memory of the spread holds adequate forecasting power over longer horizons (Baillie and Bollerslev 1994). Otherwise, another factor plays an important issue in the design of the monetary policy, i.e., the persistence of the spread, in which a greater persistence may indicate that it is more difficult for monetary policy signals to be transmitted along the money market yield curve. Additionally, if spreads are highly persistent, the lasting impact of shocks may impede the transparency of policy signals and, thus, the central bank's impact on longer-term rates, implying a gradual loss of control over interest rates by Central Banks (see Cassola and Morana 2008; Hassler and Nautz 2008; Cömert 2012 for Europe and the USA).

References

Araç, A., & Yalta, A. Y. (2015). Testing the expectations hypothesis for the Eurozone: A nonlinear cointegration analysis. *Finance Research Letters, 15,* 41–48.

Ayuso, J., & Repullo, R. (2003). A model of the open market operations of the European Central Bank. *The Economic Journal, 113*(490), 883–902.

Baillie, R. T., & Bollerslev, T. (1994). Cointegration, fractional cointegration, and exchange rate dynamics. *The Journal of Finance, 49*(2), 737–745.

Bansal, R., & Shaliastovich, I. (2012). A long-run risks explanation of predictability puzzles in bond and currency markets. *The Review of Financial Studies, 26*(1), 1–33.

Bekaert, G., & Hodrick, R. J. (2001). Expectations hypotheses tests. *The Journal of finance, 56*(4), 1357–1394.

Bekaert, G., Hodrick, R. J., & Marshall, D. A. (1997). On biases in tests of the expectations hypothesis of the term structure of interest rates. *Journal of Financial Economics, 44*(3), 309–348.

Belke, A., Beckmann, J., & Verheyen, F. (2013). Interest rate pass-through in the EMU–New evidence from nonlinear cointegration techniques for fully harmonized data. *Journal of International Money and Finance, 37,* 1–24.

Benito, F., León, Á., & Nave, J. (2007). Modeling the Euro overnight rate. *Journal of Empirical Finance, 14*(5), 756–782.

Bernanke, B. S., & Blinder, A. S. (1992). The federal funds rate and the channels of monetary transmission. *The American Economic Review,* 901–921.

Bollerslev, T., Osterrieder, D., Sizova, N., & Tauchen, G. (2013). Risk and return: Long-run relations, fractional cointegration, and return predictability. *Journal of Financial Economics, 108*(2), 409–424.

Brüggemann, R., & Lütkepohl, H. (2005). Uncovered interest rate parity and the expectations hypothesis of the term structure. Empirical results for the US and Europe. *Applied Economics Quarterly, 51,* 143–154.

Bulkley, G., Harris, R. D., & Nawosah, V. (2011). Revisiting the expectations hypothesis of the term structure of interest rates. *Journal of Banking & Finance, 35*(5), 1202–1212.

Bulkley, G., Harris, R. D., & Nawosah, V. (2015). Can behavioral biases explain the rejections of the expectation hypothesis of the term structure of interest rates? *Journal of Banking & Finance, 58,* 179–193.

Busch, U., & Nautz, D. (2010). Controllability and persistence of money market rates along the yield curve: Evidence from the euro area. *German Economic Review, 11*(3), 367–380.

Camarero, M., Ordonez, J., & Tamarit, C. (2008). The expectations hypothesis of the term structure in the Euro area. *Economic Bulletin, 3,* 1–15.

Campbell, J. Y. (1995). Some lessons from the yield curve. *Journal of Economic Perspectives, 9,* 124–152.

Campbell, J. Y., & Shiller, R. J. (1987). Cointegration and tests of present value models. *Journal of Political Economy, 95,* 1062–1088.

Campbell, J. Y., & Shiller, R. J. (1991). Yield spreads and interest rate movements: A bird's eye view. *Review of Economic Studies, 58,* 495–514.

Caporale, G. M., & Gil-Alana, L. A. (2016). Persistence and cyclical dependence in the monthly euribor rate. *Journal of Economics and Finance, 40*(1), 157–171.

Cassola, N., & Morana, C. (2008). Modeling short-term interest rate spreads in the euro money market. *International Journal of Central Bank.*

Christiansen, H., & C. Pigott. (1997). Long-term interest rates in globalised markets. OECD economics department working papers no 175.

Clarida, R. H., Sarno, L., Taylor, M. P., & Valente, G. (2006). The role of asymmetries and regime shifts in the term structure of interest rates. *The Journal of Business, 79*(3), 1193–1224.

Cömert, H. (2012). *Decoupling between the federal funds rate and long-term interest rates: Decreasing effectiveness of monetary policy in the U.S.* No. wp295, Political Economy Research Center.

Cossetti, F., & Guidi, F. (2009). *ECB monetary policy and term structure of interest rates in the euro area: an empirical analysis*. Working Papers 334, Universitá Politecnica delle Marche (I), Dipartimento di Scienze Economiche e Sociali.

Cox, J. C., Ingersoll, J. E., & Ross, S. A. (1985). A theory of the term structure of interest rates. *Econometrica, 53*(2), 385.

Dahlquist, M., & Jonsson, G. (1995). The information in Swedish short-maturity forward rates. *European Economic Review, 39*(6), 1115–1131.

Diebold, F. X., Rudebusch, G. D., & Aruoba, S. B. (2006). The macroeconomy and the yield curve: A dynamic latent factor approach. *Journal of Econometrics, 131*(1–2), 309–338.

Enders, W., & Granger, C. W. J. (1998). Unit-root tests and asymmetric adjustment with an example using the term structure of interest rates. *Journal of Business and Economic Statistics, 16*(3), 304–311.

Engle, R. F., & Granger, C. W. (1987). Cointegration and error correction: Representation, estimation and testing. *Econometrica, 55,* 251–276.

Engle, R. F., Lilien, D. M., & Robins, R. P. (1987). Estimating time varying risk premia in the term structure: The ARCH-M model. *Econometrica: Journal of the Econometric Society*, 391–407.

Engsted, T., & Tanggaard, C. (1994). Cointegration and the US term structure. *Journal of Banking & Finance, 18*(1), 167–181.

Estrella, A., & Hardouvelis, G. A. (1991). The term structure as a predictor of real economic activity. *Journal of Finance, 46,* 555–576.

Estrella, A., & Mishkin, F. S. (1997). Is there a role for monetary aggregates in the conduct of monetary policy? *Journal of Monetary Economics, 40*(2), 279–304.

Evans, C. L., & Marshall, D. A. (1998). Monetary policy and the term structure of nominal interest rates: Evidence and theory. *Carnegie-Rochester Conference Series on Public Policy, 49,* 53–111.

Fama, E. (1984). The information in the term structure. *Journal of Financial Economics, 13*(4), 509–528.

Fama, E., & Bliss, R. (1987). The information in long-maturity forward rates. *American Economic Review, 77,* 680–692.

Gerlach, S., & Smets, F. (1997). The term structure of Euro-rates: Some evidence in support of the expectations hypothesis. *Journal of International Money and Finance, 16*(2), 305–321.

Guidolin, M., & Thornton, D. L. (2010). *Predictions of short-term rates and the expectations hypothesis*. Federal Reserve Bank of St. Louis Working Paper Series 2010-013A.

Gürkaynak, R. S., & Wright, J. H. (2012). Macroeconomics and the term structure. *Journal of Economic Literature, 50*(2), 331–367.

Hafer, R. W., Kutan, A. M., & Zhou, S. (1997). Linkage in EMS term structures: Evidence from common trend and transitory components. *Journal of International Money and Finance, 16*(4), 595–607.

Hall, A. D., Anderson, H. M., & Granger, C. W. (1992). A cointegration analysis of Treasury bill yields. *The Review of Economics and Statistics*, 116–126.

Hamilton, J. D. (1988). Rational-expectations econometric analysis of changes in regime: An investigation of the term structure of interest rates. *Journal of Economic Dynamics and Control, 12,* 385–423.

Hammersland, R., & Vikøren, B. (1997). Long-term interest rates in the US and Germany, mimeo, Norges Bank.

Hansen, B. E., & Seo, B. (2002). Testing for two-regime threshold cointegration in vector error-correction models. *Journal of Econometrics, 110,* 293–318.

Hansen, P. R. (2003). Structural changes in the cointegrated vector autoregressive model. *Journal of Econometrics, 114*(2), 261–295.

Hardouvelis, G. A. (1994). The term structure spread and future changes in long and short rates in the G7 countries: Is there a puzzle? *Journal of Monetary Economics, 33,* 255–283.

Hardy, D. (1998). Anticipation and surprises in Central Bank interest rate policy: The case of the Bundesbank. *IMF Working Paper, 43*(4), 647–671.

Hassler, U., & Nautz, D. (2008). On the persistence of the Eonia spread. *Economics Letters, 101*(3), 184–187.
Holmes, M. J., Otero, J., & Panagiotidis, T. (2015). The expectations hypothesis and decoupling of short-and long-term US interest rates: A pairwise approach. *The North American Journal of Economics and Finance, 34,* 301–313.
Hurn, A. S., Moody, T., & Muscatelli, V. A. (1995). The term structure of interest rates in the London interbank market. *Oxford Economic Papers, 47*(3), 418–436.
Johansen, S. (1995). *Likelihood based inference on cointegration in the vector autoregressive model*. Oxford, UK: Oxford University press.
Johansen, S. (2008a). A representation theory for a class of vector autoregressive models for fractional processes. *Econometric Theory, 24,* 651–676.
Johansen, S. (2008b). Representation of cointegrated autoregressive processes with application to fractional processes. *Econometric Reviews, 28,* 121–145.
Johansen, S., & Nielsen, M. Ø. (2012). Likelihood inference for a fractionally cointegrated vector autoregressive model. *Econometrica, 80*(6), 2667–2732.
Junker, M., Szimayer, A., & Wagner, N. (2006). Nonlinear term structure dependence: Copula functions, empirics, and risk implications. *Journal of Banking & Finance, 30*(4), 1171–1199.
Kishor, N. K., & Marfatia, H. A. (2013). Does federal funds futures rate contain information about the treasury bill rate? *Applied Financial Economics, 23*(16), 1311–1324.
Lanne, M. (2003). Testing the expectations hypothesis of the term structure of interest rates in the presence of a potentia regime shift. *Manchester School, 71,* 54–67.
Linzert, T., & Schmidt, S. (2011). What explains the spread between the Euro overnight rate and the ECB's policy rate? *International Journal of Finance and Economics, 16*(3), 275–289.
Lutz, F. A. (1940). The structure of interest rates. *The Quarterly Journal of Economics, 55*(1), 36–63.
Macaulay, F. R. (1938). *The movement of interest rates, bond yields, and stock prices in the United States since 1856*. New York: National Bureau of Economic Research.
MacKinnon, J. G., & Nielsen, M. Ø. (2014). Numerical distribution functions of fractional unit root and cointegration tests. *Journal of Applied Econometrics, 29,* 161–171.
Mankiw, N. G. (1986). The term structure of interest rates revisited. *Brookings Papers on Economic Activity, 1,* 61–96.
Mankiw, N. G., & Summers, L. H. (1984). Do long-term interest rates overreact to short-term interest rates. *Brookings Papers Economic Activity, 1,* 223–242.
Mili, M., Sahut, J. M., & Teulon, F. (2012). New evidence of the expectation hypothesis of interest rates: A flexible nonlinear approach. *Applied Financial Economics, 22*(2), 165–176.
Nautz, D., & Offermanns, C. J. (2007). The dynamic relationship between the euro overnight rate, the ECB's policy rate and the term spread. *International Journal of Finance and Economics, 12*(3), 287–300.
Nautz, D., & Scheithauer, J. (2011). Monetary policy implementation and overnight rate persistence. *Journal of International Money and Finance, 30*(7), 1375–1386.
Nielsen, M. Ø., & Popiel, M. K. (2016). *A Matlab program and user's guide for the fractionally cointegrated VAR model*. QED working paper, 1330.
Perez-Quiros, G., & Sicilia, J. (2002). *Is the European Central Bank (and the United States Federal Reserve) predictable*? ECB Working Paper Series No. 192.
Pérez-Quirós, G., & Mendizábal, H. R. (2006). The daily market for funds in Europe: What has changed with the EMU?. *Journal of Money, Credit, and Banking, 38*(1), 91–118.
Poole, W., Rasche, R. H., & Thornton, D. L. (2002). *Market anticipations of monetary policy actions* (pp. 65–94). Federal Reserve Bank of St. Louis Review.
Roley, V. V., & Sellon, G. H., Jr. (1995). Monetary policy actions and long-term interest rates. *Economic Review-Federal Reserve Bank of Kansas City, 80*(4), 73.
Rudebusch, G. D. (1995). Federal reserve interest rate targeting, rational expectations, and the term structure. *Journal of Monetary Economics, 35*(2), 245–274.

Sarno, L., & Thornton, D. L. (2003). The dynamic relationship between the federal funds rate and the treasury bill rate: An empirical investigation. *Journal of Banking & Finance, 27*(6), 1079–1110.
Sarno, L., Thornton, D. L., & Valente, G. (2007). The empirical failure of the expectations hypothesis of the term structure of bond yields. *Journal of Financial and Quantitative Analysis, 42*(1), 81–100.
Schaefer, S. M. (1981). Measuring a tax-specific term structure of interest rates in the market for British government securities. *The Economic Journal, 91*(362), 415–438.
Seo, B. (2003). Nonlinear mean reversion in the term structure of interest rates. *Journal of Economic Dynamics and Control, 27,* 2243–2265.
Shiller, R. J. (1990). The term structure of interest rates. In B. M. Friedman & F. H. Hahn (Eds.), *Handbook of Monetary Economics, 1,* 628–722. North-Holland, Amsterdam.
Sibbertsen, P., & Kruse, R. (2009). Testing for a break in persistence under long-range dependencies. *Journal of Time Series Analysis, 30*(3), 263–285.
Strohsal, T., & Weber, E. (2014). Mean-variance cointegration and the expectations hypothesis. *Quantitative Finance, 14*(11), 1983–1997.
Tamakoshi, G., & Hamori, S. (2014). Nonlinear adjustment between the Eonia and Euribor rates: A two-regime threshold cointegration analysis. *Applied Financial Economics, 24*(2), 139–143.
Thornton, D. L. (2005). Tests of the expectations hypothesis: Resolving the anomalies when the short-term rate is the federal funds rate. *Journal of Banking & Finance, 29*(10), 2541–2556.
Tkacz, G. (2001). Estimating the fractional order of integration of interest rates using a wavelet OLS estimator. *Studies in Nonlinear Dynamics and Econometrics, 5*(1), 19–32.
Vetzal, K. R. (1994). A survey of stochastic continuous time models of the term structure of interest rates. *Insurance: Mathematics and Economics, 14*(2), 139–161.
Weber, E., & Wolters, J. (2012). The US term structure and central bank policy. *Applied Economic Letters, 19,* 41–45.
Weber, E., & Wolters, J. (2013). Risk and policy shocks on the US term structure. *Scottish Journal of Political Economy, 60,* 101–119.

Affine Type Analysis for BESQ and CIR Processes with Applications to Mathematical Finance

Luca Di Persio and Luca Prezioso

Abstract This chapter aims presents the deep relationships between the Cox-Ingersoll-Ross (CIR) type-processes, the Squared Bessel (BESQ) processes and the family of affine processes, according to specific dynamics for the dividend structure behind the market scenarios, aiming at deriving pricing formulas in individual markets as well as analytical solvable or numerical tractable, schemes for dividend processes in volatility stabilized markets.

Keywords CIR processes · BESQ processes · Affine processes · Dividend structures · Dividend processes computation · Individual markets

1 Roadmap and Main Results

CIR type processes were introduced in Cox et al. (1985a, b) and then widely used, mainly because they guarantee the *mean reversion* and the *non-negativity* properties, see, e.g. Brigo and Mercurio (2001, Sect. 3.9). In finance they are exploited, e.g., to model the stochastic volatility of stock prices and credit spread, see, e.g., Brigo and Mercurio (2001, Sect. 3.2.3). Within the insurance framework, the family of CIR processes are widely used to manage asset liabilities, when stochastic jumps have to be taken into account, see, e.g., Filipović (2001, Sect. 7) or see, e.g., Albeverio et al. (2016) for a more general setting linked to Lévy type noises and related invariant measures. It is worth to underline that a generic CIR process

$$\mathrm{d}r_t = k\,(\theta - r_t)\,\mathrm{d}t + \sigma\,\sqrt{|r_t|}\,\mathrm{d}W_t\,,$$

L. Di Persio (✉)
Department of Computer Science, Head of the HPA (High Performance Analytics) Research & Development unit, University of Verona, Verona, Italy
e-mail: luca.dipersio@univr.it

L. Prezioso
Department of Computer Science, University of Verona, Verona, Italy
e-mail: luca.prezioso@unitn.it

M. Mili et al. (eds.), *New Methods in Fixed Income Modeling*, Contributions to Management Science, https://doi.org/10.1007/978-3-319-95285-7_8

with *adjustment speed* parameter k, mean θ, and volatility σ, we will specify the model and the meaning of its parameters later on, see Sect. 2, has the advantage that can be seen as a δ-dimensional BESQ process through a transformation formula, setting $\delta = \frac{4k\theta}{\sigma^2}$. Moreover, the latter can be also derived by a δ-dimensional Brownian Motion (BM). To what concerns the BESQ type processes, they share many interesting properties, such, e.g., the scaling and the additive features. The additivity of BESQ processes allows for the stability under convolution operations. What is more, they are characterized by the *absolute continuity* which related the law of a general δ-dimensional BESQ process with a zero-dimensional BESQ. Such peculiarity is of great relevance since, for $\delta = 0$, BESQ processes are martingales, so that the absorbing property holds in 0. It is also important to recall that BESQ processes are in fact diffusions determined by a known transition probability, as well as it happens with the drifted BM and the Ornstein-Uhlenbeck processes, see, e.g., Jeanblanc et al. (2009, Chap. 6). Such rich structure, allows BESQ processes to find concrete applications both in finance and in insurance settings, e.g., to price exotic-type options, see, e.g. Geman and Yor (1993, Sect. 3), to model volatility-stabilized market dynamics, see, e.g., Fernoholz and Karatzas (2008, Sect. 12), or to solve the pricing problems for zero coupon bonds within the CIR setting, see, e.g., Cox et al. (1985a, Sects. 3–4), etc. We would like to underline that many of the financial applications concerning the use of CIR and BESQ processes are implied by the fact that their diffusion terms satisfy the affine characterization for diffusion process, see, e.g. Duffie et al. (2003), Duffie and Kan (1996), Filipović and Mayerhofer (2009), Filipović (2001). Therefore, their characteristic function, evaluated at a time $T > 0$ and conditioned up to time $t \in (0, T]$, is exponentially affine w.r.t to the process at time t. We recall that a d-dimensional diffusion process X, defined on a filtered probability space $(\Omega, \mathcal{F}, \{\mathcal{F}_t\}_{t\geq 0}, \mathbb{P})$, is called *affine* if, by definition, $\forall u \in i\mathbb{R}^d$ and $t \leq T$, it satisfies the following *affine transform formula*

$$\mathbb{E}[e^{u^T X_T} | \mathcal{F}_t] = e^{\Phi(T-t,u)+\Psi(T-t,u)^T X_t} ,$$

where Φ and Ψ are the solution of a system of Riccati equations depending on the drift and the diffusion terms of the process itself. These affine, diffusive processes are widely exploited in finance mainly because of their implications about the financial treatment of *affine short rate models*, as in the case of the Vašíček setting, see Vašíček (1977), as well as concerning the CIR type models for short rates, see Cox et al. (1985a, b). In fact, under suitable conditions, BESQ and CIR process settings, and affine processes in general, allow to derive explicit pricing formulas for given payoff functions. The latter is a particularly desired characteristic when dealing whit concrete financial settings, Indeed, the usual practitioners' goal, is to find concrete and efficient ways to price financial quantities they have to manage. This means, in particular, that in real cases it is much more important to derive approximations with explicit formula, instead of finding elegant proofs for the existence and uniqueness of the solution, e.g. of viscosity type, of a given problem. In what follows we provide a detailed characterization of affine processes on the canonical d-dimensional state space $\mathbb{R}_+^m \times \mathbb{R}^n$, with $d = m + n$, see Filipović and Mayerhofer (2009, Theorem 2.2). Furthermore, we show how one-dimensional characterization is satisfied by BESQ processes, and henceforth by CIR processes. Finally, we will derive evaluation formulas for dividend processes in volatility stabilized markets, which are revealed to be analytically solvable in several settings, being also almost always numerically tractable. We would also like to underline that a similar analysis can also be conducted for HJM-type models, see, e.g., Chiarella and Kwon (2003), Cordoni and Di Persio (2015). In fact, related forward rate curves are affine functions of the state variables. Moreover, by a suitable finite dimensional approach, latter state variables can be expressed as affine functions of a finite number of forward rates or yields.

2 Introduction to CIR and BESQ Processes

In this section we provide a connection between CIR processes and BESQ processes, through a transformation formula. An overview of the related main properties will be also given. Let us consider a standard probability space $(\Omega, \mathcal{F}, \mathbb{P})$, and an associated one-dimensional BM $\{W_t\}_{t\geq 0}$, $\mathcal{F}^W$ being its generated filtration, namely $\mathcal{F}_t = \sigma\{W_s : s \leq t\}$. We will assume that $\mathcal{F}_t^W \subseteq \mathcal{F}_t$, since this feature turns to be useful for latter financial applications. Indeed, $\mathcal{F}$ will represent the *informations provided by the market development, through time*. Let k, θ, σ, x be real numbers, then r is said to be a CIR process if it is the unique solution to the following Stochastic Differential Equation (SDE)

$$\begin{cases} \mathrm{d}r_t = k\,(\theta - r_t)\,\mathrm{d}t + \sigma\,\sqrt{|r_t|}\,\mathrm{d}W_t\,, \quad t > 0 \\ r_o = x\,, \end{cases} \tag{1}$$

where k, θ and σ are, respectively, the *speed-adjustment*, the mean and the volatility parameters, while x represents the initial condition, or the starting point, for the process. A relevant feature of Eq. (1) is that it admits a closed form solution. In particular, applying the Itô formula to $f(t, x) = x\,e^{k\,t}$, we obtain

$$d(e^{k\,t}\,r_t) = k\,\theta\,e^{k\,t}\,dt + \sigma\,e^{k\,t}\,\sqrt{|r_t|}\,dW_t\,,$$

hence, for $s \in (0, t)$

$$r_t = r_s\,e^{-k\,(t-s)} + \theta\,(1 - e^{-k\,(t-s)}) + \sigma \int_s^t e^{-k\,(t-u)}\,\sqrt{|r_u|}\,d\,W_u\,,$$

and we can easily compute both its expected value and its variance, as follows

$$\mathbb{E}[r_t|r_s] = r_s\,e^{-k\,(t-s)} + \theta\,((1 - e^{-k\,(t-s)}))\,, \tag{2}$$

$$Var(r_t|r_s) = r_s \frac{\sigma^2}{k}(e^{-k(t-s)} - e^{-2k(t-s)}) + \frac{\theta\sigma^2}{2k}(1 - e^{-k(t-s)})^2, \tag{3}$$

where (2) results from $dW_u \sim N(0, du)$, and (3) from Itô isometry. Therefore by (2) we get the mean reversion property for CIR processes, in particular $\lim_{t\to\infty} \mathbb{E}[r_t|r_s] = \theta$, in fact θ is also called long-term parameter. Together with the facts that r has a solution in closed form. Moreover, for k, θ, σ, x positive constants satisfying $\sigma^2 < 2k\theta$, we also have that r remains positive. Such feature, typically referred to as the *mean reversion property*, is the main reason why CIR processes are so used in Finance, especially to model the term structure of interest rates, see. From now on we will assume $\sigma^2 < 2k\theta$, in order to gain the positiveness of r, hence we drop out the absolute value within the square root in the dynamics of the process.

It is worth to mention that CIR processes constitute a generalization of the BESQ ones. In particular, these classes of stochastic processes are related through the following transformation formula

$$r_t = e^{-kt}\rho\Big(\frac{\sigma^2}{4k}(e^{kt} - 1)\Big), \tag{4}$$

where $\{\rho(s)\}_{s\geq 0}$ is a BESQ process with dimension $\delta = \frac{4k\theta}{\sigma^2}$, that is the unique solution to the following SDE

$$\begin{cases} \mathrm{d}\rho_t = \delta\,\mathrm{d}t + 2\sqrt{|\rho_t|}\,\mathrm{d}W_t \\ \rho_o = x \end{cases}, \tag{5}$$

often the terminology $\rho \in \mathrm{BESQ}^{\delta}_{x}$, is also used. Let us provide explicitly the aforementioned link, starting with the following proposition.

Proposition 1 (Relation with BESQ processes) *The CIR process r, solution to* (1), *also solves* (5), *with* $\delta = \frac{4k\theta}{\sigma^2}$, *provided the space-time change* (4).

Proof (Cit. (Jeanblanc et al. 2009, Prop. 6.3.1.1)) □

2.1 Construction of BESQ Processes

Let $\beta = \{\beta_t\}_{t\in\mathbb{R}} = \{(\beta_t^1, \beta_t^2, \ldots, \beta_t^n)\}_{t\in\mathbb{R}}$ be an n-dimensional BM, $n \in \mathbb{N}^+$. For the sake of completeness, let us recall that $\beta = \{(\beta_t^1, \beta_t^2, \ldots, \beta_t^n)\}_{t\in\mathbb{R}}$ is an n-dimensional BM iff $\{\beta_t^i\}_{t\geq 0}$ is a 1-dimensional BM, $\forall i = 1, \ldots, n$, and $\forall s, t \in \mathbb{R}^+$, β_t^i is a random variable with $\mathbb{E}[\beta_t^i] = 0$, $\mathrm{Cov}(\beta_t^i, \beta_s^i) = t \wedge s$, and moreover the β^i-s are independent.

Define the process R as $R_t = \|\beta_t\|$, i.e. $R_t^2 = \sum_{i=1}^n (\beta_t^i)^2$, then, exploiting the Itô-Döblin formula, we have that R_t^2 is a semimartingale, i.e. it can be decomposed in a local martingale and an adapted process with finite variation, see, e.g., He et al. (1992, Chap. 8), and

$$dR_t^2 = 0\,dt + 2\sum_{i=1}^{n} \beta_t^i\, d\beta_t^i + \frac{1}{2}\sum_{i=1}^{n} 2\,d\big[\beta_t^i, \beta_t^i\big] = 2\sum_{i=1}^{n} \beta_t^i\, d\beta_t^i + n\,dt\,,$$

in fact, for a Brownian motion $W = \{W_t\}_{t\in\mathbb{R}^+}$ we have $d\big(W^2\big)_t = dt + 2\,W_t dW_t$, the differential of the quadratic variation of a BM being $d\big[\beta_t^i, \beta_t^i\big] = dt$, see, e.g., Resnick (1992, Chap. 6.12).

Moreover $\mathbb{P}(R_t = 0) = 0$, $\forall t > 0$, therefore we can define the process W as $dW_t = \frac{1}{R_t}\sum_{i=1}^{n} \beta_t^i\, d\beta_t^i$, which is a $\mathbb{R}$-valued Brownian motion satisfying

$$d(R_t^2) = 2\,R_t\, dW_t + n\,dt\ .$$

Hence, setting $\rho_t = R_t^2$, we have

$$d\rho_t = 2\sqrt{\rho_t}\, dW_t + n\,dt\,,$$

and by Itô-Döblin formula we get the following SDE for R_t

$$dR_t = d(\sqrt{\rho_t}) = 0\,dt + \frac{1}{2\sqrt{\rho_t}}\,d\rho_t + \frac{1}{2}\Big(-\frac{1}{4\sqrt{\rho^3}}\Big)\,d[\rho_t, \rho_t] = dW_t + \frac{n-1}{2\,R_t}\,dt\,,$$

since $d[\rho_t, \rho_t] = 4\,\rho_t\,dt$.

The processes R and ρ are the so called *Bessel*, respectively *squared Bessel process*, of dimension n.

Definition 1 (BESQ_x^δ) For every $\delta \geq 0$ and every $x \geq 0$ the unique solution to

$$\rho_t = x + \delta\,t + 2\int_0^t \sqrt{\rho_s}\, dW_s\ ,$$

is called the *squared Bessel process* of dimension δ starting at x, and we denote it as $\rho \in \text{BESQ}_x^\delta$.

Definition 2 (BES_r^δ) Let $\rho \in \text{BESQ}_x^\delta$. The process $R = \sqrt{\rho}$ is called *Bessel process* of dimension δ starting at $r = \sqrt{x}$, and we denote it as $R \in \text{BES}_r^\delta$.

3 Characterization of Affine Processes

In this section we are going to analyse the properties that the parameters $a = \rho^2$ and b have to fulfill in order to have the diffusion process X_t as the unique solution to the SDE

$$\begin{cases} \mathrm{d}X_t = b(X_t)\,\mathrm{d}t + \rho(X_t)\,\mathrm{d}W_t\,, & t \in [0, T] \\ X_0 = x \in \Omega \end{cases}, \tag{6}$$

provided the *affine transform formula*

$$\mathbb{E}[e^{u^T X_T} \mid \mathcal{F}_t] = e^{\Phi(T-t,u)+\Psi(T-t,u)^T X_t}\,, \tag{7}$$

where $T \in \mathbb{R}^+$ is the *final time*, e.g. the expiration time of a certain investment, $u \in i\mathbb{R}^d$, and

- $d \geq 1$, is the dimension,
- $\Omega \subset \mathbb{R}^d$, is the state space, assumed to be closed and with non-empty interior,
- $b : \Omega \to \mathbb{R}^d$, is a continuous function,
- $\rho : \Omega \to \mathbb{R}^{d\times d}$, is a measurable function such that the diffusion matrix $a(x) = \rho(x)\,\rho(x)^T$ is continuous $\forall x \in \Omega$,
- W is a d-dimensional BM defined on the filtered probability space $\big(\Omega, \mathcal{F}^W, \{\mathcal{F}_t^W\}, \mathbb{P}\big)$.

Let us recall the definition of *Affine Processes*, namely

Definition 3 (*Affine Process*) Let $X := \{X_t\}_{t\in[0,T]}$, $T \in \mathbb{R}^+$, be a stochastic process over the filtered probability space $\big(\Omega, \mathcal{F}, \{\mathcal{F}\}_{t\in[0,T]}, \mathbb{P}\big)$. X is said to be *affine* if the $\mathcal{F}_t$-conditional characteristic function of X_T is exponentially affine in X_t, for all $t \leq T$, namely if there exist two functions $\Phi : [0, T] \times i\mathbb{R}^d \to \mathbb{C}$ and $\Psi : [0, T] \times i\mathbb{R}^d \to \mathbb{C}^d$ with continuous derivative with respect to the first argument and such that $X = X_t$ solution to (6) satisfies (7), for all $u \in i\mathbb{R}^d$, $t \leq T$, and every starting point $x \in \Omega$.

The next theorem provides a characterization for affine processes, stating sufficient as well as necessary conditions that a diffusion process has to satisfy to solve Eq. (7)

Theorem 4 (Characterization of affine processes) *For $m, n \in \mathbb{N}_0$ and $m + n = d$, let $\Omega = \mathbb{R}^m_+ \times \mathbb{R}^n$ be the canonical space and consider the following index sets notation $I := \{1, \ldots, m\}$ and $J := \{m + 1, \ldots, m + n\}$. The process X, unique solution to* (6) *on Ω, is affine, namely it satisfies Eq.* (7)*, if and only if*

1. *a and b are affine of the form*

$$a(x) = a + \sum_{i=1}^{d} x_i\,\alpha_i\,,$$

$$b(x) = b + \sum_{i=1}^{d} x_i\,\beta_i = b + B\,x\,,$$

respectively, where a, α_i are $d \times d$-matrices, $\forall i = 1, \ldots, d$, while b, β_i are in $\mathbb{R}^d$, and $B := (\beta_1, \ldots, \beta_d)$. Moreover, the parameters a, α_i, b, β_i are admissible in the following sense

- *a and α_i are symmetric positive semi-definite $d \times d$ matrices, $\forall i = 1, \dots, d$,*
- *$a_{II} = 0$, where a_{II} is the principal sub-matrix of a obtained by its first m rows and columns,*
- *$\alpha_j = 0$, for all $j \in J$,*
- *$\alpha_{i,kl} = 0$, for $k \in I \setminus \{i\}$, and for all $1 \le i, l \le d$,*
- *$b \in \mathbb{R}^m_+ \times \mathbb{R}^n$,*
- *$B_{IJ} = 0$ and B_{II} has positive off-diagonal elements;*

2. *The functions Φ and $\Psi = (\Psi_1, \dots, \Psi_d)^T$, where T denotes the transpose, solve the system of Riccati equations*

$$\begin{cases} \partial_t \Phi(t,u) = \frac{1}{2}\,\Psi_J(t,u)^T\, a_{JJ}\, \Psi_J(t,u) + b^T\, \Psi(t,u) \\ \Phi(0,u) = 0 \\ \partial_t \Psi_i(t,u) = \frac{1}{2}\,\Psi(t,u)^T\, \alpha_i\, \Psi(t,u) + \beta_i^T\, \Psi(t,u)\,, \qquad i \in I\ ; \\ \partial_t \Psi_J(t,u) = B_{JJ}^T\, \Psi_J(t,u) \\ \Psi(0,u) = u \end{cases} \tag{8}$$

3. *There exists a unique global solution to Eq. (8), namely $(\Phi(\,,u), \Psi(\,,u)) : \mathbb{R}_+ \to \mathbb{C}_- \times \mathbb{C}^m_- \times i\mathbb{R}^n$, for all $u \in \mathbb{C}^m_- \times i\mathbb{R}^n$ initial values. In particular, the equation for Ψ_J forms an autonomous linear system with unique global solution $\Psi_J(t,u) = e^{\mathcal{B}^T_{JJ}\,t}\, u_J$, for all $u_J \in \mathbb{C}^m$.*

Proof For a proof see, e.g., Filipović and Mayerhofer (2009, Sect. 3). □

As an example of application of Theorem 4, in the next paragraph we show that BESQ processes can be seen as an example of one-dimensional affine process.

BESQ are affine.

According with the notation used in Theorem 4, we set the parameters as

$$\begin{aligned} d = 1\,, &\quad I = \{1\}\,, \quad J = \emptyset\,, \\ b = 2\,\mu\,, &\quad a = 0\,, \quad \alpha_1 = 4\,, \end{aligned}$$

Hence the resulting Riccati system reads as follows

$$\begin{cases} \partial_t \Phi(t,u) = b\,\Psi(t,u) \\ \Phi(0,u) = 0 \\ \partial_t \Psi(t,u) = \frac{1}{2}\,\Psi(t,u)^2\,\alpha_1 + 0 \\ \Psi(0,u) = u \end{cases},$$

therefore, by the Ψ-equation, we get

$$\Psi(t,u) = \frac{1}{-c_1(u) - 2t} = \frac{1}{1/u - 2t}, \tag{9}$$

where (9) comes from the initial conditions, while the Φ-equation implies that $\Phi(t,u) = -\mu\,\ln(1-2\,t\,u)$. Hence, we get $\Phi(t,u) = -\mu\ln(1-2\,t\,u)$, and $\Psi(t,u) = \frac{u}{1-2\,t\,u}$.

Therefore, by Theorem 4, a BESQ process ρ is indeed an affine process, satisfying the following Affine transform formula

$$\mathbb{E}[e^{u^T\,\rho_T}|\mathcal{F}_t] = \frac{\exp\left(\frac{u}{1-2\,(T-t)\,u}\,\rho_t\right)}{(1-2\,(T-t)\,u)^{\mu}}.$$

In the next sections we see two different applications of Affine processes, in particular of Bessel processes in Sect. 5, to financial applications. Namely to the option pricing problem in Sect. 4, and to describe the dividend market structure in Sect. 5.

4 Affine Short Rate Models and Pricing Formulas

In what follow we recall some preliminary results that concern affine processes in general, and affine *short rate* models, in particular. We refer to Filipović and Mayerhofer (2009, Sect. 4) for further details.

Definition 5 (*Affine rate model*) Let us consider a stochastic *short Interest Rate Model*, IRM, see, e.g., Brigo and Mercurio (2001, Chap. 1), r_t with dynamics

$$\mathrm{d}r_t = \mu(t, r_t)\,\mathrm{d}t + \sigma(t, r_t)\,\mathrm{d}W_t\,,$$

we say that r is an *affine rate model* if there exists $c \in \mathbb{R}$, $\gamma \in \mathbb{R}^d$ constant parameters, and a process X on the canonical space $\mathbb{R}^m_+ \times \mathbb{R}^n$, with admissible parameters $a, \alpha_1, \ldots, \alpha_d, b, \beta_1, \ldots, \beta_d$ as in point 1 of Theorem 4, such that

$$r_t = c + \gamma^T\,X_t\,. \tag{10}$$

For the sake of completeness, let us consider the following result, which turns to be very useful to price financial instruments with known payoff functions.

Theorem 6 *Consider a process X satisfying* (6), *and r as in Eq.* (10). *For $\tau > 0$ the following statements are equivalent:*

1. $\mathbb{E}[e^{-\int_0^\tau r_s \, ds}] < +\infty, \forall x \in \mathbb{R}_+^m \times \mathbb{R}^n$;
2. $\exists!$ *solution* $(\Phi(\cdot, u), \Psi(\cdot, u)) : [0, \tau] \to \mathbb{C} \times \mathbb{C}^d$ *to the following system of Riccati equations*

$$\begin{cases} \partial_t \Phi(t,u) = \frac{1}{2}\, \Psi_J(t,u)^T \, a_{JJ} \, \Psi_J(t,u) + b^T \, \Psi(t,u) - c \\ \Phi(0,u) = 0 \\ \partial_t \Psi_i(t,u) = \frac{1}{2}\, \Psi(t,u)^T \, \alpha_i \, \Psi(t,u) + \beta_i^T \, \Psi(t,u) - \gamma_i \\ \partial_t \Psi_J(t,u) = B_{JJ}^T \, \Psi_J(t,u) \\ \Psi(0,u) = u \end{cases}, \tag{11}$$

with $i \in I$, for $u = 0$, and I and J as in Theorem 4.

In either cases, there exists an open convex neighborhood U of 0 in $\mathbb{R}^d$ such that the system (11) *admits a unique solution* $(\Phi(\cdot, u), \Psi(\cdot, u)) : [0, \tau] \to \mathbb{C} \times \mathbb{C}^d$, $\forall u \in \mathcal{S}(U)$, *namely, for every u belonging to the $\mathbb{C}^d$-strip, we have*

$$\mathcal{S}(U) = \{z \in \mathbb{C}^d \mid \Re z \in U\} ,$$

*moreover, the $\mathcal{F}_t$-*conditional characteristic function of *X_T under $\mathbb{Q}^T$, for $u \in i\mathbb{R}^d$, allows the following affine representation*

$$\mathbb{E}[e^{-\int_t^T r_s \, ds} \, e^{u^T X_T} \mid \mathcal{F}_t] = e^{\Phi(T-t,u)+\Psi(T-t,u)^T X_t}, \tag{12}$$

$\forall u \in \mathcal{S}(U)$, $t \leq T \leq t + \tau$ *and* $x \in \mathbb{R}_+^m \times \mathbb{R}^n$.

Proof See, e.g., Filipović and Mayerhofer (2009, Sect. 4). □

From the financial perspective, a relevant consequence of Theorem 6 is provided by the following

Corollary 1 (Bond pricing formula) *For every maturity $T \leq \tau$, with τ being as in the affine representation provided by Theorem 6, the bond price at $t \leq T$, and with maturity T, is given by*

$$P(t,T) = e^{-A(T-t)-B(T-t)^T X_t} ,$$

with $A(t) = -\Phi(t,0)$ and $B(t) = -\Psi(t,0)$. Besides, for $t \leq T \leq S \leq \tau$, the $\mathcal{F}_t$-conditional characteristic function of X_T under the S-forward measure $\mathbb{Q}^S$, is given by

$$\mathbb{E}_{\mathbb{Q}^S}[e^{u^T X_T} \mid \mathcal{F}_t] = \frac{e^{-A(S-T)+\Phi\left(T-t,u-B(S-T)\right)+\Psi\left(T-t,u-B(S-T)\right)^T X_t}}{P(t,S)}, \tag{13}$$

$\forall u \in \mathcal{S}\big(U + B(S - T)\big)$, *where* U *is the neighborhood of 0 in* $\mathbb{R}^d$ *as in Theorem 6.*

Proof See, e.g., Filipović and Mayerhofer (2009, Sect. 4). □

5 BESQ Processes Approach to the Dividend Dynamics Structure

In this section we show how BESQ processes can be efficiently applied to the analysis of the dividend open markets structure, see, e.g. Fernoholz and Karatzas (2008, Chap. 12) for further details. In particular we have two distinct individual markets; in each of them there are financial agents that can operate only on one asset, namely "asset 1" for the first market and "asset 2" for the second one. Each asset produces a continuous dividend, let us indicate them by $D_t^{(1)}$ and $D_t^{(2)}$. At a certain time $t \in [0, T]$, T being a finite horizon time, markets get open: from now on all agents can operate on both assets. The assets dynamics are as follows

$$dD_t^{(i)} = \frac{\mu}{2} D_t\, dt + \sigma \sqrt{D_t^{(i)}\, D_t}\, dW_t^{(i)}, \quad \text{for } i = 1, 2\,, \tag{14}$$

where we assume the two vector of parameters $\mu = [\mu^{(1)}, \mu^{(2)}]$ and $\sigma = [\sigma^{(1)}, \sigma^{(2)}]$ to be constant, while $W = [W^{(1)}, W^{(2)}]$ is a two-dimensional BM, and D_t is determined by $D_t = D_t^{(1)} + D_t^{(2)}$, equivalently by

$$dD_t = \mu\, dD_t + \sigma \sqrt{D_t}\,\big(\sqrt{D_t^{(1)}}\, dW_t^{(1)} + \sqrt{D_t^{(2)}}\, dW_t^{(2)}\big)\,.$$

Since D_t can be rewritten as a geometric BM, namely

$$dD_t = \mu\, dD_t + \sigma\, D_t\, d\hat{W}_t\,,$$

with $\hat{W}_t$ BM defined by

$$d\hat{W}_t = \frac{\sqrt{D_t^{(1)}}\, dW_t^{(1)} + \sqrt{D_t^{(2)}}\, dW_t^{(2)}}{\sqrt{D_t}}\,,$$

then, thanks to the Itô-Döblin formula, we have that $D_t = e^{(\mu + \sigma^2/2)t + \sigma \hat{W}_t}$. On the other hand, the dividend structures of the single assets, is less easy to derive. Indeed, supposing to act in an open market, we cannot exclude an interaction between them.

Hence, consider the stochastic processes $R^{(i)}$ and Λ, solution to

$$\mathrm{d}R_t^{(i)} = 2\,\mu\,\mathrm{d}t + 2\sqrt{R_t^{(i)}}\,\mathrm{d}\tilde{W}_t^{(i)}, \quad \text{for } i = 1, 2\,,$$
$$\mathrm{d}\Lambda_t = \frac{1}{4}\,(D_t^{(1)} + D_t^{(2)})\,\mathrm{d}t\,,$$

with $\tilde{W}_t^{(i)} = \int_0^{\Lambda_t^{-1}} \sqrt{\Lambda_s'}\,\mathrm{d}W_s^{(i)}$, we have $\tilde{W}_t^{(i)}$ is a BM. In fact, $\sqrt{\Lambda_s'}$ is adapted to the filtration generated by $\hat{W}_s$, and, at the same time, it is also adapted to the filtration generated by $W_s^{(i)}$, because $\hat{W}$ is independent from the increments of $W_s^{(i)}$. Therefore, $\tilde{W}_t^{(i)}$ is a martingale and its variance is

$$\langle \tilde{W}^{(i)}, \tilde{W}^{(i)} \rangle_t = \int_0^{\Lambda_t^{-1}} \Lambda_s'\,ds = t\,,$$

which implies that it is a standard BM.

Concerning the process $R^{(i)}$, it turns out that it is a BESQ process, moreover, exploiting the Itô-Döblin formula, $D_t^{(i)} = R^{(i)} \circ \Lambda_t$, in fact

$$\begin{aligned} \mathrm{d}\big(R^{(i)} \circ \Lambda_t\big) &= 2\,\mu\,\mathrm{d}\Lambda_t + 2\sqrt{R^{(i)} \circ \Lambda_t}\,\mathrm{d}\big(\tilde{W}^{(i)} \circ \Lambda_t\big) \\ &= \frac{\mu}{2}\,D_t\,\mathrm{d}t + \sqrt{R^{(i)} \circ \Lambda_t}\,\sqrt{D_t}\,\mathrm{d}W_t^{(i)}. \end{aligned}$$

We would like to underline that latter *time change* reveals to be particularly useful to study the asymptotic behavior of the solution to Eq. 14, see, e.g., Fernoholz and Karatzas (2008, Sect. 12.1).

References

Albeverio, S., Di Persio, L., Mastrogiacomo, E., & Smii, B. (2016). A class of Lévy driven SDEs and their explicit invariant measures. *Potential Analysis*, *45*(2), 229–259.

Brigo, D., & Mercurio, F. (2001). *Interest rate models: Theory and practice*. Berlin: Springer.

Chiarella, C., & Kwon, O. K. (2003). Finite dimensional affine realisations of HJM models in terms of forward rates and yields. *Review of Derivatives Research*, *6*(2), 129–155.

Cox, J. C., Ingersoll, J. E., & Ross, S. A. (1985a). A theory of the term structure of interest rate. *Econometrica*, *53*(2).

Cox, J. C., Ingersoll, J. E., & Ross, S. A. (1985b). An intertemporal general equilibrium model of asset prices. *Econometrica*, *53*.

Duffie, D., Filipović, D., & Schachermayer, W. (2003). Affine processes and applications in finance. *Annals of Applied Probability*, *13*, 984–1053.

Duffie, D., & Kan, R. (1996). A yield-factor model of interest rates. *Mathematical Finance*, *6*, 379–406.

Cordoni, F., & Di Persio, L. (2015). Invariant measure for the Vasicek interest rate model in the Heath-Jarrow-Morton-Musiela framework. *Infinite Dimensional Analysis, Quantum Probability and Related Topics*, *18*(3), art. no. 1550022.

Filipović, D., & Mayerhofer, E. (2009). *Affine diffusion processes: Theory and applications*. Vienna University of Economics and Business Administration.

Filipović, D. (2001). *A general characterization of one factor affine term structure model*. Zürich: Departement Mathematik.

Fernoholz, R., & Karatzas, I. (2008). *Stochastic portfolio theory: An overview*.

Geman, H., & Yor, M. (1993). Bessel Processes, Asian options and Perpetuities. *Mathematical Finance*, *3*(4), 349–375.

He, S., Wang, J., & Yan, J. (1992). *Semimartingale theory and stochastic calculus*. Science Press and CRC Press.

Jeanblanc, M., Yor, M., & Chesney, M. (2009). *Mathematical methods for financial markets*. Springer Finance.

Resnick, S. (1992). *Adventures in stochastic processes*. Birkhäuser Boston.

Vašíček, O. (1977). An equilibrium characterization of the term structure. *Journal Financial Economics*, *5*, 177–188.

Part II
New Advances in Fixed Income Management

Sensitivity Analysis and Hedging in Stochastic String Models

Alberto Bueno-Guerrero, Manuel Moreno and Javier F. Navas

Abstract We analyze certain results on the stochastic string modeling of the term structure of interest rates and we apply them to study the sensitivities and the hedging of options with payoff functions homogeneous of degree one. Under the same framework, we use an exact multi-factor extension of Jamshidian (1989) to find the sensitivities for swaptions and we prove that it cannot be applied to captions. We present a new approximate result for pricing options on coupon bonds based on the Fenton-Wilkinson method and we show that it generalizes the fast coupon bond option pricing proposed in Munk (1999). This result can be easily applied to the approximate valuation of swaptions and captions.

Keywords Stochastic string · Sensitivity analysis · Hedging · Swaption
Stochastic duration

1 Introduction

The Greeks (the sensitivities of the price of an option with respect to different parameters) and the hedging portfolios (those self-financing portfolios that replicate the option value at maturity) are two of the main tools in the risk management of equity options. Closed-form expressions for them can be obtained under the Black-Scholes

A. Bueno-Guerrero
IES Francisco Ayala, Avda. Francisco Ayala, s/n, 18014 Granada, Spain
e-mail: alberto.bueno@alu.uclm.es

M. Moreno
Department of Economic Analysis and Finance, University of Castilla La-Mancha,
Cobertizo San Pedro Mártir s/n, 45071 Toledo, Spain
e-mail: manuel.moreno@uclm.es

J. F. Navas (✉)
Department of Financial Economics and Accounting, Universidad Pablo de Olavide,
Ctra. de Utrera, Km. 1, 41013 Seville, Spain
e-mail: jnavas@upo.es

M. Mili et al. (eds.), *New Methods in Fixed Income Modeling*, Contributions
to Management Science, https://doi.org/10.1007/978-3-319-95285-7_9

(1973) framework and several extensions are widely used. Nevertheless, less attention has been paid to these risk management tools in the fixed income case.

Since the appearance of the seminal paper of Santa-Clara and Sornette (2001), the stochastic string approach has been considerably enlarged in a number of papers, see for example, Bueno-Guerrero et al. (2015a, 2016) showing that it can be considered as a unifying framework for the term structure of interest rates (TSIR). Thus, it seems natural to apply this setting to the risk management issues previously mentioned.

Another reason to consider the stochastic string framework is that sensitivities and hedging portfolios are intimately related under this setting. In fact, Bueno-Guerrero et al. (2016) obtains a pricing theorem for options whose payoffs are continuously differentiable and homogeneous of degree one. For these options, in a very general setting, Bueno-Guerrero et al. (2015c) states a result that allows to obtain hedging portfolios by delta-hedging. Therefore, we can obtain sensitivities and use them to design hedging portfolios under the same framework. This analysis is very important if we note that the class of options satisfying the previous conditions includes standard calls and puts, floorlets, caplets, swaptions, captions, and other compounded options.

As we have just mentioned, swaptions satisfy the conditions of our analysis and, then, we will also tackle with the problem of their valuation. Under Gaussian models of the TSIR, the pricing formulas for swaptions involve the calculation of multiple integrals on a domain with a non-linear boundary. Given the importance of this derivative in the industry, it would be useful to have another method (exact or approximate) to price swaptions.

One exact approach for pricing swaptions is based on the one-factor model of Jamshidian (1989). He obtains a formula for pricing options on a portfolio of bonds that can be applied to price swaptions. Eberlein and Kluge (2005) use Jamshidian's insight to obtain exact pricing formulas for swaptions in a Lévy model. Wei (1997), also in a one-factor framework, proposes an approximation to the price of an option on a coupon bond as a multiple of the price of an option on a zero-coupon bond with time to maturity equal to the stochastic duration of the coupon bond. Nevertheless, Wei does not provide any theoretical justification for his approximation.

The main limitation of these methods is that they only work under one-factor models of the TSIR. Many other approaches have been developed to overcome this drawback. Munk (1999) generalizes the Wei's approach to multi-factor models and provides an analytical justification. Singleton and Umantsev (2002) approximates the exercise boundary of coupon bond options with a hyperplane and Collin-Dufresne and Goldstein (2002) obtains an algorithm to price swaptions using an Edgeworth expansion technique. More recent papers related to this problem are Kim (2014), Vidal and Silva (2014) and Choi and Shin (2016).

Our contribution to this topic is twofold. First, we review the result in Bueno-Guerrero et al. (2015b) that generalizes Jamshidian's work to a multifactor model and we apply it to obtain sensitivities of swaptions. Second, we use the Fenton-Wilkinson approximation (Fenton 1960) to present a new approximation for the value of an option on a coupon bond which is directly applicable to swaptions. The Fenton-Wilkinson method is based on assuming that the sum of lognormal variables can be approximated by another lognormal variable whose two first moments match those of the sum of lognormals. This method is widely used in telecommunications and gives accurate approximations for large number of summands and small variances (Asmussen et al. 2016). For an application in a financial setting see, for instance, Moreno and Navas (2008).

The rest of the chapter is organized as follows. Section 2 presents the sensitivities for options whose payoffs are differentiable and homogeneous of degree one and apply them to simple cases. Section 3 recalls the delta-hedging result and illustrates it with the hedging portfolio for a discount bond call option. The multi-factor extension of Jamshidian (1989) and its application to value swaptions are reviewed and used to obtain swaptions sensitivities. It is also proved that this method cannot be applied to captions, motivating the next result of the section, i.e., the theorem for approximate valuation of coupon bond options that is obtained with the Fenton-Wilkinson method. Section 4 concludes the chapter.

2 Sensitivity Analysis

We start recalling the main result related to option pricing in Bueno-Guerrero et al. (2016).

Theorem 1 *If the following statements hold:*

(a) The final pay-off of a contingent claim at time T_0 is given by $\boldsymbol{C}\big[T_0, \boldsymbol{P}_{T_0}\big] = max\big(\Phi\big(\boldsymbol{P}_{T_0}\big), 0\big)$ where $\boldsymbol{P}_{T_0} = (P(T_0, T_0), P(T_0, T_1), \ldots, P(T_0, T_n))$, $P(T_0, T_i)$ denotes the price at time T_0 of a zero-coupon bond maturing at time T_i, $i = 0, 1, \ldots, n$, and $\Phi : \mathbb{R}^{n+1} \to \mathbb{R}$ is a continuously differentiable homogeneous function of degree one.

(b) The correlation matrix M with elements $M_{ij}(s, T_0) = \frac{\Delta_{ij}(s,T_0)}{\sqrt{\Delta_{ii}(s,T_0)}\sqrt{\Delta_{jj}(s,T_0)}}$, $i, j = 1, \ldots, n$ is deterministic and non-singular with

$$\Delta_{ij}(s, T_0) = cov\big[\ln P(T_0, T_i), \ln P\big(T_0, T_j\big)|\mathcal{F}_s\big] = \int_{t=s}^{T_0}\left[\int_{y=T_0-t}^{T_i-t}\int_{u=T_0-t}^{T_j-t} R_t(u, y)dudy\right]dt$$

where $R_t(u, y) = c(t, u, y)\sigma(t, u)\sigma(t, y)$ and $c(t, u, y)$, $\sigma(t, u)$ are, respectively, the correlation and the volatility functions of a stochastic string model.

Then, the price at time s of this contingent claim is given by

$$\boldsymbol{C}[s, \boldsymbol{P}_s] = \int_{\mathbb{R}^n} g(x_1, \ldots, x_n; M)\Big(\Phi\Big[P(s, T_0), P(s, T_1)e^{\sqrt{\Delta_{11}}x_1-\frac{1}{2}\Delta_{11}}, \ldots, P(s, T_n)e^{\sqrt{\Delta_{nn}}x_n-\frac{1}{2}\Delta_{nn}}\Big]\Big)_+ d\boldsymbol{x} \quad (1)$$

where

$$g(x_1,\ldots,x_n;M)=\frac{1}{\sqrt{(2\pi)^n|M|}}\exp\left(-\frac{1}{2}\sum_{i,j=1}^{n}x_i\left(M^{-1}\right)_{ij}x_j\right)$$

is the density function of a multivariate random variable.[1] ∎

The class of options whose payoff satisfies (a) of this Theorem is very wide and includes calls, puts, floorlets, caplets, swaptions, captions, and other compounded options. From now on, we will work with deterministic correlations and volatilities in order to satisfy (b) of this Theorem. We also consider $\Phi_i \equiv \left.\frac{\partial\Phi(y)}{\partial y_i}\right|_{y=\boldsymbol{P}_{T_0}}$.

Now we can state the main result on sensitivities of option prices.

Theorem 2 *Let $\boldsymbol{C}[t,\boldsymbol{P}_t]$ be the price at time t of an option satisfying (a) of Theorem 1 and assume that $\Phi_j>0$ for $j=1,\ldots,n$. Then, we have the following derivatives:*

$$\frac{\partial\boldsymbol{C}[t,\boldsymbol{P}_t]}{\partial P(t,T_0)}=\Phi_0\int_{x_1=-\infty}^{+\infty}\cdots\int_{x_j=\tilde{x}_j}^{+\infty}\cdots\int_{x_n=-\infty}^{+\infty}g(x_1,\ldots,x_n;M)d\boldsymbol{x}$$
$$-\frac{\Phi_0}{\sqrt{\Delta_{jj}}}\sum_{k=0}^{n}\Phi_k P(t,T_k)\times\int_{x_1=-\infty}^{+\infty}\cdots\int_{x_{j-1}=-\infty}^{+\infty}\int_{x_{j+1}=-\infty}^{+\infty}$$
$$\cdots\int_{x_n=-\infty}^{+\infty}\frac{g\left(x_1-\frac{\Delta_{1k}}{\sqrt{\Delta_{11}}},\ldots,\tilde{x}_j-\frac{\Delta_{jk}}{\sqrt{\Delta_{jj}}},\ldots,x_n-\frac{\Delta_{nk}}{\sqrt{\Delta_{nn}}};M\right)}{\sum_{\substack{m=0\\m\neq j}}^{n}\Phi_m P(t,T_m)e^{\sqrt{\Delta_{mm}}x_m-\frac{1}{2}\Delta_{mm}}}d\boldsymbol{x}_j$$

$$\frac{\partial\boldsymbol{C}[t,\boldsymbol{P}_t]}{\partial P\left(t,T_j\right)}=\Phi_j\int_{x_1=-\infty}^{+\infty}\cdots\int_{x_j=\tilde{x}_j}^{+\infty}\cdots\int_{x_n=-\infty}^{+\infty}g\left(x_1-\frac{\Delta_{1j}}{\sqrt{\Delta_{11}}},\ldots,x_n-\frac{\Delta_{nj}}{\sqrt{\Delta_{nn}}};M\right)d\boldsymbol{x}$$
$$+\frac{1}{\sqrt{\Delta_{jj}}P\left(t,T_j\right)}\sum_{k=0}^{n}\Phi_k P(t,T_k)\times\int_{x_1=-\infty}^{+\infty}\cdots\int_{x_{j-1}=-\infty}^{+\infty}\int_{x_{j+1}=-\infty}^{+\infty}\cdots$$
$$\int_{x_n=-\infty}^{+\infty}g\left(x_1-\frac{\Delta_{1k}}{\sqrt{\Delta_{11}}},\ldots,\tilde{x}_j-\frac{\Delta_{jk}}{\sqrt{\Delta_{jj}}},\ldots,x_n-\frac{\Delta_{nk}}{\sqrt{\Delta_{nn}}};M\right)d\boldsymbol{x}_j$$

[1] For the probabilistic setting and assumptions of the stochastic string framework, we refer the reader to Bueno-Guerrero et al. (2015a).

$$\frac{\partial C[t, \boldsymbol{P}_t]}{\partial \Delta_{ij}} = \sum_{k=0}^{n} \Phi_k P(t, T_k) \int_{x_1=-\infty}^{+\infty} \cdots \int_{x_l=\tilde{x}_l}^{+\infty} \cdots \int_{x_n=-\infty}^{+\infty} g\left(x_1 - \frac{\Delta_{1k}}{\sqrt{\Delta_{11}}}, \ldots, x_n - \frac{\Delta_{nk}}{\sqrt{\Delta_{nn}}}; M\right)$$
$$\times \left[\boldsymbol{D} \circ \left(\boldsymbol{M}^{-1}\boldsymbol{x}\boldsymbol{x}^T\boldsymbol{M}^{-1} - \boldsymbol{M}^{-1}\right)\right]_{ij} d\boldsymbol{x}$$

$$\frac{\partial C[t, \boldsymbol{P}_t]}{\partial \Delta_{ii}} = \frac{1}{2\sqrt{\Delta_{ii}}} \sum_{k=0}^{n} \Phi_k P(t, T_k)$$
$$\times \int_{x_1=-\infty}^{+\infty} \cdots \int_{x_{i-1}=-\infty}^{+\infty} \int_{x_{i+1}=-\infty}^{+\infty} \cdots \int_{x_n=-\infty}^{+\infty} g\left(x_1 - \frac{\Delta_{1k}}{\sqrt{\Delta_{11}}}, \ldots, \tilde{x}_i - \frac{\Delta_{ik}}{\sqrt{\Delta_{ii}}}, \ldots, x_n - \frac{\Delta_{nk}}{\sqrt{\Delta_{nn}}}; M\right)$$
$$\times \left(\frac{1}{\Delta_{ii}} \ln \left(-\frac{1}{\Phi_i P(t, T_i)} \sum_{\substack{m=0 \\ m \neq i}}^{n} \Phi_m P(t, T_m) e^{\sqrt{\Delta_{mm}} x_m - \frac{1}{2}\Delta_{mm}} \right) - \frac{1}{2} \right) d\boldsymbol{x}_i$$
$$-\frac{1}{2} \sum_{k=0}^{n} \Phi_k P(t, T_k) \int_{x_1=-\infty}^{+\infty} \cdots \int_{x_i=\tilde{x}_i}^{+\infty} \cdots \int_{x_n=-\infty}^{+\infty} g\left(x_1 - \frac{\Delta_{1k}}{\sqrt{\Delta_{11}}}, \ldots, x_n - \frac{\Delta_{nk}}{\sqrt{\Delta_{nn}}}; M\right)$$
$$\times \left(\frac{1}{\Delta_{ii}} \sum_{\substack{\boldsymbol{m}=1 \\ \boldsymbol{m} \neq i}}^{\boldsymbol{n}} \left[\boldsymbol{M} \circ \left(\boldsymbol{M}^{-1}\boldsymbol{x}\boldsymbol{x}^T\boldsymbol{M}^{-1} - \boldsymbol{M}^{-1}\right)\right]_{\boldsymbol{im}} - \delta_{\boldsymbol{ik}}\left(\frac{x_k}{\sqrt{\Delta_{kk}}} - 1\right) \right) d\boldsymbol{x}$$

for any $i, j \in \{1, \ldots, n\}$, $i \neq j$, *with* $(\boldsymbol{D})_{ij} = \frac{1}{\sqrt{\Delta_{ii}}\sqrt{\Delta_{jj}}}$,

$$\tilde{x}_j = \frac{1}{\sqrt{\Delta_{jj}}} \ln \left(-\frac{1}{\Phi_j P(t, T_j)} \sum_{\substack{k=0 \\ k \neq j}}^{n} \Phi_k P(t, T_k) e^{\sqrt{\Delta_{kk}} x_k - \frac{1}{2}\Delta_{kk}} \right) + \frac{1}{2}\sqrt{\Delta_{jj}}$$

and where the symbol $\circ$ *stands for the Hadamard product of matrices and* $d\boldsymbol{x}_i \equiv d\boldsymbol{x}_1 \ldots d\boldsymbol{x}_{i-1} d\boldsymbol{x}_{i+1} \ldots d\boldsymbol{x}_n$. ■

Proof See the Appendix. ■

Remark 1 Theorem 2 also holds when $\Phi_j < 0$ *for* $j = 1, \ldots, n$. We just need to change the integrals $\int_{\tilde{x}}^{+\infty}$ by $\int_{-\infty}^{\tilde{x}}$ everywhere and change the sign of the other term. This case is useful for put options on a bond portfolio with long positions (that is, for caps and swaptions). ■

In general, the computation of the sensitivities in Theorem 2 requires to calculate numerically high-dimensional integrals.[2] However, in the simplest case of just one bond, a straightforward application of this theorem provides closed-form expressions for these sensitivities. The following corollary states this result.

Corollary 1 *Consider an option whose payoff satisfies condition (a) of Theorem 1 with* $\Phi(\boldsymbol{P}_{T_0}) = \Phi(P(T_0, T_0), P(T_0, T_1))$ *and* $\Phi_1 > 0$*. The sensitivities of this option are given by*

$$\begin{aligned}
\frac{\partial \boldsymbol{C}[t, \boldsymbol{P}_t]}{\partial P(t, T_0)} &= \Phi_0 \phi\left(\tilde{k}_2\right) - \frac{\Phi_0}{\sqrt{\Delta}} \rho\left(\tilde{k}_2\right) - \frac{\Phi_1}{\sqrt{\Delta}} \frac{P(t, T_1)}{P(t, T_0)} \rho\left(\tilde{k}_1\right) \\
\frac{\partial \boldsymbol{C}[t, \boldsymbol{P}_t]}{\partial P(t, T_1)} &= \Phi_1 \phi\left(\tilde{k}_1\right) + \frac{\Phi_0}{\sqrt{\Delta}} \frac{P(t, T_0)}{P(t, T_1)} \rho\left(\tilde{k}_2\right) + \frac{\Phi_1}{\sqrt{\Delta}} \rho\left(\tilde{k}_1\right) \\
\frac{\partial \boldsymbol{C}[t, \boldsymbol{P}_t]}{\partial \Delta} &= -\frac{1}{2\Delta}\left[\Phi_1 P(t, T_1) \tilde{k}_2 \rho\left(\tilde{k}_1\right) + \Phi_0 P(t, T_0) \tilde{k}_1 \rho\left(\tilde{k}_2\right)\right]
\end{aligned}$$

where ρ *and* ϕ *are, respectively, the density and the distribution functions of a standard Gaussian random variable,* $\Delta \equiv \Delta_{11}$*, and*

$$\tilde{k}_1 = \frac{\ln\left(-\frac{\Phi_1 P(t, T_1)}{\Phi_0 P(t, T_0)}\right) + \frac{1}{2}\Delta}{\sqrt{\Delta}}, \quad \tilde{k}_2 = \tilde{k}_1 - \sqrt{\Delta}$$

■

This corollary can be easily applied to obtain the sensitivities of different options such as calls or floorlets. We illustrate it with an example.

Example 1 Consider a call option with maturity T_0 and strike K written on a zero-coupon bond maturing at T_1 . Its payoff can be written as $\boldsymbol{Call}_K(T_0, T_0, T_1) = [\mathrm{P}(T_0, T_1) - K P(T_0, T_0)]_+$ and then $\Phi(x_0, x_1) = x_1 - K x_0$, $\Phi_0 = -K$, $\Phi_1 = 1$. By Corollary 1 we have

$$\begin{aligned}
\frac{\partial \boldsymbol{Call}_K[t, T_0, T_1]}{\partial P(t, T_0)} &= -K \phi(d_2) + \frac{K}{\sqrt{\Delta}} \rho(d_2) - \frac{1}{\sqrt{\Delta}} \frac{P(t, T_1)}{P(t, T_0)} \rho(d_1) \\
\frac{\partial \boldsymbol{Call}_K[t, T_0, T_1]}{\partial P(t, T_1)} &= \phi(d_1) - \frac{K}{\sqrt{\Delta}} \frac{P(t, T_0)}{P(t, T_1)} \rho(d_2) + \frac{1}{\sqrt{\Delta}} \rho(d_1) \\
\frac{\partial \boldsymbol{Call}_K[t, T_0, T_1]}{\partial \Delta} &= -\frac{1}{2\Delta}[P(t, T_1) d_2 \rho(d_1) - K P(t, T_o) d_1 \rho(d_2)]
\end{aligned}$$

with

$$d_1 = \frac{\ln\left(\frac{P(t, T_1)}{K P(t, T_0)}\right) + \frac{1}{2}\Delta}{\sqrt{\Delta}}, \quad d_2 = d_1 - \sqrt{\Delta}$$

■

[2]We will return to this problem in the next section.

3 Hedging

We consider now that Φ_i is $\mathcal{F}_t$-measurable for $i = 0, 1, \ldots, n$.[3] This assumption jointly with deterministic correlations and volatilities conform the Gauss-Markov setting of Bueno-Guerrero et al. (2015c). Under this framework, the hedging portfolios for the options treated in Sect. 1 have no bank account part and are given by delta hedging. We reproduce here Theorem 4 of Bueno-Guerrero et al. (2015c).

Theorem 3 *Under the Gauss-Markov setting, if the conditions of Theorem 2 are satisfied, we can hedge* $\boldsymbol{C}\left[T_0, \boldsymbol{P}_{T_0}\right]$ *with the portfolio composed by* $\left.\frac{\partial\phi(y)}{\partial y_i}\right|_{y=\overline{\boldsymbol{P}}_{T_0}}$ *units of the bond maturing at* T_i, $i = 0, 1, \ldots, n$, *zero units of bank account and discounted portfolio value* $\overline{V}_t = \nabla\phi\left(\overline{\boldsymbol{P}}_t\right)\cdot\overline{\boldsymbol{P}}_t$, *where* ϕ *is given by* $\overline{V}_t = \phi\left(\overline{\boldsymbol{P}}_t\right)$ *and the overline means discounting with respect to the bank account.* ■

Note that $\overline{V}_t = \boldsymbol{C}\left[t, \overline{\boldsymbol{P}}_t\right]$ and then we can obtain explicit expressions for the hedging portfolios by using the two first sensitivities of Theorem 2. Moreover, as these portfolios have no bank account part, the expression for the hedging portfolio is also a pricing formula for the option.

Example 2 Applying Theorem 3 to the call option of Example 1 we have

$$\begin{aligned}\overline{V}_t = \boldsymbol{C}\left[t, \overline{\boldsymbol{P}}_t\right] &= \frac{\partial \boldsymbol{Call}_K[t, T_0, T_1]}{\partial P(t, T_0)}\overline{P}(t, T_0) + \frac{\partial \boldsymbol{Call}_K[t, T_0, T_1]}{\partial P(t, T_1)}\overline{P}(t, T_1)\\ &= \overline{P}(t, T_1)\phi(d_1) - K\overline{P}(t, T_0)\phi(d_2)\end{aligned}$$

This is the same expression (in discounted terms) obtained in Bueno-Guerrero et al. (2015a) for the value of a call option in the stochastic string framework. ■

As we have just seen, it is not difficult to obtain sensitivities or hedging portfolios for options with only one bond underlying. Nevertheless, there are other options, such as caps or swaptions, whose payoffs depend typically on a higher number of bonds. For one-factor models, one way to circumvent this problem is the so-called *Jamshidian's trick* (Jamshidian 1989), that allows to price an option on a portfolio of bonds as a portfolio of options with appropriate strikes. Bueno-Guerrero et al. (2015b) presented a multi-factor extension of this *trick* and applied it to price swaptions. We recall here the basis of this approach for completeness. Consider that the function $R_t(x, y)$, defined in (b) of Theorem 1, is time-homogeneous and given by

$$R^{*(m)}(x, y) = e^{-\tau(x+y)}\sum_{k=0}^{m}\lambda_k L_k(x)L_k(y), \quad 0 < \tau < \frac{1}{2}$$

with $\lambda_k > \lambda_{k+1} > 0$ and where L_k are Laguerre polynomials, $L_k = \frac{1}{k!}e^x\frac{d^k}{dx^k}(e^{-x}x^k)$. Bueno-Guerrero et al. (2015b) showed that, with this choice and under the equivalent

[3] Usually this assumption is satisfied with constant Φ_i, $i = 0, 1, \ldots, n$.

martingale measure, the short-term interest rate $r(t)$ follows the dynamics proposed in Hull and White (1990), that is,

$$dr(t) = [\omega(t) - \kappa r(t)]dt + \sigma d\widetilde{W}(t)$$

with

$$\omega(t) = \kappa \left[\int_{u=0}^{t} \left. \frac{\partial f(u,x)}{\partial x} \right|_{x=0} du + r(0) \right] + \left. \frac{\partial f(0,x)}{\partial x} \right|_{x=0} + \int_{u=0}^{t} \left. \frac{\partial^2 f(u,x)}{\partial x^2} \right|_{x=0} du + \sigma^2 t$$

where

$$\kappa = \frac{1}{\sigma^2} \sum_{k=0}^{m} \lambda_k (k+\tau), \quad \sigma = \sqrt{\sum_{k=0}^{m} \lambda_k}$$

and the bond price has the form

$$P(r(t), t, T) = e^{A(t,T) - B(t,T) r(t)} \tag{2}$$

with

$$A(t,T) = \frac{1}{2}\sigma^2 \int_{s=t}^{T} B^2(s,T) ds - \int_{s=t}^{T} \omega(s) B(s,T) ds$$

$$B(t,T) = \frac{1 - e^{-\kappa(T-t)}}{\kappa}$$

Under this multi-factor framework we have the following hedging and pricing result for a swaption (Bueno-Guerrero et al. (2015b), Proposition 3.9).

Proposition 1 *If we consider $R_t(x,y) = R^{*(m)}(x,y)$, the price at time t of a fixed payer swaption, with strike K and \$1 principal that matures at time T_0, is given by*

$$\boldsymbol{Swn}^{*(m)}(t) = \left[\sum_{i=1}^{n} C_i K_i \phi\left(-d_{2,i}^{*(m)}\right) \right] P(t, T_0) - \sum_{i=1}^{n} C_i \phi\left(-d_{1,i}^{*(m)}\right) P(t, T_i) \tag{3}$$

where $C_j = \delta K$ for $j = 1, \cdots, n-1$, $C_n = 1 + \delta K$, $\{T_j\}_{j=1}^{n}$ are the settlement times of the swap, $\delta = T_j - T_{j-1}$, and

$$d_{1,i}^{*(m)} = \frac{\ln\left(\frac{P(t,T_i)}{K_i P(t,T_0)}\right) + \frac{1}{2}\Delta_{ii}^{*(m)}}{\sqrt{\Delta_{ii}^{*(m)}}}, \quad d_{2,i}^{*(m)} = d_{1,i}^{*(m)} - \sqrt{\Delta_{ii}^{*(m)}}$$

with

$$\Delta_{ii}^{*(m)}(t, T_0, T_i) = \int_{v=t}^{T_0} \left[\int_{y=T_0-v}^{T_i-v} \int_{w=T_0-v}^{T_i-v} R^{*(m)}(w, y) dw dy \right] dv$$

where $K_i = P(r^*, T_0, T_i)$ *and* r^* *solves* $\sum_{i=1}^{n} C_i P(r^*, T_0, T_i) = 1$, *for P(r, t, T) given by (2).* ■

Taking partial derivatives in (3) we obtain immediately the sensitivities of the swaption.

Corollary 2 *Under the conditions of Proposition 1, the sensitivities of the fixed payer swaption are given, for* $i, j \in \{1, \cdots n\}, i \neq j$, *by*

$$\frac{\partial \boldsymbol{Swn}^{*(m)}(t)}{\partial P(t, T_0)} = \sum_{i=1}^{n} C_i \left[\frac{K_i}{\sqrt{\Delta_{ii}^{*(m)}}} \rho\left(-d_{2,i}^{*(m)}\right) + K_i \phi\left(-d_{2,i}^{*(m)}\right) - \frac{1}{\sqrt{\Delta_{ii}^{*(m)}}} \rho\left(-d_{1,i}^{*(m)}\right) \frac{P(t, T_i)}{P(t, T_0)} \right]$$

$$\frac{\partial \boldsymbol{Swn}^{*(m)}(t)}{\partial P(t, T_j)} = -C_j \left[\phi\left(-d_{1,j}^{*(m)}\right) + \frac{1}{\sqrt{\Delta_{jj}^{*(m)}}} \rho\left(-d_{1,j}^{*(m)}\right) - \frac{K_j}{\sqrt{\Delta_{jj}^{*(m)}}} \rho\left(-d_{2,j}^{*(m)}\right) \frac{P(t, T_0)}{P(t, T_j)} \right]$$

$$\frac{\partial \boldsymbol{Swn}^{*(m)}(t)}{\partial \Delta_{ii}^{*(m)}} = -\frac{C_i}{2\Delta_{ii}^{*(m)}} \left[d_{2,i}^{*(m)} \rho\left(d_{1,i}^{*(m)}\right) P(t, T_i) - K_i d_{1,i}^{*(m)} \rho\left(d_{2,i}^{*(m)}\right) P(t, T_0) \right]$$

$$\frac{\partial \boldsymbol{Swn}^{*(m)}(t)}{\partial \Delta_{ij}^{*(m)}} = 0$$

■

It is not difficult to show that the payoff of the fixed payer swaption in Proposition 1 can be written as

$$\boldsymbol{Swn}(T_0) = \left[P(T_0, T_0) - \sum_{i=1}^{n} P(T_0, T_i) C_i \right]_+$$

with $C_i = \delta K, i = 1, \cdots, n-1$ and $C_n = 1 + \delta K$. Then, it satisfies the conditions of Theorem 2 (with $\Phi_j < 0$, see Remark 1) and by Theorem 3 it can be hedged by delta-hedging. Thus, expression (3) gives a hedging portfolio for the payer swaption.

To apply the Jamshidian's trick to coupon bond options, as with the previous swaptions, we need all the payments to be positive. Unfortunately, not all the options in our framework satisfy this condition. For instance, consider a caption with strike S, maturing at T_0, cap rate K and principal \$1. The underlying cap has settlement dates $\{T_j\}_{j=1}^{n}$ with $T_j - T_{j-1} = \delta$. The expression for the price of the cap in the Gaussian stochastic string framework (see Bueno-Guerrero et al. 2015b) is given by

$$\boldsymbol{Cap}(t) = \sum_{j=1}^{n} P\left(t, T_{j-1}\right) \phi\left(h_{j,1}(t)\right) - (1 + \delta K) P\left(t, T_j\right) \phi\left(h_{j,2}(t)\right)$$

with

$$h_{j,1}(t)=\frac{-\ln\left(\frac{(1+\delta K)P(t,T_j)}{P(t,T_{j-1})}\right)+\frac{1}{2}\Delta_{jj}\left(t,T_{j-1}\right)}{\sqrt{\Delta_{jj}\left(t,T_{j-1}\right)}},\quad h_{j,2}(t)=h_{j,1}(t)-\sqrt{\Delta_{jj}\left(t,T_{j-1}\right)}$$

Then, the payoff of the caption is given by

$$\left[\boldsymbol{Cap}(T_0)-S\right]_+=\left[\sum_{j=1}^{n-1}\left[\phi\left(h_{j+1,1}(T_0)\right)-(1+\delta K)\phi\left(h_{j,2}(T_0)\right)\right]P\left(T_0,T_j\right)\right.$$
$$\left.-(1+\delta K)\phi\left(h_{n,2}(T_0)\right)P(T_0,T_n)-\left[S-\phi\left(h_{1,1}(T_0)\right)\right]P(T_0,T_0)\right]_+$$

that corresponds to a call option with strike $S-\phi\left(h_{1,1}(T_0)\right)$ on a coupon bond with payments $C_j=\phi\left(h_{j+1,1}(T_0)\right)-(1+\delta K)\phi\left(h_{j,2}(T_0)\right),\ j=1,\cdots,n-1; C_n=-(1+\delta K)\phi\left(h_{n,2}(T_0)\right)$. As some payments are negative, we cannot apply the Jamshidian's trick to captions. A possible solution could be to develop an approximate method for pricing coupon bond options.

The next result is based on the Fenton-Wilkinson method and states that the price of a coupon bond option can be approximated by a multiple of the price of an option on a zero-coupon bond with an appropriate maturity.[4]

Theorem 4 *Consider a call option with strike K maturing at T_0 on a coupon bond with payments $C_1,\cdots,C_n$ at times $T_1,\cdots,T_n$, and such that $\sum_{i=1}^n C_iP(t,T_i)>0$. Then, an approximation for the price of the option at time $t<T_0$, $\boldsymbol{Call}_K[t,\boldsymbol{P}_t]$, is given by*

$$\boldsymbol{Call}_K[t,\boldsymbol{P}_t]\cong\frac{\sum_{i=1}^n C_iP(t,T_i)}{P\left(t,t+\widehat{D}(t)\right)}\boldsymbol{Call}_{\widehat{K}}\left(T,t_0,t+\widehat{D}(t)\right)\tag{4}$$

where $\boldsymbol{Call}_{\widehat{K}}\left(t,t_0,t+\widehat{D}(t)\right)$ is the price of a call option with strike $\widehat{K}=\frac{KP\left(t,t+\widehat{D}(t)\right)}{\sum_{i=1}^n C_iP(t,T_i)}$ and maturity T_0 on a zero-coupon bond maturing at $t+\widehat{D}(t)$, where $\widehat{D}(t)$ is the solution of

$$\int_{t=s}^{T_0}\left[\int_{y=T_0-t}^{\widehat{D}(t)}\int_{u=T_0-t}^{\widehat{D}(t)}R_t(u,y)\,du\,dy\right]dt=\ln\sum_{i,j=1}^n w_iw_je^{\Delta_{ij}(s,T_0)}\tag{5}$$

and $w_i=\frac{C_iP(t,T_i)}{\sum_{i=1}^n C_iP(t,T_i)}$. ■

Proof See the Appendix. ■

[4]We state the result for calls as the put case can be obtained in a similar way.

Theorem 4 generalizes the fast coupon bond option pricing of Munk (1999) to the stochastic string framework as we state in the following corollary. Moreover, we show that this approximation works because it comes from the Fenton-Wilkinson method.

Corollary 3 *Under the multifactor HJM stochastic string case (Bueno-Guerrero et al. 2016) $\widehat{D}(s)$ coincides with the stochastic duration of Munk (1999) in the HJM case.* ■

Proof See the Appendix. ■

Theorem 4 can be easily applied to obtain approximate formulas for the price of swaptions or captions simply by replacing the corresponding parameters of each derivative. For swaptions, note that an expression analogous to (4) also holds for put options.

4 Conclusions

In this chapter we have analyzed some results of the stochastic string framework and we have applied them to obtain sensitivities and hedging portfolios for options with payoffs homogeneous of degree one. Moreover, we have applied the Fenton-Wilkinson method to obtain a new result for the approximate valuation of coupon bond options, generalizing the fast coupon bond option pricing proposed in Munk (1999). An extension of this work is to apply the Malliavin calculus techniques to obtain sensitivities and hedging portfolios for path-dependent options. This issue is object of ongoing research.

Acknowledgements Moreno and Navas gratefully acknowledge financial support by grants ECO2017-85927-P and P12-SEJ-1733. The usual caveat applies.

Appendix of Proofs

Proof of Theorem 2 Using the Euler Theorem for homogeneous functions, the equality

$$g(x_1, \ldots, x_n; M)e^{\sqrt{\Delta_{kk}}x_k - \frac{1}{2}\Delta_{kk}} = g\left(x_1 - \frac{\Delta_{1k}}{\sqrt{\Delta_{11}}}, \ldots, x_n - \frac{\Delta_{nk}}{\sqrt{\Delta_{nn}}}; M\right) \quad (6)$$

and $\Delta_{00} = 0$, expression (1) can be rewritten as

$$C[t, \boldsymbol{P}_t] = \sum_{k=0}^{n} \Phi_k P(t, T_k) \int_{x_1=-\infty}^{+\infty} \cdots \int_{x_j=\tilde{x}_j}^{+\infty} \cdots \int_{x_n=-\infty}^{+\infty} g\left(x_1 - \frac{\Delta_{1k}}{\sqrt{\Delta_{11}}}, \ldots, x_n - \frac{\Delta_{nk}}{\sqrt{\Delta_{nn}}}; M\right) d\boldsymbol{x}$$

Taking derivative with respect to $P(t, T_0)$ and using $\Delta_{i0} = 0$ we obtain the first sensitivity. The second one follows from a similar calculation.

For the next derivative, taking into account that $\tilde{x}_l$ is independent of Δ_{ij} , we have

$$\frac{\partial \boldsymbol{C}[t, \boldsymbol{P}_t]}{\partial \Delta_{ij}} = \sum_{k=0}^{n} \Phi_k P(t, T_k) \int_{x_1=-\infty}^{+\infty} \cdots \int_{x_l=\tilde{x}_l}^{+\infty} \cdots \int_{x_n=-\infty}^{+\infty} \frac{\partial g(x_1, \ldots, x_n; M)}{\partial \Delta_{ij}} e^{\sqrt{\Delta_{kk}} x_k - \frac{1}{2}\Delta_{kk}} d\boldsymbol{x} \tag{7}$$

The partial derivative of the Gaussian density can be obtained as

$$\begin{aligned}
\frac{\partial g(x_1, \ldots, x_n; M)}{\partial \Delta_{ij}} &= g(x_1, \ldots, x_n; M) \frac{\partial \ln g(x_1, \ldots, x_n; M)}{\partial \Delta_{ij}} \\
&= -\frac{1}{2} g(x_1, \ldots, x_n; M) \left[\frac{\partial \ln |\boldsymbol{M}|}{\partial \Delta_{ij}} + \frac{\partial}{\partial \Delta_{ij}} \left(\boldsymbol{x}^T \boldsymbol{M}^{-1} \boldsymbol{x} \right) \right] \\
&= g(x_1, \ldots, x_n; M) \frac{1}{\sqrt{\Delta_{ii}} \sqrt{\Delta_{jj}}} \left[\left(\boldsymbol{M}^{-1} \boldsymbol{x} \boldsymbol{x}^T \boldsymbol{M}^{-1} \right)_{ij} - \left(\boldsymbol{M}^{-1} \right)_{ij} \right] \\
&= g(x_1, \ldots, x_n; M) \left[\boldsymbol{D} \circ \left(\boldsymbol{M}^{-1} \boldsymbol{x} \boldsymbol{x}^T \boldsymbol{M}^{-1} - \boldsymbol{M}^{-1} \right) \right]_{ij}
\end{aligned} \tag{8}$$

where we have used the definition of M_{ij}, the chain rule, the symmetry of $\boldsymbol{M}$ and the equalities $\frac{\partial \ln|\boldsymbol{M}|}{\partial M_{ij}} = 2(\boldsymbol{M}^{-1})_{ij}$ and $\frac{\partial}{\partial M_{ij}} \left(\boldsymbol{x}^T \boldsymbol{M}^{-1} \boldsymbol{x} \right) = -2 \left(\boldsymbol{M}^{-1} \boldsymbol{x} \boldsymbol{x}^T \boldsymbol{M}^{-1} \right)_{ij}$. Replacing (8) in (7) we arrive at the expression for $\frac{\partial \boldsymbol{C}[t, \boldsymbol{P}_t]}{\partial \Delta_{ij}}$.

For the last sensitivity we have

$$\begin{aligned}
\frac{\partial \boldsymbol{C}[t, \boldsymbol{P}_t]}{\partial \Delta_{ii}} &= \frac{\partial}{\partial \Delta_{ii}} \sum_{k=0}^{n} \Phi_k P(t, T_k) \int_{x_1=-\infty}^{+\infty} \cdots \int_{x_i=\tilde{x}_i}^{+\infty} \cdots \\
&\int_{x_n=-\infty}^{+\infty} g\left(x_1 - \frac{\Delta_{1k}}{\sqrt{\Delta_{11}}}, \ldots, x_n - \frac{\Delta_{nk}}{\sqrt{\Delta_{nn}}}; M \right) d\boldsymbol{x} = -\sum_{k=0}^{n} \Phi_k P(t, T_k) \\
&\times \int_{x_1=-\infty}^{+\infty} \cdots \int_{x_{i-1}=-\infty}^{+\infty} \int_{x_{i+1}=-\infty}^{+\infty} \cdots \int_{x_n=-\infty}^{+\infty} g\left(x_1 - \frac{\Delta_{1k}}{\sqrt{\Delta_{11}}}, \ldots, \tilde{x}_i \right. \\
&\left. - \frac{\Delta_{ik}}{\sqrt{\Delta_{ii}}}, \ldots, x_n - \frac{\Delta_{nk}}{\sqrt{\Delta_{nn}}}; M \right) \times \frac{\partial \tilde{x}_i}{\partial \Delta_{ii}} d\boldsymbol{x}_i + \sum_{k=0}^{n} \Phi_k P(t, T_k) \int_{x_1=-\infty}^{+\infty} \cdots \\
&\int_{x_i=\tilde{x}_i}^{+\infty} \cdots \int_{x_n=-\infty}^{+\infty} \frac{\partial}{\partial \Delta_{ii}} g\left(x_1 - \frac{\Delta_{1k}}{\sqrt{\Delta_{11}}}, \ldots, x_n - \frac{\Delta_{nk}}{\sqrt{\Delta_{nn}}}; M \right) d\boldsymbol{x}
\end{aligned} \tag{9}$$

Moreover, we have

$$\frac{\partial \tilde{x}_i}{\partial \Delta_{ii}} = -\frac{1}{2\sqrt{\Delta_{ii}}}\left(\frac{1}{\Delta_{ii}}\ln\left(-\frac{1}{\Phi_i P(t,T_i)}\sum_{\substack{m=0\\ m\neq i}}^{n}\Phi_m P(t,T_m)e^{\sqrt{\Delta_{mm}}x_m-\frac{1}{2}\Delta_{mm}}\right)-\frac{1}{2}\right) \tag{10}$$

For the second one, by the chain rule we have

$$\begin{aligned}\frac{\partial g(x_1,\ldots,x_n;M)}{\partial \Delta_{ii}} &= \sum_{\substack{m=1\\ m\neq i}}^{n}\frac{\partial M_{im}}{\partial \Delta_{ii}}\frac{\partial g(x_1,\ldots,x_n;M)}{\partial M_{im}}\\ &= -\frac{1}{2\Delta_{ii}\sqrt{\Delta_{ii}}}\sum_{\substack{m=1\\ m\neq i}}^{n}\frac{\Delta_{im}}{\sqrt{\Delta_{mm}}}\frac{\partial g(x_1,\ldots,x_n;M)}{\partial M_{im}}\\ &= -\frac{1}{2\Delta_{ii}}\sum_{\substack{m=1\\ m\neq i}}^{n}\Delta_{im}\frac{\partial g(x_1,\ldots,x_n;M)}{\partial \Delta_{im}}\end{aligned}$$

Using expressions (6) and (8) and the definition of M_{ij} , we get

$$\begin{aligned}&\frac{\partial}{\partial \Delta_{ii}}g\left(x_1-\frac{\Delta_{1k}}{\sqrt{\Delta_{11}}},\ldots,x_n-\frac{\Delta_{nk}}{\sqrt{\Delta_{nn}}};M\right)\\ &=\frac{1}{2}g\left(x_1-\frac{\Delta_{1k}}{\sqrt{\Delta_{11}}},\ldots,x_n-\frac{\Delta_{nk}}{\sqrt{\Delta_{nn}}};M\right)\\ &\times\left(\frac{1}{\Delta_{ii}}\sum_{\substack{m=1\\ m\neq i}}^{n}\left[\boldsymbol{M}\circ\left(\boldsymbol{M}^{-1}\boldsymbol{x}\boldsymbol{x}^{T}\boldsymbol{M}^{-1}-\boldsymbol{M}^{-1}\right)\right]_{im}-\delta_{ik}\left(\frac{x_k}{\sqrt{\Delta_{kk}}}-1\right)\right)\end{aligned} \tag{11}$$

Replacing (10) and (11) in (9) we obtain the last sensitivity. ■

Proof of Theorem 4 Bueno-Guerrero et al. (2015b) showed that the exact price of a coupon bond call in the Gaussian stochastic string framework is given by

$$\boldsymbol{Call}_K[t, \boldsymbol{P}_t] = \int_{\Omega_t} g(x_1, \ldots, x_n; M) \sum_{i=1}^{n} C_i P(t, T_i) e^{\sqrt{\Delta_{ii}} x_i - \frac{1}{2}\Delta_{ii}} d\boldsymbol{x}$$

$$- K P(t, T_0) \int_{\Omega_t} g(x_1, \ldots, x_n; M) d\boldsymbol{x} \tag{12}$$

where

$$\Omega_t = \left\{ \boldsymbol{x} \in \mathbb{R}^n : \sum_{i=1}^{n} C_i P(t, T_i) e^{\sqrt{\Delta_{ii}} x_i - \frac{1}{2}\Delta_{ii}} - K P(t, T_0) > 0 \right\}$$

Define $\mathrm{y}_i = \sqrt{\Delta_{ii}} x_i - \frac{1}{2}\Delta_{ii}$. As x_i has a standard normal distribution, $\mathrm{y}_i \sim N\left(-\frac{1}{2}\Delta_{ii}, \Delta_{ii}\right)$ and then

$\mathrm{w}_i e^{y_i} \sim \log N\left(-\frac{1}{2}\Delta_{ii} + \ln \mathrm{w}_i, \Delta_{ii}\right)$.

Define $\mathrm{l}_i = \log N\left(m_{li}, \sigma_{\mathrm{l}_i}^2\right)$.The Fenton-Wilkinson method allows to approximate $\sum_{i=1}^{n} l_i$ by another lognormal variable $Z \sim \log N\left(m_z, \sigma_z^2\right)$ with

$$m_z = 2 \ln \mathrm{u}_1 - \frac{1}{2} \ln \mathrm{u}_2, \quad \sigma_z^2 = \ln \mathrm{u}_2 - 2 \ln \mathrm{u}_1$$

where

$$\mathrm{u}_1 = \sum_{i=1}^{n} e^{m_{l_i} + \frac{1}{2}\sigma_{\mathrm{l}_i}^2}$$

$$\mathrm{u}_2 = \sum_{i=1}^{n} e^{2\left(m_{l_i} + \sigma_{\mathrm{l}_i}^2\right)} + 2 \sum_{i=1}^{n-1} \sum_{j=i+1}^{n} e^{m_{l_i} + m_{lj}} e^{\frac{1}{2}\left(\sigma_{\mathrm{l}_i}^2 + \sigma_{\mathrm{l}_j}^2 + 2 r_{ij} \sigma_{\mathrm{l}_i} \sigma_{\mathrm{l}_j}\right)}$$

and $r_{ij} = \mathrm{corr}\left(l_i, l_j\right)$ (see Pirinen 2003).

In our case, we have $\mathrm{l}_i = \mathrm{w}_i e^{y_i}$, $r_{ij} = M_{ij}$, $u_1 = 1$ and $\mathrm{u}_2 = \sum_{i,j=1}^{n} w_i w_j e^{\Delta_{ij}}$. Then,

$$m_z = -\frac{1}{2} \ln \sum_{i,j=1}^{n} w_i w_j e^{\Delta_{ij}}, \quad \sigma_z^2 = -2\, m_z \tag{13}$$

Defining Δ by $\mathrm{Z} = \sqrt{\Delta} x_1 - \frac{1}{2}\Delta$ we have $m_z = -\frac{1}{2}\Delta$ and $\sigma_z^2 = \Delta$, which replaced in (13) give

$$\Delta = \ln \sum_{i,j=1}^{n} w_i w_j e^{\Delta_{ij}}$$

Thus we can make $\sum_{i,j=1}^{n} \mathrm{w}_i e^{y_i} \simeq e^{\sqrt{\Delta} x_1 - \frac{1}{2}\Delta}$ and rewrite (12) approximately as

$$\boldsymbol{Call}_K[t, \boldsymbol{P}_t] \simeq \sum_{i=1}^{n} C_i P(t, T_i) \int\limits_{\widehat{\Omega}_t} g(x_1, \ldots, x_n; M) e^{\sqrt{\Delta} x_1 - \frac{1}{2}\Delta} d\boldsymbol{x}$$

$$- K P(t, T_0) \int\limits_{\widehat{\Omega}_t} g(x_1, \ldots, x_n; M) d\boldsymbol{x}$$

with

$$\widehat{\Omega}_t = \left\{ \boldsymbol{x} \in \mathbb{R}^n : \sum_{i=1}^{n} C_i P(t, T_i) e^{\sqrt{\Delta} x_1 - \frac{1}{2}\Delta} - K P(t, T_0) > 0 \right\}$$

Integrating in $x_2, \ldots, x_n$, we get

$$\boldsymbol{Call}_K[t, \boldsymbol{P}_t] \simeq \sum_{i=1}^{n} C_i P(t, T_i) \int\limits_{\frac{1}{\sqrt{\Delta}}\left[\ln\left(\frac{\hat{K} P(t,T_0)}{P(t,t+\hat{D}(t))}\right) + \frac{1}{2}\Delta\right]}^{+\infty} g\left(x - \sqrt{\Delta}; 1\right) dx - K P(t, T_0)$$

$$\int\limits_{\ln\left(\frac{\hat{K} P(t,T_0)}{P(t,t+\hat{D}(t))}\right) + \frac{1}{2}\Delta}^{+\infty} g(x; 1) dx = \frac{\sum_{i=1}^{n} C_i P(t, T_i)}{P\left(t, t + \hat{D}(t)\right)} \left[P\left(t, t + \hat{D}(t)\right) \phi \left(\frac{\ln\left(\frac{P\left(t,t+\hat{D}(t)\right)}{\hat{K} P(t,T_0)}\right) + \frac{1}{2}\Delta}{\sqrt{\Delta}} \right) \right.$$

$$\left. - \hat{K} P(t, T_0) \phi \left(\frac{\ln\left(\frac{P\left(t,t+\hat{D}(t)\right)}{\hat{K} P(t,T_0)}\right) - \frac{1}{2}\Delta}{\sqrt{\Delta}} \right) \right] = \frac{\sum_{i=1}^{n} C_i P(t, T_i)}{P\left(t, t + \hat{D}(t)\right)} \boldsymbol{Call}_{\hat{K}}\left(t, T_0, t + \hat{D}(t)\right)$$

■

Proof of Corollary 3 We write expression (5) under the multifactor HJM stochastic string case (see Bueno-Guerrero et al. 2016) given by $R_t(u, y) = \sum_{k=0}^{m} \sigma_{HJM}^{(k)}(t, u) \sigma_{HJM}^{(k)}(t, y)$ where $\sigma_{HJM}^{(k)}(t, u), k = 0, 1, \ldots, m$ are the HJM volatilities in the Musiela parameterization. Expression (5) becomes

$$\int\limits_{t=s}^{T_0} \sum_{k=0}^{m} \left[\int\limits_{y=T_0-t}^{\widehat{D}(t)} \sigma_{HJM}^{(k)}(t, y) dy \right]^2 dt$$

$$= \ln \sum_{i,j=1}^{n} w_i w_j \exp \left\{ \int\limits_{t=s}^{T_0} \sum_{k=0}^{m} \left[\int\limits_{u=T_0-t}^{T_i-t} \sigma_{HJM}^{(k)}(t, u) du \right] \left[\int\limits_{y=T_0-t}^{T_j-t} \sigma_{HJM}^{(k)}(t, y) dy \right] dt \right\}$$

Taking derivative with respect to s provides

$$\sum_{k=0}^{m}\left[\int_{y=T_0-s}^{\widehat{D}(s)} \sigma_{HJM}^{(k)}(s,y)dy\right]^2 = \sum_{k=0}^{m}\sum_{i,j=1}^{n} w_{ij}\left[\int_{u=T_0-s}^{T_i-s} \sigma_{HJM}^{(k)}(s,u)du\right]\left[\int_{y=T_0-s}^{T_j-s} \sigma_{HJM}^{(k)}(s,y)dy\right] \tag{14}$$

with

$$w_{ij}(s,T_0) = \frac{w_i w_j \exp\left\{\int_{t=s}^{T_0}\sum_{k=0}^{m}\left[\int_{u=T_0-t}^{T_i-t}\sigma_{HJM}^{(k)}(t,u)du\right]\left[\int_{y=T_0-t}^{T_j-t}\sigma_{HJM}^{(k)}(t,y)dy\right]dt\right\}}{\sum_{i,j=1}^{n} w_i w_j \exp\left\{\int_{t=s}^{T_0}\sum_{k=0}^{m}\left[\int_{u=T_0-t}^{T_i-t}\sigma_{HJM}^{(k)}(t,u)du\right]\left[\int_{y=T_0-t}^{T_j-t}\sigma_{HJM}^{(k)}(t,y)dy\right]dt\right\}} \tag{15}$$

and $\sum_{i,j=1}^{n} w_{ij}(s,T_0) = 1$. Replacing T_0 by s in expressions (14) and (15), we get

$$\sum_{k=0}^{m}\left[\int_{y=0}^{\widehat{D}(s)} \sigma_{HJM}^{(k)}(s,y)dy\right]^2 = \sum_{k=0}^{m}\left[\sum_{i=1}^{n} w_i \int_{u=0}^{T_i-s} \sigma_{HJM}^{(k)}(s,u)du\right]^2$$

that is exactly the definition of the stochastic duration in Munk (1999) for the HJM case. ∎

References

Asmussen, S., Jensen, J. L., & Rojas-Nandayapa, L. (2016). Exponential family techniques for the lognormal left tail. *Scandinavian Journal of Statistics, 43*(3), 774–787.

Black, F., & Scholes, M. (1973). The pricing of options and corporate liabilities. *Journal of Political Economy, 81*(3), 637–654.

Bueno-Guerrero, A., Moreno, M., & Navas, J. F. (2015a). Stochastic string models with continuous semimartingales. *Physica A: Statistical Mechanics and its Applications, 433*(1), 229–246.

Bueno-Guerrero, A., Moreno, M., & Navas, J. F. (2015b). Valuation of caps and swaptions under a stochastic string model. Working paper available online at http://ssrn.com/abstract=2438678.

Bueno-Guerrero, A., Moreno, M., & Navas, J. F. (2015c). Bond market completeness under stochastic strings with distribution-valued strategies. Working paper available online at http://ssrn.com/abstract=2598403.

Bueno-Guerrero, A., Moreno, M., & Navas, J. F. (2016). The stochastic string model as a unifying theory of the term structure of interest rates. *Physica A: Statistical Mechanics and its Applications, 461*(1), 217–237.

Choi, J., & Shin, S. (2016). Fast swaption pricing in gaussian term structure models. *Mathematical Finance, 26*(4), 962–982.

Collin-Dufresne, P., & Goldstein, R. S. (2002). Pricing swaptions within an affine framework. *Journal of Derivatives, 10*(1), 9–26.

Eberlein, E., & Kluge, W. (2005). Exact pricing formulae for caps and swaptions in a Lévy term structure model. *Journal of Computational Finance, 9*(2), 99–125.
Fenton, L. F. (1960). The sum of log-normal probability distributions in scatter transmission systems. *IRE Transactions on Communication Systems, 8*(1), 57–67.
Hull, J. C., & White, A. (1990). Pricing interest rate derivative securities. *Review of Financial Studies, 3*(4), 573–592.
Jamshidian, F. (1989). An exact bond option formula. *Journal of Finance, 44*(1), 205–209.
Kim, D. H. (2014). Swaption pricing in affine and other models. *Mathematical Finance, 24*(4), 790–820.
Moreno, M., & Navas, J. F. (2008). Australian options. *Australian Journal of Management, 33*(1), 69–93.
Munk, C. (1999). Stochastic duration and fast coupon bond option pricing in multi-factor models. *Review of Derivatives Research, 3*(2), 157–181.
Pirinen, P. (2003). Statistical power sum analysis for nonidentically distributed correlated lognormal signals. In *Proceedings of the Finnish Signal Processing Symposium (FINSIG)*, pp. 254–258.
Santa-Clara, P., & Sornette, D. (2001). The dynamics of the forward interest rate curve with stochastic string shocks. *Review of Financial Studies, 14*(1), 149–185.
Singleton, K. J., & Umantsev, L. (2002). Pricing coupon-bond options and swaptions in affine term structure models. *Mathematical Finance, 12*(4), 427–446.
Vidal, J. P., & Silva, P. M. (2014). Pricing swaptions under multifactor Gaussian HJM Models. *Mathematical Finance, 24*(4), 762–789.
Wei, J. Z. (1997). A simple approach to bond option pricing. *Journal of Futures Markets, 17*(2), 131–160.

Hedging Asian Bond Options with Malliavin Calculus Under Stochastic String Models

Alberto Bueno-Guerrero, Manuel Moreno and Javier F. Navas

Abstract In this chapter we use some recent hedging results for bond options, obtained with Malliavin calculus in the context of the stochastic string framework, to hedge different types of Asian options. In all the cases, we show that the hedging portfolio has no bank account part.

Keywords Stochastic string · Hedging · Malliavin calculus · Asian option

1 Introduction

Asian options provide a payoff that depends on a certain average of the past prices of their underlying asset. This feature brings some advantages for Asian options with respect to their non-averaged counterparts.[1] First, Asian options protect their holders against price manipulation of the underlying at the maturity date. Second, stock call Asian options are always cheaper than their standard versions (see Kemna and Vorst 1990).

Different kinds of Asian options arise depending on the type of average used. Thus, we can have discrete/continuous arithmetic/geometric Asian options. The study of

[1] As in the case of standard options, Asian options can be issued in American or European style. In this chapter we will focus on the European case.

A. Bueno-Guerrero (✉)
IES Francisco Ayala, Avda. Francisco Ayala, S/N, 18014 Granada, Spain
e-mail: alberto.bueno@alu.uclm.es

M. Moreno
Department of Economic Analysis and Finance, University of Castilla La-Mancha, Cobertizo San Pedro Mártir S/N, 45071 Toledo, Spain
e-mail: manuel.moreno@uclm.es

J. F. Navas
Department of Financial Economics and Accounting,
Universidad Pablo de Olavide, Ctra. de Utrera, Km. 1, 41013 Seville, Spain
e-mail: jnavas@upo.es

M. Mili et al. (eds.), *New Methods in Fixed Income Modeling*, Contributions to Management Science, https://doi.org/10.1007/978-3-319-95285-7_10

Asian stock options has been developed in a number of papers. Kemna and Vorst (1990) price analytically geometric Asian options and price continuous arithmetic Asian options with Monte Carlo simulations. Geman and Yor (1993) use a result for the Laplace transform of the integral of geometric Brownian motion to obtain a closed-form expression for continuous arithmetic in-the-money Asian options. Jacques (1996) obtains hedging portfolios for discrete arithmetic Asian options by lognormal and inverse Gaussian approximations. Vecer (2002) suggests a numerical pricing method for arithmetic (discrete and continuous) Asian options. Dufresne (2005) provides a review of the application of Bessel processes to price Asian options. Using also the relationship between arithmetic and geometric means, Roman (2010) applies the Curran (1994) model to price arithmetic Asian options. Wiklund (2012) also focuses on arithmetic Asian options and assumes that the price of the underlying asset follows a Geometric Brownian motion. Finally, Ewald et al. (2017) price Asian commodity options assuming an stochastic convenience yield and jumps in the underlying asset price.

Much less work has been done for Asian bond options. To the best of our knowledge the first contribution related to this issue is Baaquie (2010), in which coupon bond Asian options are priced with quantum finance techniques. Recently, Landa-Fernández and Moreno (2017) have priced Asian and Australian[2] bond options by using the term structure models introduced and developed in Moreno and Platania (2015) and Moreno et al. (2018). For both models, these authors provide analytical expressions for geometric (discrete and continuous) options and apply several numerical techniques for arithmetic options. Nevertheless, nothing has been done with respect to the problem of hedging Asian bond options. The present chapter tries to fill this gap.

Recently, a new method for obtaining bond option hedging portfolios has been proposed under the stochastic string framework (Bueno-Guerrero et al. 2015b). This approach has been improved with the use of Malliavin calculus techniques in a subsequent work (Bueno-Guerrero et al. 2017), where the procedure is illustrated by hedging a bond barrier option.

The main objective of this chapter is to apply the method previously mentioned to obtain the hedging portfolio for other types of exotic, path-dependent options, namely, the arithmetic and geometric versions of Asian call options with discrete and continuous average and fixed strike on a zero-coupon bond. To the best of our knowledge, this is the first result related to this issue.

The rest of the chapter is organized as follows. Section 2 presents preliminary results on hedging options under the stochastic string modeling and their formulation with Malliavin calculus. Section 3 states the main hedging result for all the types of Asian bond options considered. Section 4 concludes the chapter.

---

[2]For a description, valuation, and hedging of Australian options, see Moreno and Navas (2008).

2 Preliminary Results

In this section we review those results from the stochastic string framework that we will need in what follows. We present them without proofs, that can be found in the original papers.

The stochastic string model is based on the following dynamics for the instantaneous forward interest rate, $f(t, x)$, (Santa-Clara and Sornette 2001)

$$df(t,x) = \alpha(t,x)dt + \sigma(t,x)dZ(t,x)$$

where t is calendar time, x is time to maturity, $\alpha(t,x)$ is the drift, $\sigma(t,x)$ is the volatility, and $Z(t, x)$ is the stochastic string process under the physical probability measure $\mathbb{P}$.[3]

Bueno-Guerrero et al. (2015b) presented a new framework for bond portfolios under which the market is complete. We recall here the main definitions and results.

Definition 1 A **portfolio** in the bond market is a pair $\{g_t, h(t,\cdot)\}$ where

(a) g is a predictable process.
(b) For each ω, t, $h(\omega, t, \cdot)$ is a generalized function in (t, ∞).
(c) For each T, the process h(t, T) is predictable. ■

The process g_t can be interpreted as the number of units of the risk-free asset in the portfolio at time t, whereas $h(t, T)dT$ is the "number" of bonds with maturities between T and $T+dT$ hold at time t in the same portfolio.

Definition 2 The **value process**, V, of a portfolio {g, h} is defined by

$$V_t = g_t B_t + \int_{T=t}^{\infty} h(t,T)P(t,T)dT \tag{1}$$

where B_t is the risk-free asset or bank account process and P(t, T) is the price, at time t, of a zero-coupon bond maturing at T. ■

Definition 3 A portfolio is **self-financing** if its value process satisfies

$$dV_t = g_t dB_t + \int_{T=t}^{\infty} h(t,T)dP(t,T)dT$$

From now on we will consider the overline symbol "$\overline{}$" meaning discounting with respect to the risk-free asset B_t.

[3]We will not need any property of $Z(t,x)$ in this chapter. For a detailed study of the stochastic string process and the probabilistic framework, we refer the reader to Bueno-Guerrero et al. (2015a).

Definition 4 Consider a discounted contingent claim $\overline{X} \in L^{\infty}(\mathcal{F}_{T0})$. We say that $\overline{X}$ can be **replicated** or that we can **hedge** against $\overline{X}$ if there exists a self-financing portfolio with bounded, discounted value process $\overline{V}$, such that $\overline{V}_{T0} = \overline{X}$. ■

The main theorem in Bueno-Guerrero et al. (2015b), stated as follows, gives the explicit expression for the bond part of the hedging portfolio.

Theorem 1 *In the stochastic string model of Bueno-Guerrero et al. (2015b), the market is complete and the generalized function $h(t,\cdot)$ in the hedging portfolio is given by*

$$h(t,T) = \frac{1}{\overline{P}(t,T)}\left[\frac{j(t,T-t)}{\sigma(t,T-t)}\right]' \tag{2}$$

where the symbol $'$ means derivative with respect to T in the sense of distributions, $j(t,\cdot)$ is given by the martingale representation of $\overline{V}_t$

$$d\overline{V}_t = \int_{u=0}^{\infty} d\tilde{Z}(t,u)j(t,u)du \tag{3}$$

and $\tilde{Z}(t,u)$ is the stochastic string process with respect to the equivalent martingale measure. ■

The problem that arises with the application of Theorem 1 is that usually we do not know the process $j(t,u)$ in the martingale representation of $\overline{V}_t$. In order to solve this problem, Bueno-Guerrero et al. (2015b) develops a Malliavin Calculus valid for stochastic string models that allows to obtain the martingale representation (3) in terms of the Malliavin derivative of the payoff.[4] Applying this result, Theorem 2 can be rewritten as follows [Theorem 5 of Bueno-Guerrero et al. (2017)].

Theorem 2 *In the Gaussian stochastic string model, the generalized function $h(t,\cdot)$ in the hedging portfolio is given by*

$$h(t,T) = \frac{1}{\overline{P}(t,T)}\left[\frac{\mathbb{E}^{\mathbb{Q}}\left[D_{t,T-t}\overline{X}|\mathcal{F}_t\right]}{\sigma(t,T-t)}\right]' \tag{4}$$

whenever the discounted payoff $\overline{X}$ is Malliavin differentiable and where $D_{t,T-t}$ is the Malliavin derivative for stochastic strings and $\mathbb{Q}$ is the equivalent martingale measure. ■

All the general results from the Malliavin calculus that do not rely on the Brownian motion framework can be applied to our framework. To apply expression (4) to Asian options, we need to know the Malliavin derivative of bond prices. The following result correspond to Propositions 6 and 9 of Bueno-Guerrero et al. (2017).

[4]We refer the reader to Nualart (2006) for the general theory of Malliavin Calculus and to Bueno-Guerrero et al. (2017) for the specific issues related to stochastic strings.

Proposition 1 *In the Gaussian stochastic string framework, and working under the equivalent martingale measure, $\overline{P}(\upsilon, T_1)$ and $P(\upsilon, T_1)$ are Malliavin differentiable and*

$$D_{t,T-t}\overline{P}(\upsilon, T_1) = -\overline{P}(\upsilon, T_1)\sigma(t, T-t)1_{T<T_1} \tag{5}$$

$$D_{t,T-t}P(\upsilon, T_1) = -P(\upsilon, T_1)\sigma(t, T-t)1_{t<\upsilon<T}1_{T<T_1} \tag{6}$$

■

3 Asian Bond Options

We consider fixed strike Asian call options on zero-coupon bonds. The payoff of these options can be written as

$$X_{T_0} = [A(T_0) - K]\, 1_{A(T_0)>K}$$

where T_0 and K denote, respectively, maturity and strike of the option, **1** is the indicator function and $A(T_0)$ is the average of bond prices. Four types of Asian options arise depending on whether (a) bond prices are treated as discrete or continuous random variables and (b) arithmetic or geometric averaging is used. In the discrete case we consider the partition $0 = t_1 < \cdots < t_n = T_0$.

- Discrete Arithmetic:

$$A(T_0) = \frac{1}{n}\sum_{i=1}^{n} P(t_i, T_1) \tag{7}$$

- Discrete Geometric:

$$A(T_0) = \left[\prod_{i=1}^{n} P(t_i, T_1)\right]^{1/n} \tag{8}$$

- Continuous Arithmetic:

$$A(T_0) = \frac{1}{T_0}\int_0^{T_0} P(\nu, T_1)d\nu \tag{9}$$

- Continuous Geometric:

$$A(T_0) = \exp\left\{\frac{1}{T_0}\int_0^{T_0} \ln(P(\nu, T_1))d\nu\right\} \tag{10}$$

The following theorem shows the hedging portfolios for the different types of Asian bond options under the stochastic string framework. In all the cases, these hedging portfolios have no bank account part.

Theorem 3 *Under the framework of the previous section, the hedging portfolios for the different types of Asian bond options are given by*

(a) Discrete arithmetic Asian option:

$$\frac{C_{DA}(t,T_0)}{P(t,T_0)} \quad \text{units of bond maturing a } T_0$$

$$\frac{1}{nP(t,T_1)} \sum_{i=1}^{n} b(t,t_i,T_1)\, 1_{t<t_i<T_0} \quad \text{units of bond maturing a } T_1$$

$$-\frac{b(t,t_i,T_1)}{nP(t,t_i)} \quad \text{units of bond maturing a } t_i,\ t<t_i<T_0$$

where $C_{DA}(t, T_0)$ is the price at time t of the discrete arithmetic Asian option that matures at time T_0 and $b(t, t_i, T_1)$ is the price at time t of a binary option that pays $P(t_i, T_1)$ if the Asian option ends up in-the-money.

(b) Discrete geometric Asian option:

$$\frac{C_{DG}(t,T_0)}{P(t,T_0)} \quad \text{units of bond maturing a } T_0$$

$$\frac{1}{nP(t,T_1)} C_{DG}(t,T_0) + Kb(t,T_0,T_0) \sum_{i=1}^{n} 1_{t<t_i<T_0} \quad \text{units of bond maturing a } T_1$$

$$-\frac{1}{nP(t,t_i)} (C_{DG}(t,T_0) + Kb(t,T_0,T_0)) \quad \text{units of bond maturing a } t_i,\ t<t_i<T_0$$

where $C_{DG}(t, T_0)$ is the price at time t of the discrete geometric Asian option that matures at time T_0 and $b(t, T_0, T_0)$ is the price at time t of a binary option that pays $P(T_0, T_0) = \$1$ if the Asian option ends up in-the-money.

(c) Continuous arithmetic Asian option:

$$\frac{C_{CA}(t,T_0)}{P(t,T_0)} \quad \text{units of bond maturing a } T_0$$

$$\frac{1}{P(t,T_1)} \left(C_{CA}(t,T_0) + \left(K - \frac{t}{T_0} A(t) \right) b(t,T_0,T_0) \right) \quad \text{units of bond maturing a } T_1$$

$$-\frac{b(t,T,T_1)}{T_0 P(t,T)} \quad \text{units of bond maturing a } T,\ T \in [t,T_0)$$

where $C_{CA}(t, T_0)$ is the price at time t of the continuous arithmetic Asian option that matures at time T_0, A(t) is the continuous arithmetic average up to time t, and b(t, x, y) is the price at time t of a binary option that pays P(x, y), x = T, T_0 and $y = T_0, T_1$ if the Asian option ends up in-the-money.

(d) Continuous geometric Asian option:

$\frac{C_{CG}(t,T_0)}{P(t,T_0)}$ *units of bond maturing a* T_0

$\frac{T_0-t}{T_0 P(t,T_1)}(C_{CG}(t,T_0)+Kb(t,T_0,T_0))$ *units of bond maturing a* T_1

$-\frac{C_{CG}(t,T_0)+Kb(t,T_0,T_0)}{T_0 P(t,T)}$ *units of bond maturing a* $T,\ T \in [t, T_0)$

where $C_{CG}(t, T_0)$ *is the price at time t of the continuous geometric Asian option that matures at time* T_0 *and* $b(t, T_0, T_0)$ *is the price at time t of a binary option that pays* $P(T_0, T_0) = \$1$ *if the Asian option ends up in-the-money.* ∎

Proof Taking Malliavin derivative on the discounted payoff $\overline{X}_{T_0}$ we get

$$D_{t,T-t}\overline{X}_{T_0} = X_{T_0} D_{t,T-t} B^{-1}(T_0) + B^{-1}(T_0) D_{t,T-t} X_{T_0} \tag{11}$$

Applying (5) and $P(T_0, T_0) = 1$, the first Malliavin derivative in the right-hand side of this equation becomes

$$D_{t,T-t} B^{-1}(T_0) = -B^{-1}(T_0)\sigma(t, T-t)\mathbf{1}_{T<T_0} \tag{12}$$

Applying the multiplication rule for Malliavin derivatives and the properties of the Dirac delta, the second Malliavin derivative in the right-hand side of Eq. (11) becomes

$$\begin{aligned} D_{t,T-t} X_{T_0} &= D_{t,T-t}\left\{[A(T_0) - K]\mathbf{1}_{A(T_0)K}\right\} \\ &= \mathbf{1}_{A(T_0)>K} D_{t,T-t} A(T_0) \end{aligned} \tag{13}$$

Replacing (12) and (13) in (11) we obtain

$$D_{t,T-t}\overline{X}_{T_0} = -\overline{X}_{T_0}\sigma(t, T-t)\mathbf{1}_{T<T_0} + B^{-1}(T_0)\mathbf{1}_{A(T_0)>K} D_{t,T-t} A(T_0) \tag{14}$$

Now we will prove each of the cases in the theorem

(a) Taking Malliavin derivative in (7) and applying (6) we get

$$D_{t,T-t} A(T_0) = -\frac{1}{n}\sigma(t, T-t)\mathbf{1}_{T<T_1}\sum_{i=1}^{n} P(t_i, T_1)\mathbf{1}_{t<t_i<(T\wedge T_0)}$$

that replaced in (14) gives

$$D_{t,T-t}\overline{X}_{T_0} = -\sigma(t, T-t)\left[\overline{X}_{T_0}\mathbf{1}_{T<T_0} + \frac{1}{n}B^{-1}(T_0)\mathbf{1}_{A(T_0)>K}\mathbf{1}_{T<T_1}\sum_{i=1}^{n} P(t_i, T_1)\mathbf{1}_{t<t_i<(T\wedge T_0)}\right]$$

and by (4)

$$h(t, T) = \frac{1}{\bar{P}(t, T)}\left[\frac{\boldsymbol{E}^{Q}[D_{t,T-t}\overline{X}]|\mathcal{F}_t}{\sigma(t, T-t)}\right]'$$

$$
\begin{aligned}
&= \frac{1}{\overline{P}(t,T)}\left[\frac{\mathbb{E}^{\mathbb{Q}}\left[D_{t,T-t}\overline{X}|\mathcal{F}_t\right]}{\sigma(t,T-t)}\right]'. \\
&\quad + \frac{1}{n}\delta(T-T_1)\boldsymbol{E}^{Q}\left[B^{-1}(T_0)\mathbf{1}_{A(T_0)>K}\sum_{i=1}^{n}P(t_i,T_1)\mathbf{1}_{t<t_i<(T\wedge T_0)}|\mathcal{F}_t\right] \\
&\quad - \frac{1}{n}\mathbf{1}_{T<T_1}\boldsymbol{E}^{Q}\left[B^{-1}(T_0)\mathbf{1}_{A(T_0)>K}\sum_{i=1}^{n}P(t_i,T_1)1_{t<t_i<T_0}\delta(T-T_i)|\mathcal{F}_t\right]\Bigg\} \\
&= \frac{1}{\overline{P}(t,T)}\overline{C}_{DA}(t,T_0)\delta(T-T_0) \\
&\quad + \frac{1}{n\overline{P}(t,T_1)}\sum_{i=1}^{n}\bar{b}(t,t_i,T_1)\mathbf{1}_{t<t_i<T_0}\delta(T-T_1) \\
&\quad - \frac{1}{n}\sum_{i=1}^{n}\frac{\bar{b}(t,t_i,T_1)}{\bar{P}(t,t_i)}\mathbf{1}_{t<t_i<T_0}\delta(T-T_i)
\end{aligned}
$$

where we have used that $\left(1_{T<T_0}\right)' = -\delta(T-T_0)$. The value of the bond part of the portfolio is given by

$$
\begin{aligned}
V_t - g_t B_t &= \int_{T=t}^{\infty} h(t,T)P(t,T)\mathrm{d}T \\
&= C_{DA}(t,T_0) + \frac{1}{n}\sum_{i=1}^{n}b(t,t_i,T_1)1_{t<t_i<T_0} - \frac{1}{n}\sum_{i=1}^{n}b(t,t_i,T_1)1_{t<t_i<T_0} \\
&= C_{DA}(t,T_0)
\end{aligned}
$$

(b) From (8) we have

$$
\begin{aligned}
D_{t,T-t}A(T_0) &= \frac{1}{n}[A(T_0)]^{1-n}D_{t,T-t}\left[\prod_{i=1}^{n}P(t_i,T_1)\right] \\
&= \frac{1}{n}[A(T_0)]^{1-n}\sum_{j=1}^{n}\prod_{\substack{i=1\\ i\neq j}}^{n}P(t_i,T_1)D_{t,T-t}P\left(t_j,T_1\right) \\
&= -\frac{1}{n}A(T_0)\sigma(t,T-t)\mathbf{1}_{T<T_1}\sum_{j=1}^{n}\mathbf{1}_{t<t_j<(T\wedge T_0)}
\end{aligned}
$$

Replacing in (14) we obtain

$$D_{t,T-t}\overline{X}_{T_0} = -\sigma(t, T-t)\left[\overline{X}_{T_0}\mathbf{1}_{T<T_0} + \frac{1}{n}B^{-1}(T_0)A(T_0)\mathbf{1}_{A(T_0)>K}\mathbf{1}_{T<T_1}\sum_{j=1}^{n}\mathbf{1}_{t<t_j<(T\wedge T_0)}\right]$$
$$= -\sigma(t, T-t)\left[\overline{X}_{T_0}\mathbf{1}_{T<T_0} + \frac{1}{n}B^{-1}(T_0)\big(X_{T_0} + K\mathbf{1}_{A(T_0)>K}\big)\mathbf{1}_{T<T_1}\sum_{j=1}^{n}\mathbf{1}_{t<t_j<(T\wedge T_0)}\right]$$

and then, by (4)

$$h(t, T) = \frac{1}{\overline{P}(t, T_0)}\overline{C}_{DG}(t, T_0)\delta(T - T_0)$$
$$+ \frac{1}{n\overline{P}(t, T_1)}\big(\overline{C}_{DG}(t, T_0) + K\bar{b}(t, T_0, T_0)\big)\left(\sum_{j=1}^{n}\mathbf{1}_{t<t_j<T_0}\right)\delta(T - T_1)$$
$$- \frac{1}{n}\big(\overline{C}_{DG}(t, T_0) + K\bar{b}(t, T_0, T_0)\big)\sum_{j=1}^{n}\frac{\mathbf{1}_{t<t_j<T_0}}{\overline{P}\big(t, T_j\big)}\delta\big(T - T_j\big)$$

The value of the bond part of the hedging portfolio is now

$$V_t - g_t B_t = \int_{T=t}^{\infty} h(t, T)P(t, T)dT$$
$$= C_{DG}(t, T_0) + \frac{1}{n}(C_{DG}(t, T_0) + Kb(t, T_0, T_0))\sum_{j=1}^{n}\mathbf{1}_{t<t_j<T_0}$$
$$- \frac{1}{n}(C_{DG}(t, T_0) + Kb(t, T_0, T_0))\sum_{j=1}^{n}\mathbf{1}_{t<t_j<T_0}$$
$$= C_{DG}(t, T_0)$$

(c) In this case, from (6) and (9) we have

$$D_{t,T-t}A(T_0) = -\frac{1}{T_0}\sigma(t, T-t)\mathbf{1}_{T<T_1}\int_{t}^{T\Lambda T_0} P(v, T_1)\mathrm{d}v$$

from where

$$D_{t,T-t}\overline{X}_{T_0} = -\sigma(t, T-t) + \left[\overline{X}_{T_0}\mathbf{1}_{T<T_0} + \frac{1}{T_0}B^{-1}(T_0)\mathbf{1}_{A(T_0)>K}\mathbf{1}_{T<T_1}\int_{t}^{T\Lambda T_0} P(v, T_1)dv]\right]$$

and

$$h(t, T) = \frac{\overline{C}_{CA}(t, T_0)}{\bar{P}(t, T_0)}\delta(T - T_0)$$

$$
\begin{aligned}
&- \frac{1}{T_0 \overline{P}(t,T)} \mathbf{E}^{\mathbf{Q}}\left[B^{-1}(T_0)\mathbf{1}_{A(T_0)>K} P(T,T_1)|\mathcal{F}_t\right]\mathbf{1}_{T<T_0} \\
&+ \frac{1}{T_0 \overline{P}(t,T_1)} \mathbf{E}^{\mathbf{Q}}\left[B^{-1}(T_0)\mathbf{1}_{A(T_0)>K} \int_t^{T_0} P(v,T_1)dv|\mathcal{F}_t\right]\delta(T-T_1) \\
&= \frac{\overline{C}_{CA}(t,T_0)}{\overline{P}(t,T_0)}\delta(T-T_0) - \frac{\bar{b}(t,T,T_1)}{T_0\bar{P}(t,T)}\mathbf{1}_{T<T_0} \\
&+ \frac{1}{\overline{P}(t,T_1)}\mathbf{E}^{\mathbf{Q}}\left[\overline{X}_{T_0} + KB^{-1}(T_0)\mathbf{1}_{A(T_0)>K}|\mathcal{F}_t\right]\delta(T-T_1) \\
&- \frac{1}{T_0\overline{P}(t,T_1)}A(t)\mathbf{E}^{\mathbf{Q}}\left[B^{-1}(T_0)\mathbf{1}_{A(T_0)>K}|\mathcal{F}_t\right]\delta(T-T_1) \\
&= \frac{\overline{C}_{CA}(t,T_0)}{\overline{P}(t,T_0)}\delta(T-T_0) - \frac{\bar{b}(t,T,T_1)}{T_0\overline{P}(t,T)}\mathbf{1}_{T<T_0} \\
&+ \frac{1}{\overline{P}(t,T_1)}\left[\overline{C}_{CA}(t,T_0) + \left(K - \frac{t}{T_0}A(t)\right)\bar{b}(t,T_0,T_0)\right]\delta(T-T_1)
\end{aligned}
$$

For the absence of bank account part we get

$$
\begin{aligned}
V_t - g_t B_t &= 2C_{CA}(t,T_0) + \left(K - \frac{t}{T_0}A(t)\right)b(t,T_0,T_0) - \frac{1}{T_0}\int_t^{T_0} b(t,T,T_1)dT \\
&= 2C_{CA}(t,T_0) + \left(K - \frac{t}{T_0}A(t)\right)b(t,T_0,T_0) \\
&- \frac{B(t)}{T_0}\int_t^{T_0}\mathbf{E}^{\mathbf{Q}}\left[B^{-1}(T_0)\mathbf{1}_{A(T_0)>K}P(T,T_1)|\mathcal{F}_t\right]dT \\
&= 2C_{CA}(t,T_0) + \left(K - \frac{t}{T_0}A(t)\right)b(t,T_0,T_0) \\
&- B(t)\mathbf{E}^{\mathbf{Q}}\left[B^{-1}(T_0)\left(X_{T_0} + K\mathbf{1}_{A(T_0)>K}\right)|\mathcal{F}_t\right] \\
&+ \frac{1}{T_0}\mathbf{E}^{\mathbf{Q}}\left[B^{-1}(T_0)\mathbf{1}_{A(T_0)>K}|\mathcal{F}_t\right]\int_0^t P(T,T_1)dT \\
&= 2C_{CA}(t,T_0) + \left(K - \frac{t}{T_0}A(t)\right)b(t,T_0,T_0) - C_{CA}(t,T_0) - Kb(t,T_0,T_0) \\
&+ \frac{b(t,T_0,T_0)}{T_0}\int_0^t P(T,T_1)dT \\
&= C_{CA}(t,T_0)
\end{aligned}
$$

(d) Working as in the previous cases we have the following sequence of expressions that concludes the proof.

$$D_{t,T-t}A(T_0) = \frac{(T\wedge T_0)-t}{T_0}A(T_0)\sigma(t,T-t)\mathbf{1}_{T<T_1}$$

$$D_{t,T-t}\overline{X}_{T_0} = -\sigma(t,T-t)\left[\overline{X}_{T_0}\mathbf{1}_{T<T_0} + \frac{(T\wedge T_0)-t}{T_0}B^{-1}(T_0)A(T_0)\mathbf{1}_{A(T_0)>K}\mathbf{1}_{T<T_1}\right]$$

$$\begin{aligned} h(t,T-t) &= \frac{\overline{C}_{CG}(t,T_0)}{\overline{P}(t,T_0)}\delta(T-T_0) - \frac{1}{T_0\overline{P}(t,T)}\left[\overline{C}_{CG}(t,T_0)+K\bar{b}(t,T_0,T_0)\right]\mathbf{1}_{T<T_0} \\ &+ \frac{T_0-t}{T_0\overline{P}(t,T_1)}\left[\overline{C}_{CG}(t,T_0)+K\bar{b}(t,T_0,T_0)\right]\delta(T-T_1) \\ V_t - g_tB_t &= C_{CG}(t,T_0) - \frac{T_0-t}{T_0}\left[\overline{C}_{CG}(t,T_0)+K\bar{b}(t,T_0,T_0)\right] \\ &+ \frac{T_0-t}{T_0}\left[\overline{C}_{CG}(t,T_0)+K\bar{b}(t,T_0,T_0)\right] \\ &= C_{CG}(t,T_0) \end{aligned}$$

■

4 Conclusions

We have reviewed recent hedging results on the stochastic string framework and we have applied them to Asian bond options, showing the power and the generality of this approach. A possible extension would be to obtain explicit expressions for the prices of the different options involved in the hedging portfolios. These expressions could be used in the cases in which the options are not traded or their prices cannot be obtained from the market.

Acknowledgements Moreno and Navas gratefully acknowledge financial support by grants ECO2017-85927-P and P12-SEJ-1733. The usual caveat applies.

References

Baaquie, B. E. (2010). *Interest rates and coupon bonds in quantum finance*. Cambridge: Cambridge University Press.

Bueno-Guerrero, A., Moreno, M., & Navas, J. F. (2015a). Stochastic string models with continuous semimartingales. *Physica A: Statistical Mechanics and its Applications, 433*, 229–246.

Bueno-Guerrero, A., Moreno, M., & Navas, J.F. (2015b). *Bond market completeness under stochastic strings with distribution-valued strategies*. Working paper available online at http://ssrn.com/abstract=2598403.

Bueno-Guerrero, A., Moreno, M., & Navas, J.F. (2017). *Malliavin calculus for stochastic strings with applications to barrier options and optimal portfolios*. Working paper available online at https://ssrn.com/abstract=2935579.

Curran, M. (1994). Valuing Asian and Portfolio options by conditioning on the geometric mean price. *Management Science, 40*(12), 1705–1711.

Dufresne, D. (2005). Bessel Processes and Asian Options. In Breton, M., & Ben-Ameur, H. (Eds.) *Numerical methods in finance* (pp. 35–57). Springer, Berlin.

Ewald, C., Wu, Y., & Zhang, A. (2017). *Pricing Asian options with stochastic convenience yield and jumps*. Working paper available online at https://ssrn.com/abstract=2984865.

Geman, H., & Yor, M. (1993). Bessel processes, Asian options and perpetuities. *Mathematical Finance, 3*(4), 349–375.

Jacques, M. (1996). On the hedging Portfolio of Asian options. *Astin Bulletin, 26*(2), 165–183.

Kemna, A. G. Z., & Vorst, A. C. F. (1990). A pricing method for options based on an average asset value. *Journal of Banking & Finance, 14,* 113–129.

Landa-Fernández, M., & Moreno, M. (2017). *Pricing and hedging of fixed income Asian and Australian options: An approach based on seasonal models*. Working paper available upon request.

Moreno, M., & Navas, J. F. (2008). Australian options. *Australian Journal of Management, 33*(1), 69–93.

Moreno, M., Novales, A., & Platania, F. (2018). A term structure representation with cyclical mean reversion. *Economic Modelling, 72*, 140–150.

Moreno, M., & Platania, F. (2015). A cyclical square-root model for the term structure of interest rates. *European Journal of Operational Research, 241,* 109–121.

Nualart, D. (2006). *The Malliavin calculus and related topics* (2nd ed.). Springer, Berlin.

Roman, J.R.M. (2010). *Curran model for pricing asian options*. New York City: Mimeo.

Santa-Clara, P., & Sornette, D. (2001). The dynamics of the forward interest rate curve with stochastic string shocks. *Review of Financial Studies, 14,* 149–185.

Vecer, J. (2002). *Unified pricing of Asian options*. Working paper available online at www.stat.columbia.edu/$\sim$~vecer/asian-vecer.pdf.

Wiklund, E. (2012). *Asian option pricing and volatility*. Working paper available online at https://www.math.kth.se/matstat/seminarier/reports/M-exjobb12/120412a.pdf.

Stochastic Recovery Rate: Impact of Pricing Measure's Choice and Financial Consequences on Single-Name Products

Paolo Gambetti, Geneviève Gauthier and Frédéric Vrins

Abstract The ISDA CDS pricer is the market-standard model to value credit default swaps (CDS). Since the Big Bang protocol moreover, it became a central quotation tool: just like options prices are quoted as implied vols with the help of the Black-Scholes formula, CDSs are quoted as running (conventional) spreads. The ISDA model sets the procedure to convert the latter to an upfront amount that compensates for the fact that the actual premia are now based on a standardized coupon rate. Finally, it naturally offers an easy way to extract a risk-neutral default probability measure from market quotes. However, this model relies on unrealistic assumptions, in particular about the deterministic nature of the recovery rate. In this paper, we compare the default probability curve implied by the ISDA model to that obtained from a simple variant accounting for stochastic recovery rate. We show that the former typically leads to *underestimating* the reference entity's credit risk compared to the latter. We illustrate our views by assessing the gap in terms of implied default probabilities as well as on credit value adjustments (CVA) figures and pricing mismatches of financial products like deep in-/out-of-the-money standard CDSs and digital CDSs (main building block of credit linked notes, CLNs).

Keywords Credit default swap · Bootstrapping · Default probability curve
Stochastic recovery rate

P. Gambetti · F. Vrins (✉)
Louvain Finance Center and CORE, Université catholique de Louvain,
Voie du Roman Pays 34, 1348 Louvain-la-Neuve, Belgium
e-mail: frederic.vrins@uclouvain.be

P. Gambetti
e-mail: paolo.gambetti@uclouvain.be

G. Gauthier
HEC Montréal and GERAD, Montreal, Canada
e-mail: genevieve.gauthier@hec.ca

M. Mili et al. (eds.), *New Methods in Fixed Income Modeling*, Contributions to Management Science, https://doi.org/10.1007/978-3-319-95285-7_11

1 Introduction

Credit Default Swaps (CDSs) are credit derivatives allowing one party (*protection buyer*) to buy protection on a given *reference entity* from another party (*protection seller*) Fabozzi (2003). These instruments act as insurance contracts between the two parties, who trade the default risk of the reference entity for a given notional N up to a maturity date T. Before the crisis, the protection buyer had to make quarterly payments (defined by a specific calendar, called *IMM dates*) with amounts determined by the coupon rate c (called *running* spread), the contract's notional N and some day-count convention. This spread was the premium that makes the deal be worth zero at inception. Hence, the running spread agreed at inception with the prevailing *par* (also called *break-even spread*). In exchange of those premia, the protection seller agrees to make a payment (called *contingent flow*) to the protection buyer in case the reference entity effectively defaults prior to the contract's maturity. In the case of standard CDS contracts, the amount of the contingent flow depends on the contract notional as well as on the actual *recovery rate* of the firm that will be determined either by the residual value of issued Bonds or via an auction. The ISDA (International Swap and Derivatives Association) proposed a model to value CDS contracts based on a particular specification of the general no-arbitrage pricing equations. This model became the standard on the market. It is for example the default model in Bloomberg (the other alternatives being just variants of the latter), but also inspired most of the other pricing platforms like Markit or Summit, that differ from the ISDA model only by minor specificities.

But this way of trading CDSs is not really convenient from a Treasury management perspective. For instance, if a trader wants to close the position afterwards (i.e. at a different running spread) by entering in a reverse trade, she is left with a stream of residual payments. In order to facilitate back office operations, a standardization was needed. The fundamental reviewing of CDS trading conventions is called the Big Bang protocol Markit (March 13, 2009). While traders still quote CDS contracts in the form of a running spread (now called *conventional spread*), the mechanism of the quarterly payments made by the protection buyer has been standardized. The premium payments are decomposed in two parts: the first part consists in quarterly payments but, in contrast with the former convention, the coupon rate is no longer the quoted (*running*) spread. Instead, it is a standard coupon rate (k, say), being either 100 or 500 bps depending on the credit risk of the counterparty at inception.[1] As there is in general no reason that the credit risk premium of a reference entity (i.e. the quoted spread c) matches with the standard coupon rate k, a financial adjustment needs to take place. This correction takes the form of an upfront payment. For instance, if $c > k$, the quarterly payments made by the protection buyer to the protection seller do not properly compensate the value of the protection leg, and the trade would have a positive Mark-to-Market (MtM) to the buyer. To compensate the protection seller, the buyer has to pay (if positive, receive if negative) an additional (upfront)

[1]On the European contracts, a standard coupon rate of 25 bps is also considered for a couple of entities having a very high creditworthiness.

amount that precisely compensates the MtM. With this additional upfront amount (corresponding to the difference between (i) the present value of the payments made if the coupon rate were c and (ii) the present value of those made according to the standard rate k), the contract is *at par*. Therefore, the Big Bang protocol is just another way to schedule the premium flows. In such a setup, a trader willing to close her trade with another party only needs to make an upfront payment. In particular, all the quarterly flows will cancel each other, independently of the prevailing levels of quoted spreads. In this context, the ISDA model also plays the role of *converter*. Just like Black-Scholes formula allows to convert "quoted vols" to "cash amounts", the ISDA CDS pricer (also called *ISDA converter* since then) allows to quote deals on a running basis rather than on upfront amounts, which is much more intuitive and convenient for traders (ISDA CDS Standard Model 2004). The ISDA converter wipes out any ambiguity about how to convert quoted spreads to upfront amounts by making the present value of quarterly payments in the new setup (upfront + premia based on the standard coupon rate k) equal to the present value of the payments in the old setup assuming the quoted spread is the running spread c (i.e. without upfront, but replacing k by c in the premia).

On the top of being a pricer and a converter, it also became the central tool to select a pricing measure in (incomplete) credit markets. In the classical no-arbitrage setup, the value of a contract is the present value of the future cashflows. In the CDS case, this means that at inception, the value of the premium leg (upfront + present value of payments based on the standard coupon rate) agrees with the protection leg (present value of the contingent cashflow). The ISDA pricer consists in simplifying the pricing equations to have a simple form for both legs. Therefore, inverting the ISDA pricing equations, starting from quoted spreads of a set of CDS contracts, provides an easy way to pick up a pricing probability measure and, in particular, to back out a parametric risk-neutral default probability curve. This measure could then be used to compute other quantities, like credit valuation adjustments (CVA) or prices of non-standard credit-sensitive deals. Whereas most of these simplifications have little impact and are quite realistic, one of them may have substantial financial consequences: the recovery rate that determines the contingent payment in case of default is assumed to be known, and is typically set to 40% (or 20% for sovereigns). While 40% is indeed close to the average of observed recovery rates from past defaults, fixing the recovery rate to that level contradicts empirical evidence. Since the '90s in fact, researchers started to analyze recovery rates information and build evidence about their time-series and cross-sectional variation. It became clear that recovery rates exhibit significant differences across seniority levels of the defaulted bonds and sector to which issuers belong (Altman and Kishore 1996): interactions among these two features can also play an important role. While it is straightforward to think that lack of collateralization and higher degrees of subordination lead to lower recoveries, industry effects have been justified by asset redeployability and/or anticipated government support considerations[2] (Shleifer and Vishny 1992; Acharya et al. 2007; Sarbu et al. 2013). Moreover, there is clear evidence that recovery rates are nega-

[2]This is particularly true in the banking sector, even though there are famous counter-examples.

tively correlated with the business cycle and in particular with default probabilities: on average, the higher the default rate, the lower the recovery rate (Frye 2000; Hu and Perraudin 2002; Altman et al. 2005; Boudreault et al. 2013). It is worth noting that the interest about recovery rates uncertainty is not only limited to academia, but it is also shared among the industry under the form of research reports, mainly by rating agencies (Hamilton et al. 2001; Cantor and Varma 2004; Keisman and Van de Castle 1999; Van de Castle and Keisman 2000). The main message of all these studies is again that recovery rates are unknown before default, and they should be considered as random variables. Surprisingly, none of the standard approaches account for this crucial point.

The goal of this paper is to show that in some circumstances, disregarding this reality may have important consequences in terms of pricing and risk-management of financial products. We first recall in Sect. 2 the no-arbitrage pricing equations associated to CDS contracts and show that the ISDA model essentially is an approximation where the recovery rate assumption plays a central role. We then introduce a simple CDS model in Sect. 3. The purpose of this model is to provide some intuition about the possible consequence of the recovery rate's uncertainty. The model is used in Sect. 4 as an external benchmark to identify the potential model risk embedded in the standard pricers. We start with an empirical analysis of recovery rates based on the Moody's Default and Recovery Database. This will provide guidelines to adjust the recovery rate parameters in our "in-house" CDS model. Then, the two models are compared from various points of view: implied default probabilities, CVA figures and mispricing of instruments like digital CDSs [customized CDSs that strike the recovery rate to 0% contractually, they are the cornerstone of credit linked notes, CLNs Fabozzi et al. (2007)] as well as on existing standard CDSs that are not explicitly quoted.

2 Pricing Equations

We derive the pricing equations under the assumption that we are valuing deals at time t coming prior to maturity T and that the firm did not default yet. All contracts are valued from the standpoint of the protection buyer. In order to ease the notations the valuation date is used as a refence time ($t = 0$). Therefore, depending on the context, a same symbol s can be used to loosely denote either a date or the remaing time from the valuation date ($t = 0$) to date s. For instance, T represents both the contract maturity and the time-to-maturity. This is a common abuse of notation that drastically eases the mathematical exposition.

2.1 General No-Arbitrage CDS Pricing Equations

The price of a CDS is given by the difference between the risk-neutral expectation of the discounted contingent cashflow with that of the discounted flows of the premium leg. Let us start with the protection leg. The default of the reference entity triggers the payment of the contingent flow at default's time, should the latter comes prior to the contract maturity T. The payment is a fraction $L = (1 - R)$ of the notional N where R stands for the recovery rate of the firm:

$$(1 - R)N\ \mathbb{I}_{\{\tau \le T\}}\ ,$$

where $\mathbb{I}_{\{\omega\}}$ stands for the indicator function defined as 1 if ω is true and 0 otherwise. The corresponding present value in a non-arbitrage setup is

$$\mathrm{Prot}(T) := N\mathbb{E}\left[\mathbb{I}_{\{\tau \le T\}}\frac{(1-R)}{\beta_\tau}\right],$$

where $\mathbb{E}$ stands for the expectation operator under a chosen risk-neutral probability measure $\mathbb{Q}$ and β denotes the (risk-free) bank account numéraire.

Seen from the valuation time $t = 0$, the premium leg consists in a stream of (future) periodic flows paid at specific dates $t_1 < \ldots < t_m = T$. We denote by t_0 either the last payment date (prior to t) or the inception date (if none premium payment took place before t). The cashflows associated to the quarterly payments can be split in two parts. The first term deals with the protection between the valuation date and the next coupon date. The part of the coupon associated to the period (t, t_1) is due if $\tau \ge t_1$. Otherwise, only a part of it (corresponding to the (t, τ) period) has to be paid. More precisely, let $\Delta(s, t)$ be the fraction of year between the dates (s, t) according to the specific day count convention. Then, the discounted flow associated to the credit protection for the period $(t, t_1]$ is[3]

$$kN\ \mathbb{I}_{\{\tau \ge t_1\}}\frac{\Delta(t, t_1)}{\beta_{t_1}} + kN\ \mathbb{I}_{\{\tau < t_1\}}\frac{\Delta(t, \tau)}{\beta_\tau}\ .$$

The remaining terms deal with the following payments. As above, each of them can be split in two terms. A first term that deals with the case where there is no default up to the payment date (in which case the full coupon is due) and a second term (called *rebate*) accounting for the fact that whenever the default takes place between two payment dates, only a part of the coupon (corresponding to the period up to default) has to be paid. More specifically, the discounted cashflow is

---

[3]Note that the period between the last payment date (or inception) t_0 and the valuation date does not enter the picture as it deals with past protection, and there is no reason to pay for it. The corresponding amount $kN\Delta(t_0, t)$ is called the *accrued* and explains the difference between clean and dirty prices as for Bonds.

$$kN\left(\sum_{i=2}^{m} \mathbb{1}_{\{\tau \geq t_i\}} \frac{\Delta(t_{i-1}, t_i)}{\beta_{t_i}} + \sum_{i=2}^{m} \mathbb{1}_{\{t_{i-1} < \tau < t_i\}} \frac{\Delta(t_{i-1}, \tau)}{\beta_\tau}\right) .$$

Hence, the risk-neutral present value of the premium flows is given by

$$\mathrm{Prem}(T) = \mathrm{up} + k\mathrm{PV01}(T) ,$$

where $\mathrm{PV01}(T)$ is called the *risky duration* of the deal and is given by the present value of the premium payments based on a unitary coupon rate. Denoting $t_i^+ := \max(0, t_i)$ for conciseness,[4] it comes

$$\mathrm{PV01}(T) := N\sum_{i=1}^{m}\left(\Delta(t_{i-1}^+, t_i)\mathbb{E}\left[\frac{\mathbb{1}_{\{\tau \geq t_i\}}}{\beta_{t_i}}\right] + \mathbb{E}\left[\mathbb{1}_{\{t_{i-1} < \tau < t_i\}} \frac{\Delta(t_{i-1}^+, \tau)}{\beta_\tau}\right]\right) .$$

Consider the special case where the risk-free rate underlying the numéraire β is independent from both the default time τ and recovery rate R. Then,

$$\mathrm{Prot}(T) = -N\int_0^T (1 - \mathbb{E}[R|\tau = u])P(u)dG(u), \tag{1}$$

$$\mathrm{PV01}(T) = N\sum_{i=1}^{m}\left(\Delta(t_{i-1}^+, t_i)P(t_i)G(t_i) - \int_{t_{i-1}^+}^{t_i} \Delta(t_{i-1}^+, u)P(u)dG(u)\right), \tag{2}$$

where

$$P(s) := \mathbb{E}\left[\frac{1}{\beta_s}\right] , \quad s \geq 0$$

is the price of a risk-free zero-coupon bond paying 1 unit of currency at time S and

$$G(s) := \mathbb{E}\left[\mathbb{1}_{\{\tau > s\}}\right] = \mathbb{Q}(\tau > s) = \mathbb{Q}(\tau \geq s) , \quad s \geq 0$$

is the risk-neutral survival probability of the reference entity.[5] Observe that the assumption that the reference entity did not default by the valuation time ($\tau > 0$) leads G to be non-increasing after t with $G(0) = 1$.

[4]This notation is needed to deal with the term $i = 1$, to make sure that $t_{i-1}^+ = t_0^+ = t$ and not t_0.

[5]In this paper, we assume that τ admits a density. In particular, $\mathbb{Q}(\tau = s) = 0$ for all $s \in \mathbb{R}^+$.

2.2 ISDA Pricing Equations

The ISDA model derives from JP Morgan routines. Except little subtleties that are negligible for our purposes, the protection leg and risky duration are given by[6]

$$\overline{\text{Prot}}(T) := (1-x)N\sum_{i=1}^{m}\bar{P}(t_i)\left(\bar{G}(t_{i-1}^+)-\bar{G}(t_i)\right)$$

$$\overline{\text{PV01}}(T) := N\sum_{i=1}^{m}\Delta(t_{i-1}^+,t_i)\bar{P}(t_i)\frac{\bar{G}(t_{i-1}^+)+\bar{G}(t_i)}{2}$$

$$= N\sum_{i=1}^{m}\Delta(t_{i-1}^+,t_i)\bar{P}(t_i)\left(\bar{G}(t_i)+\frac{\bar{G}(t_{i-1}^+)-\bar{G}(t_i)}{2}\right).$$

A key assumption is that it assumes a fixed loss-given-default, i.e. it requires the knowledge of the recovery rate x. In the standard ISDA pricer, the discount curve $\bar{P}(.)$ is the risk-free discount curve built from the prices of specific instruments taken from the previous day. The curve $\bar{G}$ represents the reference entity's survival probability, and is built in agreement with market quotes (see Sect. 2.3).

These equations resemble those obtained by no-arbitrage provided in (1) and (2). Indeed, replacing $\bar{P} \leftarrow P$ and $\bar{G} \leftarrow G$ and assuming independence between default time and recovery (setting $x \leftarrow \mathbb{E}[R|\tau = u] = \mathbb{E}[R]$) the protection leg expressions agree up to the discretization of the integral. Same applies to the risky duration, provided that one assumes that the payments always take place at payment date (even in case of default) and that if the default happens between two payment dates, half of the coupon is due. Most of these assumptions are known to have little impact. The impact of discretization is limited because the only effect is to discount from slightly different dates. For the same reason, the impact of choosing $\bar{P}$ instead of P is most of the time negligible. The parametric curve $\bar{G}$ allows to obtain a continuous survival probability curve from a limited number of quotes, but due to calibration constraints, its impact is minimal. All in all, the fundamental difference between the ISDA model and the no-arbitrage CDS equations is arguably the assumption regarding the dependence between variables and processes. The credit-rate independence assumption is known to be acceptable (Brigo and Alfonsi 2005). Eventually, the fundamental specificity of the ISDA *versus* the general CDS equations has to be found in the protection leg, and with the treatment of the recovery rate in particular.

[6]In particular, we adopt a discretization scheme in line with the payment schedule (i.e. quarterly), and assume that all payments impacted by the occurence of the reference entity's default take place at the first payment date following the default event.

2.3 Calibration of the ISDA Model to Observed Mark-to-Market Values

It is clear from above that $\bar{G}$ aims at representing the survival function under the pricing measure. Credit models are incomplete. Hence several risk-neutral measures exist. However, in order to avoid arbitrage opportunities, any chosen measure must be *calibrated to the market*. In other words, the chosen measure has to comply with market quotes. We now show how such quotes uniquely determine the parametric curve $\bar{G}$ in the ISDA model when a set of CDS calibration instruments is provided.

We start from a set of n (say) liquid CDS quotes of the reference entity with various maturities ($T_1 < T_2 < \ldots < T_n$, typically 1, 3, 5 and sometimes 10 years) called *calibration* instruments. As we have only n constraints, the curve $\bar{G}(.)$ will have n degrees of freedom. In the ISDA model, it is parametrized via a *hazard rate* function h as

$$\bar{G}(s) = \mathrm{e}^{-\int_0^s h(u)du} \ , \quad s \geq 0 \ .$$

The function h is piecewise constant between the maturities of the calibration CDSs with flat extrapolation beyond the last martutity. In order to *calibrate* the ISDA model, we make sure that the model and market prices of the calibration instruments agree. In particular, we make sure that the protection and premium legs of the ISDA model match when the coupon rate is set to the quoted spread $c(T_1), \ldots, c(T_n)$ for the assumed recovery rates $\vec{x} = (x_1, \ldots, x_n)$.[7] Hence, this curve (i.e. the piecewise constant levels of h) is constructed iteratively (in the order given by the products' maturities) so as to ensure

$$c(T_i) = \frac{\overline{\mathrm{Prot}}(T_i)}{\overline{\mathrm{PV01}}(T_i)} \ , \quad i \in \{1, 2, \ldots, n\} \ . \tag{3}$$

These calibration constraints simultaneously provide the function $\bar{G}$ as well as the legs $\overline{\mathrm{Prot}}$ and $\overline{\mathrm{PV01}}$. As stressed before, the coupon rate actually used in quarterly payments is not the quoted spread $c(T)$ but instead the standard coupon rate k, and the MtM difference is compensated via the upfront. The ISDA model is the market standard procedure to determine the actual upfront amounts from quoted (conventional) spreads:

$$\mathrm{up}(k, T_i) = \overline{\mathrm{Prot}}(T_i) - k\overline{\mathrm{PV01}}(T_i) = (c(T_i) - k)\overline{\mathrm{PV01}}(T_i) \ . \tag{4}$$

The way $\bar{G}$ is constructed combined with the way c is converted to an upfront amount implies that all calibration instruments have, by construction, an ISDA-model price that is inline with the market.

[7] Usually, the term-structure of recovery rate is flat, i.e. $x_1 = \cdots = x_n = x$.

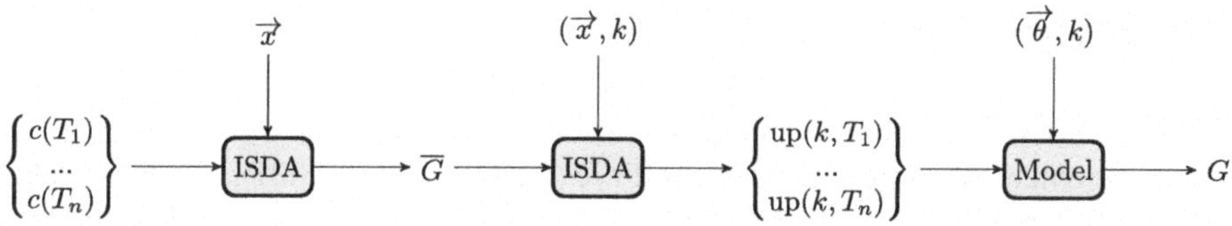

Fig. 1 Methodology to extract the model implied default probability curve G from n quoted spreads. The procedure works as follows: (1) quoted spreads are first used to compute the ISDA's curve $\bar{G}$, (2) the ISDA model fed with $\bar{G}$ is then used to convert quoted spreads to upfront amounts and (3) the survival probability curve G of the in-house model is tuned such that it implies the same prices (upfront) for the instruments used in the calibration procedure

2.4 Calibration of In-house Model to Observed Mark-to-Market Values

Let us now consider an alternative (*in-house*) model controlled by a set $\vec{\theta}$ of exogenous parameters. As before, the model price of the calibration instruments must agree with those implied by the model. There is a fundamental difference, however. Whereas the ISDA model is calibrated from quoted (conventional) spreads, the calibration of the in-house model has to be done on the *actual prices*. In particular, the target that we have to meet is that, for all calibration instruments, the difference between the model-implied protection leg $\text{Prot}(T)$ and the risky duration $\text{PV01}(T)$ needs to agree with the upfront resulting from the application of the ISDA converter to the quoted spreads [as per Eq. (4)]:

$$\text{up}(k, T_i) = \text{Prot}(T_i) - k\text{PV01}(T_i) \ . \tag{5}$$

This puts some constraints on the protection leg and the risky duration and so on the survival probability curve G. The procedure is illustrated on Fig. 1. The key point here is to notice that the differences between the two models will change the selected risk-neutral measure. In particular, there is no reason that the survival probability in the ISDA model (computed with the help of the measure selected during the ISDA calibration procedure) agrees with that of the in-house model (that is based on another measure, the one picked up from the in-house model calibration step).

As explained above, the survival probability curve $\bar{G}$ obtained by inverting the ISDA CDS pricer equations provide a rather fair estimation of the risk-neutral default probability curve G *when the recovery rate is known in advance* (and agrees with the value plugged in the ISDA model). The other discrepancies have indeed very little impact. However, the curve $\bar{G}$ is obtained by calibrating the ISDA model (that assumes a fixed recovery) to prices of standard (i.e. floating recovery) CDS. This inconsistency suggests that one makes an error when assuming that the curve $\bar{G}$ is a valid proxy for G, depending on the stochastic properties of the recovery rate. In the next section, we provide a simple model to illustrate the potential impact of this way of working.

3 A Simple Model

The first models of stochastic recovery rates have been developed in the context of risk management. We refer the reader to Frye (2000), Jarrow (2001), Jokivuolle and Peura (2000) and Pykthin (2003), just to name a few. To the best of our knowledge, the Pioneering work with regards to random recovery rates in the context of credit derivatives pricing is due to Andersen and Sidenius. In Andersen and Sidenius (2004), the authors extend the One Factor Gaussian Copula model for CDO pricing [also known as Li's model, Li (2016)], which was the market standard at that time. The credit crisis triggered a specific interest for stochastic recovery models. At some point, super senior CDO tranches (with attachment point higher than 60%) started to trade. However, such tranches are not worth anything under a fixed 40% recovery rate assumption. This is the best evidence that recovery rates have either to be decreased or, more realistically, have to be made stochastic. Since then, many stochastic recovery rate models have been introduced for the sake of pricing CDOs. We can mention for instance Gaspar and Slinko (2008), Ech-Chatbi (2008), Krekel (2010), Amraoui et al. (2012).

Surprisingly however, the impact of stochastic recovery rates on single-name CDS did not receive much attention. Yet, some authors tackled this point. For instance, Boudreault et al. (2013, 2015) and Bégin et al. (2017) propose a credit risk model in which the default intensity and the recovery rate are a non-linear function of the firm leverage ratio. This approach allows to capture the negative relationship between default probability and recovery rate observed in the empirical studies. The model is estimated using a filtering approach on a sample of CDS premiums. They show that the interrelation between recovery rate and default probability modifies the term structure of zero-coupon yield to maturity and impacts significantly the standard risk measures such as the VaR and the expected shortfall. While these methods can be adapted for this purpose, we prefer to introduce a simple model that is tailored to illustrate the potential impact of disregarding recovery rates uncertainty. Our goal here is indeed to put forward the impact of the uncertainty of recovery rate as well as its correlation with default rate on single-name credit derivatives. Therefore, it is more convenient to work with a model based on the standard approach, but adjusted for the effect we want to analyze.

In the sequel, we analyze "how far" can the default probability curve $\bar{G}$ extracted from the ISDA pricer be from the actual default probability curve G extracted from a CDS pricing model that would account for the stochastic nature of recovery rate. To concentrate on the effect we are effectively interested in, we disregard some technicalities that are known to have a minor impact on CDS valuation. First, we assume a unit notional ($N = 1$) and that the discount curve P and $\bar{P}$ agree and are parametrized by a constant and known risk-free rate r so that

$$P(s) = \bar{P}(s) = \mathrm{e}^{-rs}, \quad s \geq 0 .$$

Second, we make some simplifications in terms of payment schedule; the premiums are paid on a continuous-time basis. Finally, we assume that we have only one calibration instrument ($n = 1$). All these assumptions can be easily relaxed but significantly simplify the exposition. For the ISDA pricing equations, we assume that a single calibration CDS instrument is used, so that $h(s) = h$ and $\bar{G}(s) = \mathrm{e}^{-hs}$ for $s \geq 0$. Setting the ISDA recovery rate level to the value x, the corresponding protection leg and risky duration become

$$\begin{aligned}\overline{\mathrm{Prot}}(T) &= -(1-x)\int_0^T P(u)d\bar{G}(u) = (1-x)h\int_0^T \mathrm{e}^{-(r+h)u}du \\ &= (1-x)\frac{h}{r+h}(1-\mathrm{e}^{-(r+h)T}),\\ \overline{\mathrm{PV01}}(T) &= \int_0^T P(u)\bar{G}(u)du = \int_0^T \mathrm{e}^{-(r+h)u}du = \frac{1-\mathrm{e}^{-(r+h)T}}{r+h}.\end{aligned}$$

From (3), we find the *credit triangle* $h = c(T)/(1-x)$.

We now introduce our simple in-house model. A standard approach in credit risk modeling consists in representing the default time τ as the first jump of a Poisson process with intensity λ under $\mathbb{Q}$. By doing so, $G(T) = \mathbb{Q}(\tau > T) = \mathrm{e}^{-\lambda T}$. Accounting for the possible dependency of R and τ, one gets

$$\begin{aligned}\mathrm{Prot}(T) &= \lambda\int_0^T (1-\mathbb{E}[R|\tau = u])\mathrm{e}^{-(r+\lambda)u}du\,,\\ \mathrm{PV01}(T) &= \int_0^T P(u)G(u)du = \int_0^T \mathrm{e}^{-(r+\lambda)u}du = \frac{1-\mathrm{e}^{-(r+\lambda)T}}{r+\lambda}.\end{aligned}$$

Interestingly, we observe that the risky duration in both the ISDA and the in-house models reads $\mathrm{PV01}(h, T)$ and $\mathrm{PV01}(\lambda, T)$ where

$$\mathrm{PV01}(z, T) := \frac{1-\mathrm{e}^{-(r+z)T}}{r+z}.$$

As explained above, the protection leg needs to be modeled with care. One directly observes that when R and τ are independent, we have $h = \lambda$ (i.e. $G = \bar{G}$) whenever $\mathbb{E}[R] = x$. Generally speaking however, credit events and recovery rates are not independent. The most famous study supporting that claim is undoublty that of Altman et al (Altman et al. 2005). This is why an alternative to ISDA, accounting for that effect, needs to be considered.

An easy choice consists in adopting a static copula setup. The idea is similar to the resampling approach used in counterparty risk applications to compute the credit value adjustment (CVA) under wrong-way or right-way risk (Gregory 2010; Sokol 2011; Vrins 2017). One models the conditional expectation $\mathbb{E}[R_\tau|\tau = s]$ using a function $f(s)$ that collapses to $\mu_R := \mathbb{E}[R]$ when there is no dependency between τ and R. We start by fixing the marginal laws of R and τ. The survival probability function of τ is implied by our Poisson model and is denoted by $G(s)$. On the other hand, R is modeled with a Beta distribution with shape parameters α and β whose distribution function is denoted by $F_R(.; \alpha, \beta)$. Then, R and τ are coupled with a copula C with a fixed dependency parameter ρ, so that one can sample (R, τ) using independent uniform random variables (U, V):

$$(R, \tau) \sim C\left(F_R^{-1}(U; \alpha(\tau), \beta(\tau)), F_\tau^{-1}(V); \rho\right) .$$

Choosing the Gaussian copula with correlation ρ, the conditional distribution of R given $\tau \in ds$ is

$$f(s; \rho, Z)ds := F_R^{-1}\left(\Phi\left(\rho\Phi^{-1}(G(s)) + \sqrt{1-\rho^2}Z\right); \alpha, \beta\right) ds$$

where $Z \sim \mathcal{N}(0, 1)$ and Φ its cumulative distribution function. Hence,

$$\mathbb{E}[R|\tau \in ds]/ds = \mathbb{E}\left[f(s; \rho, Z)\right] =: f(s; \rho)$$

so that

$$\mathbb{E}\left[\frac{R\, \mathbb{I}_{\{\tau \leq T\}}}{\beta_\tau}\right] = -\int_0^T f(u; \rho)P(u)dG(u) .$$

Eventually, the protection leg takes the following form:

$$\begin{aligned}
\text{Prot}(T) &= \frac{\lambda\left(1 - \mathrm{e}^{-(r+\lambda)T}\right)}{r+\lambda} - \lambda \int_0^T f(u; \rho)\mathrm{e}^{-(r+\lambda)u}du \\
&= \lambda\left(\text{PV01}(\lambda, T) - \int_0^T f(u; \rho)\mathrm{e}^{-(r+\lambda)u}du\right) .
\end{aligned}$$

In order to correctly price the CDS contract with this model, the difference between the protection leg and the present value of the periodic premium payments must correspond to the upfront amount (the MtM of the deal without upfront); this is summarized in Fig. 2. This in turns means that λ needs to satisfy Eq. (5) which in this case reads as

$$\text{up}(k, T) = (\lambda - k)\text{PV01}(\lambda, T) - \lambda \int_0^T f(u; \rho)\mathrm{e}^{-(r+\lambda)u}du .$$

Interestingly, $f(u; 0) = \mu_R$ whenever $\rho = 0$ or when the variance of R vanishes. This means that if the recovery rate is deterministic (and equal to the constant used in the ISDA model) or if it is stochastic but independent from default's time (but whose risk-neutral expected value agrees with the constant used in the ISDA model), then $\lambda = h$. On the other hand, one can see that when $\rho > 0$, a high default probability (lower G) implies on average a low recovery rate. Hence, the observation in Altman

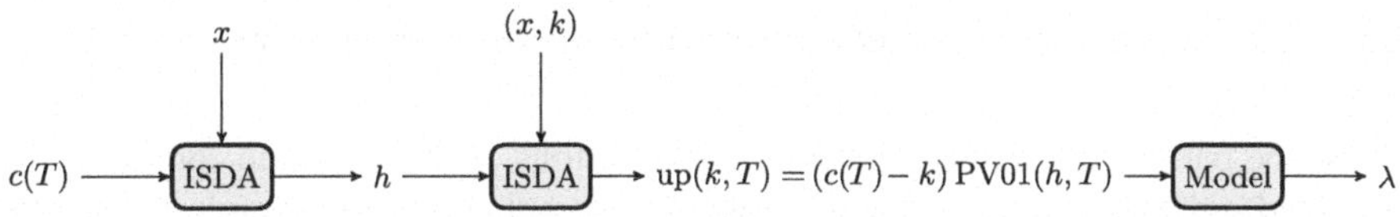

Fig. 2 Methodology to calibrate the model's parameters to correctly price standard instruments. For simplicity, we assume here that only one CDS is quoted on the market so that only one conventional spread (with maturity T) is available for the reference entity. Hence, the hazard rate curve h collapses to a constant, and same for the default rate of the in-house model. The procedure works as follows: (1) quoted spreads are first used to compute the ISDA's parameters, (2) the ISDA model is used to convert quoted spread to upfront amounts and (3) the in-house model is calibrated such that it implies the same upfronts for the standard instruments

et al. (2005) suggests that one should use $\rho > 0$.[8] If we believe in our in-house model, the correct default rate λ can be obtained by looking at the hazard rate h implied by the ISDA model. In the other cases however, estimating the default rate λ (and equivalently the default probability curve G) from the hazard rate h leads to an error. We provide some order of magnitudes in the next section.

4 Numerical Experiments

We take the α and β shape coefficients so as to satisfy two constraints about the mean μ_R and variance σ_R^2 of the recovery rate. First, we want the conditional expectation of R given $\tau \in ds$ to be a given value μ_R when $\rho = 0$, i.e. $f(t; 0) = \mu_R$. Second, we would like the variance σ_R^2 to be "valid". As the variance of a distribution whose support is $[0, 1]$ and mean is π is upperbounded by that of a Bernoulli with parameter π, we choose:

$$\sigma_R = a\sqrt{\mu_R(1-\mu_R)} \ , \quad a \in [0, 1] \ .$$

In this expression, a controls the uncertainty we have about the recovery rate around the mean. These two constraints yield the shape and scale parameters

$$\alpha := \mu_R \left(\frac{\mu_R(1-\mu_R)}{\sigma_R^2} - 1 \right) \ , \quad \beta := \frac{1-\mu_R}{\mu_R}\alpha \ .$$

In the sequel, we consider the "ideal case" where the deterministic recovery rate value x used in the ISDA pricer is 40% and does indeed correspond to $\mathbb{E}[R] = \mu_R$ (the alternatives will of course lead to more significant errors).

---

[8]The reason why a negative relationship between recovery rate and default intensity corresponds to a positive ρ stems from the fact that the default-time is negatively correlated with the default probability: the higher the default intensity, the sooner the default time, on average.

4.1 Maximum Likelihood Estimation of the Parameters

The shape parameters of the conditional distributions of recovery rate can be easily estimated either by moment matching techniques (plugging the sample mean and sample variance into the above formulas) or via maximum likelihood. The latter method works as follows. Given the probability density function of the Beta distribution parametrized by $\alpha, \beta > 0$

$$f(x;\alpha,\beta) = \frac{1}{B(\alpha,\beta)} x^{\alpha-1}(1-x)^{\beta-1} = \frac{\Gamma(\alpha+\beta)}{\Gamma(\alpha)\Gamma(\beta)} x^{\alpha-1}(1-x)^{\beta-1},$$

where $B(\cdot,\cdot)$ and $\Gamma(\cdot)$ denote the beta and gamma functions respectively, the likelihood function, given the sample of independent recovery rate observations $X = (x_1, ..., x_n)$, has the following form:

$$\mathcal{L}_n(\alpha,\beta|X) = \prod_{i=1}^{n} f(x_i;\alpha,\beta) = \prod_{i=1}^{n} \frac{\Gamma(\alpha+\beta)}{\Gamma(\alpha)\Gamma(\beta)} x_i^{\alpha-1}(1-x_i)^{\beta-1}.$$

The maximum likelihood estimator of α and β is then given by:

$$\frac{\partial \ell_n(\alpha,\beta|X)}{\partial \alpha} = n\Psi(\alpha+\beta) - n\Psi(\alpha) + \sum_{i=1}^{n} \log(x_i) = 0 \tag{6}$$

$$\frac{\partial \ell_n(\alpha,\beta|X)}{\partial \beta} = n\Psi(\alpha+\beta) - n\Psi(\beta) + \sum_{i=1}^{n} \log(1-x_i) = 0 \tag{7}$$

where $\Psi(\cdot)$ denotes the digamma function. In Table 1, we report summary statistics of recovery rate empirical distributions taken from Moody's Default and Recovery Database, together with the mean μ^{ML} and standard deviation σ^{ML} of the theoretical Beta conditional distributions obtained by maximum likelihood estimation with respect to the shape parameters α and β.[9] Conditioning is made with respect to the seniority of the defaulted Bond or the industry of the issuing company.

These results support the empirical findings discussed in Sect. 1 and confirm that the assumption of a constant recovery rate of 40% is indeed misleading. We observe important differences both in expected recovery rates μ^{ML} of different Bond seniorities (with mean recovery rates increasing with the level of seniority) and in their variability σ^{ML}. We also document a large intra-class variability (i.e. considerable values of σ^{ML} once we have conditioned for a specific seniority). In particular, we point out the higher variability of recoveries on Senior Secured Bonds: given the mean being around 50% and the standard deviation being nearly $1/\sqrt{12} \approx 28.87\%$, one should notice that this conditional distribution is in fact close to be Uniform in

[9]Note that, in the log-likelihood function we have to subtract a machine-precision quantity from full recoveries (i.e. when the recovery rate is equal to 100%).

Table 1 Summary statistics of empirical conditional distributions of recovery rates where recoveries are expressed as percentages of the bond face value. The last two columns display the mean and the standard deviation of the theoretical distribution with shape parameters α and β obtained via maximum likelihood. Data for the analysis are taken from Moody's Default and Recovery Database: the sample includes 2035 American bond defaults covering the period 1st January 1912–23rd January 2017. Regarding the optimization, we adopted a quasi-Newton method (L-BFGS algorithm with lower-bound constraints)

	Min.	1st Qu.	Median	μ	3rd Qu.	Max.	σ	μ^{ML}	σ^{ML}
Junior subordinated	0.63	8.00	13.99	20.34	33.00	74.00	18.98	20.68	17.74
Senior subordinated	0.01	10.31	25.00	30.39	44.14	100.00	24.04	31.46	25.22
Senior unsecured	0.01	15.00	31.94	36.63	56.06	100.00	26.03	37.85	26.80
Senior secured	0.01	20.00	40.00	44.57	65.36	100.00	27.73	46.81	28.81
Banking	0.01	3.94	18.00	23.55	37.00	92.08	22.70	23.91	22.57
Capital industries	0.01	14.43	30.14	36.35	57.00	100.00	26.18	37.05	26.81
Consumer Industries	0.01	15.00	31.67	36.95	55.00	100.00	25.73	38.60	26.48
Energy & environment	0.01	19.00	36.75	38.86	52.82	100.00	25.70	39.89	27.93
FIRE	0.13	10.00	25.00	32.48	46.63	100.00	25.36	35.47	26.50
Media & publishing	0.01	16.00	33.50	38.45	54.00	99.00	26.99	38.74	27.84
Retail & distribution	0.50	13.62	29.00	33.52	48.62	99.50	25.63	35.97	26.98
Technology	0.25	10.00	23.75	30.32	45.00	100.00	25.36	32.91	26.18
Transportation	1.75	16.00	25.38	32.76	45.88	99.88	21.99	34.69	22.19
Utilities	13.99	43.63	67.18	62.87	84.65	100.00	26.46	65.27	27.26

[0, 1].[10] Similarly to Bond seniorities, we document sensible differences in mean and standard deviation of recovery rates conditional distributions when conditioning is made on the industrial sector. Also for this type of conditioning the recovery rates intra-class variability is high. In the sequel, we run our experiments as if the expected recovery rate μ^{ML} were exactly equal to the fixed one of 40% used as input in the ISDA pricer but with different levels of uncertainty. However, simulations could be run also by taking as inputs the last two columns of Table 1: this would lead to an amplification of the results. The time-evolution of recovery rate using sliding window is depicted on Fig. 3.

4.2 *Impact on Implied Default Probability*

We look at the impact of the correlation parameter (ρ) for two uncertainty levels about the recovery rate value: a small uncertainty ($a = 10\%$ leading to $\sigma_R = 5\%$) and a large

[10] Uniform distribution is a particular case of the Beta distribution when both the shape parameters α and β are equal to one (indeed our estimation of this conditional distribution yields $\alpha = 0.9$ and $\beta = 1$). The flat distribution for Senior Secured bonds is in accordance with the findings of Schuermann (2004).

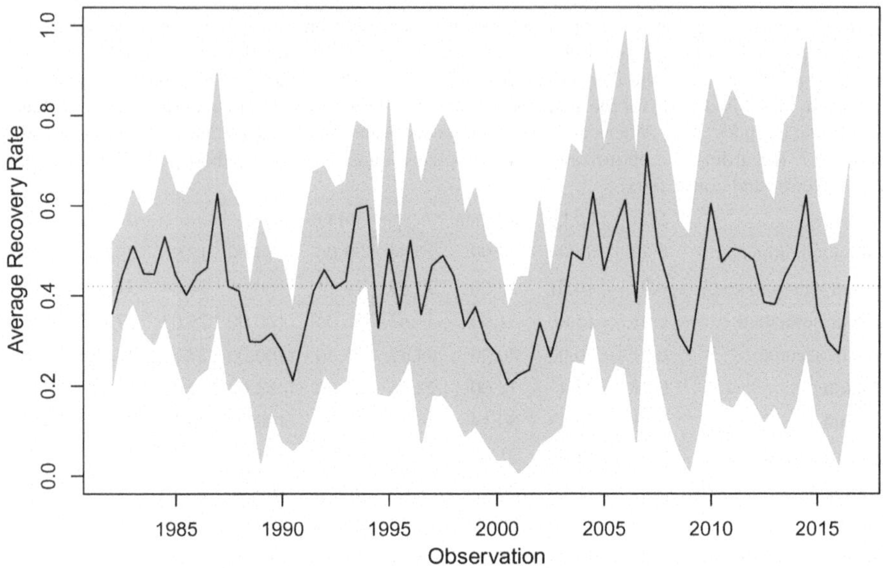

Fig. 3 Half-yearly averages of recovery rates black solid line) with the ±1 standard deviation envelope (grey area) for American bond defaults in the period 5th January 1982–31st December 2016. The red dotted line corresponds to the average recovery rate observed in this period (Color figure online)

uncertainty ($a = 50\%$ leading to $\sigma_R = 25\%$, which is indeed close to the average σ^{ML} of Table 1). We consider a single calibration instrument which is a $T = 5Y$ CDS with running spread of 250 bps. The standard coupon is set to $k = 100$ bps. Figure 4 shows the mismatch between λ and h as well as a similar information but translated into survival probability at maturity to ease the interpretation. The ISDA model always returns the same constant value located at the intersection of the dotted lines. One can see that for $\rho > 0$ (suggested by empirical evidences), the survival probability implied by the "in-house" model is lower than that of the ISDA model. To put it differently, the ISDA model underestimates the (risk-neutral) default likelihood in such cases.

4.3 *Impact on Deep In-/Out-of-the-Money Standard CDSs*

Suppose that a trader wants to price a CDS with outstanding maturity of 3Y and that the market only quotes a 5Y contract. The ISDA procedure would be to first extract the survival probability curve which, in the simplified setup depicted above, is given by the credit triangle: $h = c(5Y)/(1 - x)$ (where x is the recovery rate assumed). This in turns indicates that the break-even spread for the 3Y contract is, $c(3Y) = c(5Y)$. In our simple in-house model, the procedure is slightly different.

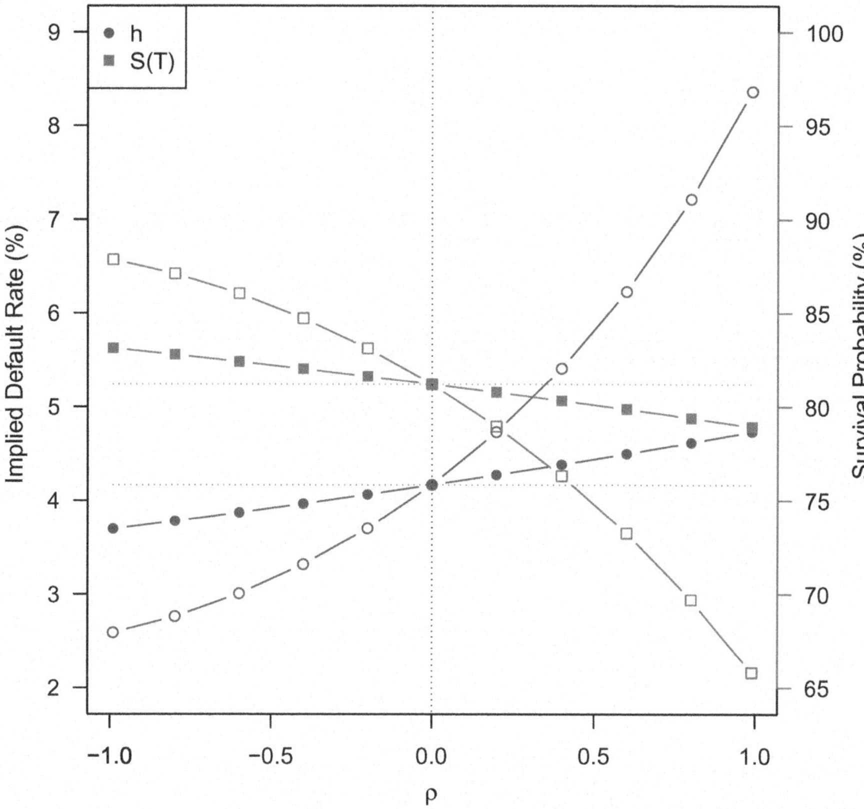

Fig. 4 Impact of correlation on the implied default rate (blue, solid+dotted markers, left axis) and survival probability at maturity (red, solid+red squares, right axis) for two different values of recovery rate volatility: $a = 1/10$ (filled markers) and $a = 1/2$ (empty markers). Parameters: $x = \mu_R = 40\%$, calibration instrument: $T = 5Y$ maturity swap with $c(T) = 250$ bps, $k = 100$ bps (Color figure online)

We still assume $\mu_R = x = 40\%$ but the implied default rate λ depends on ρ and a. From these assumptions, one can extract the break-even spread of the 3Y contract. Eventually the upfront is given by scaling the difference between the latter spread and the standard coupon rate k with the risky duration of the deal computed according to the model. As above, one needs to convert the model spread $c(3Y)$ to a quoted spread $\hat{c}(3Y)$ in agreement to the quoting convention. A similar development as above yields

$$\hat{c}(3Y) = k + (c(3Y) - k)\frac{\text{PV01}(\lambda, 3Y)}{\text{PV01}(h, 3Y)} .$$

Figure 5 summarizes the methodology and Fig. 6 illustrates the impact in terms of conventional spreads and MtM.

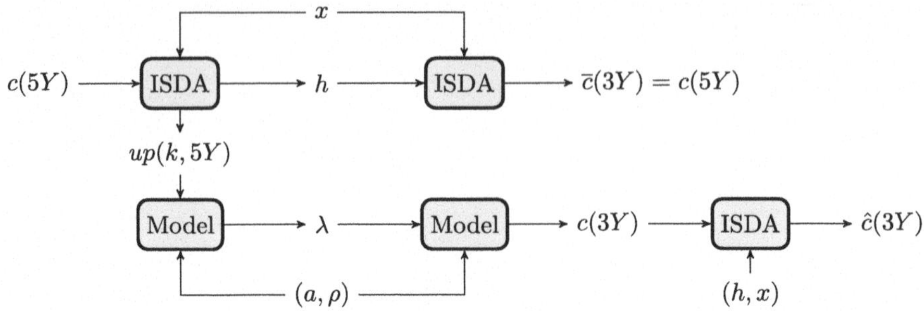

Fig. 5 Methodology to analyze the model impact on the 3Y spread of a standard CDS starting from a 5Y spread of a standard CDS

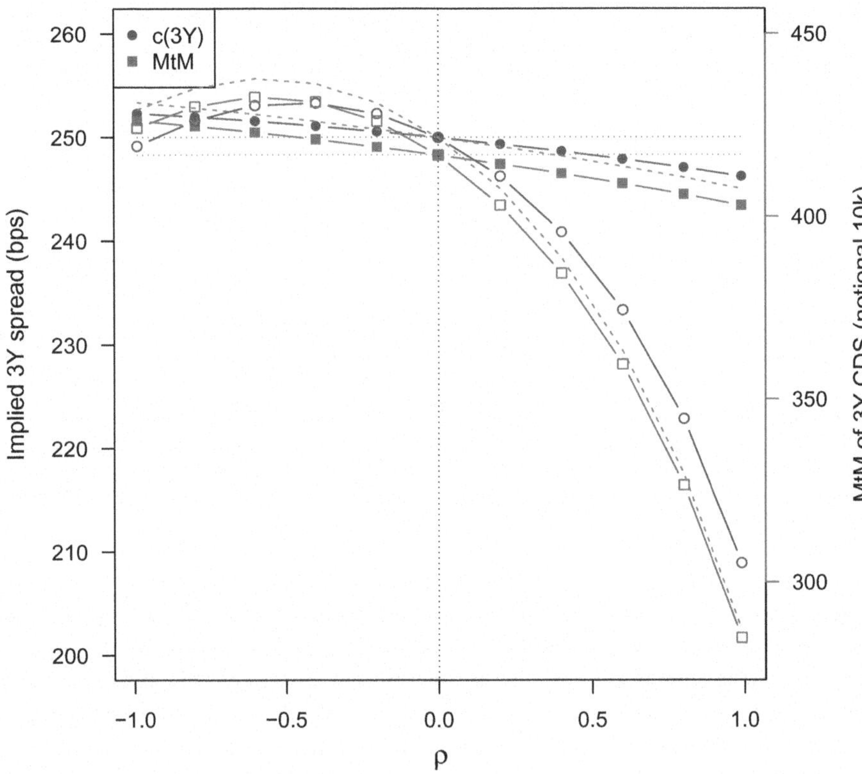

Fig. 6 Impact of correlation on the implied break-even CDS spread with maturity 3Y (blue, dot markers, left axis) and corresponding MtM with 10k notional (red, square markers, right axis) for two different values of recovery rate volatility: $a = 1/10$ (filled markers) and $a = 1/2$ (empty markers). Parameters: $x = \mu_R = 40\%$, calibration instrument: 5Y maturity swap with $c = 250$ bps, $k = 100$ bps (Color figure online)

4.4 Impact on CVA of a Call

Credit value adjustment (CVA) is the current market price corresponding to the option—implicitly given to our counterparty—to default during the life of a trade. Let us consider a deal where we trade a call option with maturity T and strike K on a stock S (with volatility σ) with a counterparty whose default time τ is independent from S. Then, letting $\tilde{C}_s := C_s/\beta_s$ be the time-s discounted price of the call and G is the survival probability of the counterparty, one gets, from Brigo et al. (2018)

$$\mathrm{CVA} = \mathbb{E}\left[\mathbb{I}_{\{\tau<T\}}(1-R)\frac{C_\tau^+}{\beta_\tau}\right] = -\int_0^T (1-\mathbb{E}[R|\tau=s])\mathbb{E}[\tilde{C}_s^+]dG(s)\ .$$

But the price of a call and the numéraire β are always non-negative, so that $\tilde{C}_s^+ = \tilde{C}_s$ and $\mathbb{E}[\tilde{C}_s] = \tilde{C}_0 = C_0$ as $\tilde{C}$ is a $\mathbb{Q}$-martingale. It becomes clear that in this context, CVA is nothing but the protection leg of a CDS whose reference entity is the counterparty, with zero risk-free rate and notional C_0:

$$\mathrm{CVA} = -C_0\int_0^T (1-\mathbb{E}[R|\tau=s])dG(s)\ .$$

In Fig. 7a we compare two types of CVAs. The first one is computed by assuming an independent recovery rate whose expected value is equal to the ISDA level, $\mathbb{E}[R|\tau=s] = x = 40\%$ and used $\lambda \leftarrow h$ to be consistent:

$$\overline{\mathrm{CVA}} = h\int_0^T (1-x)\mathbb{E}[\tilde{C}_s^+]\mathrm{e}^{-hs}du = (1-x)C_0(1-\mathrm{e}^{-hT})\ ,$$

so that the CVA on a Call option with maturity T per unit of option premium is the ISDA protection leg $\overline{\mathrm{Prot}}(T)$ with $r = 0$. This is the horizontal black curve on Fig. 7a.

The second way to compute CVA is by extracting λ with our in-house model assuming a given pair (a, ρ) and then price CVA consistently. Hence, we are consistent with the ways recovery rate is considered in both the calibration and in the CVA pricing steps:

$$\mathrm{CVA} = \lambda\int_0^T (1-f(s,\rho))\mathbb{E}[\tilde{C}_s^+]\mathrm{e}^{-\lambda s}ds = \lambda C_0\int_0^T (1-f(s,\rho))\mathrm{e}^{-\lambda s}ds\ .$$

As before, the "in-house" CVA, expressed per unit of option premium, is the "in-house" protection leg $\mathrm{Prot}(T)$ with $r = 0$. This is the blue curve on Fig. 7a. One can see that if we are consistent in the way we extract default probabilities and price CVA, the error is limited (the blue and black curves do not deviate too much from each other). If by contrast we lack consistency, i.e. use the internal model to compute the default probability curve but assume a fixed recovery in the CVA's payoff as below,

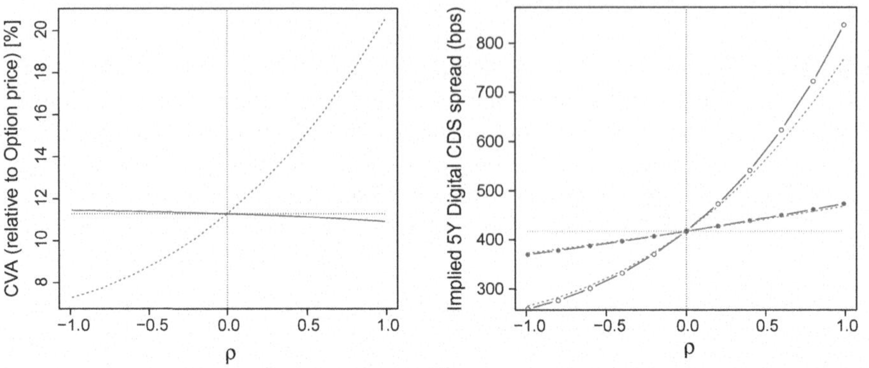

(a) The in-house model CVA ρ (blue), the "mixed-CVA" $\widetilde{\mathrm{CVA}}$ (using G but $x = 40\%$ in the CVA payoff, red) and $\overline{\mathrm{CVA}}$ obtained by assuming a fixed (here, $x = 40\%$) recovery rate everywhere (horizontal black). Parameters: $a = 0.5$, $S_0 = 50, K = 45, \sigma = 30\%$.

(b) Implied spread $\bar{d}$ (horizontal dots), d (blue) and $\hat{d}$ (magenta) of a digital CDS for two different values of recovery rate volatility: $a = 1/10$ (filled markers) and $a = 1/2$ (empty markers).

Fig. 7 Impact of stochastic recovery rate on CVA (left) and implied par digital CDS spread (right) with respect to ρ. Calibration instrument: 5Y maturity swap with $c = 250$ bps and $k = 100$ bps (Color figure online)

$$\widetilde{\mathrm{CVA}} = \lambda(1-x)\int_0^T \mathbb{E}[\tilde{C}_s^+]\mathrm{e}^{-\lambda s}ds = (1-x)C_0(1-\mathrm{e}^{-\lambda T})\,.$$

This is similar to computing CVA assuming a deterministic recovery rate of x but assuming a stochastic recovery rate when extracting the default probability (by using λ instead of h). The inconsistency is striking by comparing this CVA (red curve) with the others on Fig. 7a.

4.5 *Impact on Par Spread of Digital CDSs*

As a final experiment we focus on the price of digital CDS, i.e. a CDS contract where the recovery rate is *contractually* set to 0, so that the contingent flow is either 0 if $\tau > T$ or N otherwise. Applying the ISDA procedure, the hazard rate h is extracted from standard CDS, and the valuation of the digital CDS with the ISDA approach naturally leads to the spread $\bar{d}(5Y) = h = c(5Y)/(1-x)$. Similarly, once the default rate λ of our in-house model has been extracted, it is straightforward to check that the break-even spread implied by the in-house model is $d(5Y) = \lambda$. Note that a proper comparison would require not to compare $\bar{d}$ and d but to compare $\bar{d}$ with the "ISDA-equivalent" spread $\hat{d}(5Y)$. Indeed, from the in-house model, the trader would not quote $d(5Y)$ but would quote a spread $\hat{d}(5Y)$ such that, when converting this spread according to the ISDA converter, she would find the same MtM (i.e. upfront amount)

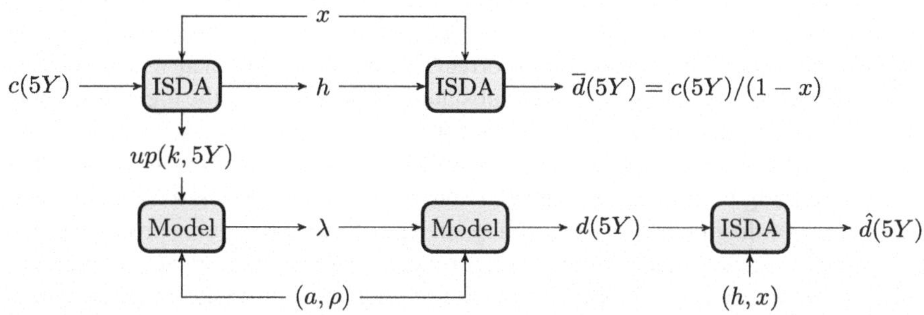

Fig. 8 Methodology to analyze the model impact on the spread of a 5Y digital CDS starting from the spread of a 5Y standard CDS

as the one implied by the in-house model when the actual (standard) coupon rate k is used. The spread $\hat{d}(5Y)$ is computed such that the "in-house model" price (correct to the trader) and "ISDA model" (conventional quotation) price agree:

$$(d(5Y) - k)\text{PV01}(\lambda, T) = \text{up} = (\hat{d}(5Y) - k)\text{PV01}(h, T) \Rightarrow$$
$$\hat{d}(5Y) = k + (d(5Y) - k)\frac{\text{PV01}(\lambda, T)}{\overline{\text{PV01}}(h, T)} .$$

Figure 8 summarizes the methodology and Fig. 7b shows the impact. From the inputs given above, we find $\bar{d}(5Y) = c(5Y)/0.6 \approx 417$ bps (horizontal dotted blue line). By contrast, the break-even spread $d(5y)$ (blue) or its ISDA-equivalent $\hat{d}(5Y)$ (magenta) substantially depends on (a, ρ) but are relatively close to each other. Accounting for a positive dependency between ρ and λ leads to a digital CDS conventional spread larger than the one given by the simple rescaling of the conventional spread $c(5Y)/(1 - 0.4)$ suggested by the ISDA model.

5 Conclusion

In this paper, we have stressed that the recovery rate uncertainty (which had been emphasized in many studies and again emphasized above) is a key driver of CDS prices. Moreover, quite a few empirical analyses suggest a negative dependency between recovery rate and default probability. Based on our simple in-house model, we have shown that the common way of extracting risk-neutral probabilities (i.e. inverting the equations derived from the ISDA CDS pricer, the standard pricer in the market) leads to *underestimate* the actual market-implied default probabilities in standard situations. This can have serious consequences when assessing credit risk of firms. Moreover, it can also impact the valuation of less liquid instruments (like digital CDSs) or existing standard deals. This type of mispricing is of the same nature as those that led to introduce multi-curve pricing (*OIS discounting*): if one bootstraps the Libor curve from new trades, these trades will, by construction, price at par. By contrast, the impact of adopting the (correct) multi-curve approach will be clearly visible for swaps where the fixed rates substantially deviate from the prevailing swap rates. Therefore, just like Black-Scholes formula is a handy tool to communicate prices in terms of implied vols, the ISDA CDS model is a nice way to quote CDS prices as running premiums. Nevertheless, as pricer, it must be used with care as it neglects one of the major risks underlying such products. The figures produced with the simple model introduced above suggest that traders and risk managers may not make the economy of developing their own model when it comes to extract risk-neutral probabilities or price non standard deals. To the best of our knowledge, there is no standard alternative to the ISDA CDS pricer involving stochastic recovery rate. Developing such alternatives is the purpose of active research in the field.

Acknowledgements This work was supported by the Fonds de la Recherche Scientifique - FNRS under Grant J.0037.18.

References

Acharya, V., Bharath, S., & Srinivasan, A. (2007). Does industry-wide distress affect defaulted firms? Evidence from creditor recoveries. *Journal of Financial Economics*, *85*, 787–821.

Altman, E., & Kishore, V. (1996). Almost everything you wanted to know about recoveries on defaulted bonds. *Financial Analysts Journal*, *52*, 57–64.

Altman, E., Brady, B., Resti, A., & Sironi, A. (2005). The link between default and recovery rates: Theory, empirical evidence, and implications. *Journal of Business*, *78*(6), 2203–2227.

Amraoui, S., Cousot, L., Hitier, S., & Laurent, J.-P. (2012). Pricing CDOs with state-dependent stochastic recovery rates. *Quantitative Finance*, *12*(8), 1219–1240.

Andersen, L., & Sidenius, J. (2004). Extensions to the gaussian copula: Random recovery and random factor loadings. *Journal of Credit Risk*, *1*(1), 29–70.

Bégin, J.-F., Boudreault, M., & Gauthier, G. (2017). Firm-specific credit risk estimation in the presence of regimes and noisy prices. *Accepted in Finance Research Letter*.

Boudreault, M., Gauthier, G., & Thomassin, T. (2015). Estimation of correlations in portfolio credit risk models based on noisy security prices. *Journal of Economic Dynamics and Control*, *61*, 334–349. https://doi.org/10.1016/j.jedc.2015.10.001.

Boudreault, M., Gauthier, G., & Thomassin, T. (2013). Recovery rate risk and credit spreads in a hybrid credit risk model. *Journal of Credit Risk*, *9*(3), 3–39.

Brigo, D., Hvolby, T., & Vrins, F. (2018). *Wrong-way risk adjusted exposure: Analytical approximations for options in default intensity models*. In: Innovations in insurance, risk and asset management, Springer (in press).

Brigo, D., & Alfonsi, A. (2005). Credit default swaps calibration and option pricing with the SSRD stochastic intensity and interest rate model. *Finance and Stochastics*, *9*, 29–42.

Cantor, R., & Varma, P. (December 2004). *Determinants of recovery rates on defaulted bonds and loans for north american corporate issuers: 1983–2003*. Technical Report.

Ech-Chatbi, C. (2008). *CDS and CDO pricing with stochastic recovery*. NA: Technical report.

Fabozzi, F. (Ed.). (2003). *The handbook of fixed income securities*, 8th ed. New York City: Mc Graw-Hill.

Fabozzi, F., Davis, H., & Choudhry, M. (2007). Credit-linked notes: A product primer. *Journal of Structured Finance*, *12*(4), 67–77.

Frye, J. (April 2000). Collateral damage. *Risk*, pp. 91–94.

Gaspar, R., & Slinko, I. (2008). On recovery and intensity's correlationa new class of credit models. *Journal of Credit Risk*, *4*(2).

Gregory, J. (2010). *Counterparty credit risk*. Wiley Finance.

Hamilton, D., Gupton, G., & Berthault, A. (February 2001). *Default and recovery rates of corporate bond issuers: 2000*. Technical Report.

Hu, Y.-T., & Perraudin, W. (2002). *The dependence of recovery rates and defaults*. Technical report.

ISDA CDS Standard Model. Technical report, 2004. http://www.cdsmodel.com/cdsmodel/.

Jarrow, R. (2001). Default parameter estimation using market prices. *Financial Analysts Journal*, *57*(5), 75–92.

Jokivuolle, E., & Peura, S. (2000). *A model for estimating recovery rates and collateral haircuts for bank loans*. Bank of Finland: Technical report.

Keisman, D., & Van de Castle, K. (June 1999). *Recovering your money: Insights into losses from defaults*. Technical Report June, 1999.

Krekel, M. (2010). *Pricing distressed cdos with base correlation and stochastic recoveries*. Technical report.

Li, D. (2016). *On default correlation: A copula function approach*. Technical report, 2016.

Pykthin, M. (August 2003). Unexpected recovery risk. *Risk*, pp. 74–78.

Sarbu, S., Schmitt, C., & Uhrig-Homburg, M. (2013). *Market expectations of recovery rates*. Working Paper, pp. 1–42.

Schuermann, T. (2004). What do we know about loss given default? *Credit Risk Models and Management*, *2*(1).

Shleifer, A., & Vishny, R. (1996). Liquidation values and debt capacity: A market equilibrium approach. *Journal of Finance*, *47*(4), 1343–1366. Reprinted in Brennan, M. J. (Ed.) The Theory of Corporate Finance, Edward Elgar Publishing Company.

Sokol, A. (2011). *A practical guide to Monte-Carlo CVA*. Risks Books.

The CDS Big-Bang: Understanding the changes to the global CDS contract and North American conventions. Technical report, March 13, 2009, Markit. http://www.markit.com/cds/announcements/resource/cds_big_bang.pdf.

Van de Castle, K., & Keisman, D. (May 2000). *Suddenly structure mattered: Insights into recoveries of defaulted debt*. Technical Report May, 2000.

Vrins, F. (2017). Wrong-way risk CVA models with analytical EPE profiles under Gaussian exposure dynamics. *International Journal of Theoretical and Applied Finance*, *20*(7), 1750045.

Dynamic Linkages Across Country Yield Curves: The Effects of Global and Local Yield Curve Factors on US, UK and German Yields

Laura Coroneo, Ian Garrett and Javier Sanhueza

Abstract We analyze the relationship between the yield curves of the USA, the UK and Germany using global and local factors. Our focus is on dynamic linkages across and between yield curves and factors. We disentangle the latent global and local factors contained in country factors, based on the Diebold and Li (J Econometrics 130:337–364, 2006) parametrization of Nelson and Siegel's (1987) three factor model and a quasi-maximum likelihood approach. The results indicate that global factors explain on average 55% of the variance of yields. Using impulse response analysis, we examine the effects of shocks to the factors on yields. We find that the response of yields to shocks to global factors is larger and longer-lasting than the response to shocks to local factors.

Keywords Yields · Global factors · Local factors · Nelson-Siegel model
Dynamic factor model

1 Introduction and Literature Review

Modeling the term structure of interest rates is important for both policymakers and market practitioners since it conveys important information about where the market expects interest rates to be in the future. It is also of relevance to the valuation of securities and portfolios. Most modeling of the yield curve has been conducted in isolation at the country level (Diebold et al. 2008). The importance of studying yields at a multi-country level has been highlighted by the most recent finan-

L. Coroneo
Department of Economics and Related Studies,
University of York, Heslington, York YO10 5DD, UK
e-mail: laura.coroneo@york.ac.uk

I. Garrett (✉)
Alliance Manchester Business School, University of Manchester,
Booth Street East, Manchester M13 9SS, UK
e-mail: ian.garrett@manchester.ac.uk

J. Sanhueza
Banco Central de Chile, 1180 Agustinas, Santiago, Chile
e-mail: sanhuezj@gmail.com

M. Mili et al. (eds.), *New Methods in Fixed Income Modeling*, Contributions to Management Science, https://doi.org/10.1007/978-3-319-95285-7_12

cial crisis, which has shown that financial markets are globally interconnected and move together. Therefore, it is important to understand the economic linkages across financial markets and in particular the mechanisms by which interest rate shocks are transmitted. Indeed, the Bank for International Settlements (BIS) indicates in its 2009 Report that the financial crisis shows the immense complexity of the modern financial system and the intricate linkage between financial markets. This highlights the need for a good understanding of the links between yield curves across different countries as they might provide important information for regulators and market participants. In particular, regulators and market participants could benefit from knowledge of the direction of the movements of global interest rate factors which could adversely affect domestic interest rates in order to take actions to counteract these effects. The main contribution of this paper is to isolate the forces of global comovements from idiosyncratic components for yield curves of different countries. We develop a model that identifies global and local factors for the yield curves of three countries: the USA, Germany and the UK.

Although there are different approaches to estimating yield curves, Bank for International Settlements (2005) and De Pooter (2007) report that the methodological approach developed by Nelson and Siegel (1987) and its extension by Svensson (1994), is widely used among practitioners and central banks. The model developed by Nelson and Siegel (1987), NS hereafter, relies on a set of predefined functions which depend on maturity and a decay factor, in order to create a fit which is flexible enough to capture the differing shapes of yield curves. This is achieved based on factor loadings predefined according to the term to maturity (short, medium and long). Diebold and Li (2006) propose a reparameterization of the model developed by NS where the coefficients are redefined in terms of the level, slope and curvature of the yield curve.[1]

More recent papers (Diebold et al. 2008; Modugno and Nikolaou 2009, for example) have focused on links between yield curves across countries. Diebold et al. (2008) use a modified version of the NS model in order to estimate level and slope factors for four countries: the UK, the USA, Japan and Germany. The yield curves for these countries are explained by a yield curve factor model which comprises orthogonal factors of two types: global and country-specific factors. Modugno and Nikolaou (2009) evaluate the forecasting power of international yield curve linkages for three countries: the UK, the USA and Germany. The methodological approach Modugno and Nikolaou (2009) use is based on the NS model's factors and a restricted vector autoregressive (VAR) process estimated by maximum likelihood, where only the same factors for different countries interact with each other. Dahlquist and Hasseltoft (2013) extend the factor model developed by Cochrane and Piazzesi (2005)

[1]Criticisms of the NS class of models are that they are not supported by a theoretical framework and are not necessarily arbitrage free. However, Coroneo et al. (2011) provide a detailed discussion about how arbitrage-free the NS model actually is. Their conclusions indicate that from a statistical point of view, the factors in the NS model are not different from those of arbitrage-free models at the 99% confidence level. Additionally, Christensen et al. (2011) develop a theoretical framework in order to estimate an affine arbitrage-free NS model (AFNS) maintaining the factor loadings of the NS model.

to an international context, estimating global and local bond factors for the UK, USA, Germany and Switzerland. The global factor is a weighted average of the Cochrane and Piazzesi (2005) factors for each country, where the weights are based on GDP growth. The global factor is closely related to bond risk premia and global macroeconomic conditions.

Although studies of international linkages across country yield curves have produced advances in knowledge of the relationship between the yield curves of different countries, they have limitations. First, studies that estimate global yield curve factors typically focus on the level and slope, ignoring curvature. The curvature factor, however, could contain important information. The evidence in Moench (2012) suggests that unanticipated movements in the curvature factor contain important information on the future evolution of the yield curve, output, prices and inflation. Second, the studies discussed above typically do not consider interactions between different factors for different countries, that is, they do not consider the links between, say, the level and slope factors across different countries. Research on single-country yield curves (see Diebold and Li 2006 and Moench 2012, for example) shows that there are interactions between different factors for the same country. In this regard, our results indicate that there are also important interactions between different factors and different countries.

In this paper we propose and examine a model based on global and local level, slope and curvature factors, in order to deepen our understanding of the mechanism by which shocks to yield curve factors are transmitted. In particular, we build a global and local factor model which explains the yields of three countries: the USA, Germany and the UK. Our global factor model includes level, slope and curvature factors allowing for interactions between both the different factors and countries.[2] Our model differs from the framework proposed by Modugno and Nikolaou (2009) since we estimate a global factor model which allows cross-interactions between factors of different countries. Our model is also different from the model proposed by Diebold et al. (2008), whose factors are orthogonal. Additionally, we use a quasi-maximum likelihood approach which overcomes the difficulties in estimating global factor models (Diebold et al. 2008).[3]

We estimate our global and local factor model using the quasi-maximum likelihood approach of Doz et al. (2012). This approach is developed for estimating dynamic factor models with a large sample size, utilizing the EM algorithm and the Kalman filter. The implementation of our model is based on Ghahramani and Hinton (1996) who provide a detailed description of the methodological procedures and steps involved in the estimation of parameters of linear dynamical systems (LDS) using the EM algorithm. This approach in turn is based on the methodological ap-

[2]Specifically, the cross-factor interaction is between level and slope, level and curvature, and slope and curvature for the USA, Germany and the UK.

[3]Diebold et al. (2008, p. 355) indicate that while under normality assumptions the estimation of the model for a single country is straightforward, in a multi-country framework estimation by maximum likelihood is "particularly difficult to implement" given the "large number of parameters to be estimated." For this reason they use a Bayesian approach.

proach developed by Shumway and Stoffer (1982) to estimate the state-space model using the EM algorithm in conjunction with the Kalman smoother.

Our results show that global factors matter in explaining yields across countries. We find that global factors explain on average 55% of the total variance of yields, with much of the explanatory power coming from the global level factor which explains, on average, 40% of the total variance. We track the effects of shocks to both local and global factors on yields using impulse response functions. Our findings indicate that effects of shocks to the local and global factors have worked their way through the system after 42 and 72 months, respectively. In addition, the range of the response of the yields to shocks on global factors is larger than the response of yields to local factor shocks. Our results suggest that given both the size and the length of time the shocks have an effect for, global factors have a larger and more permanent effect on country yields.

The rest of the paper is organized as follows. Section 2 discusses the NS model, its reparameterization by Diebold and Li (2006) and takes a first look at the data. Section 3 describes our model while Sect. 4 discusses the results. Section 5 offers some concluding remarks.

2 The Nelson-Siegel (NS) Model and Preliminary Analysis

In this section, we introduce the NS model reparameterized by Diebold and Li (2006) and provide a generalization of this model to estimate simultaneously the yield curve for three countries: the USA, Germany and the UK. We also present evidence that the NS factors for these countries contain common components, suggesting that there is a global dimension to yield curve factors. The NS model for each country i is

$$y_{i,t}(\tau) = l_{i,t} + s_{i,t}\left(\frac{1-e^{-\lambda\tau}}{\lambda\tau}\right) + c_{i,t}\left(\frac{1-e^{-\lambda\tau}}{\lambda\tau} - e^{-\lambda\tau}\right) + e_{i,t}(\tau), \tag{1}$$

where $y_{i,t}(\tau)$ is the yield for country i at time t with maturity τ, λ is the decay factor and $e_{i,t}(\tau)$ is the estimated error of the respective yield.[4] The loadings are 1 for the level, which is a long-term factor, $\left(\frac{1-e^{-\lambda\tau}}{\lambda\tau}\right)$ for the slope, which is a short-term factor, and $\left(\frac{1-e^{-\lambda\tau}}{\lambda\tau} - e^{-\lambda\tau}\right)$ for the curvature, which is a medium-term factor.

The matrix representation of this model is

$$Y_{i,t} = \Gamma_i F_{i,t} + \varepsilon_{i,t}, \tag{2}$$

where $Y_{i,t}$ is the matrix that stacks the yields of country i for n maturities, Γ_i is the matrix of the NS factor loadings, whose jth row, $\Gamma_{i,j}$, contains the NS factor loadings

[4] As suggested by Diebold and Li (2006), the decay factor λ is fixed at 0.0609 in order to maximize the curvature loadings and to aid the numerical optimization process.

$\Gamma_{i,j} = \left[1 \ \frac{1-e^{-\lambda\tau_j}}{\lambda\tau_j} \ \frac{1-e^{-\lambda\tau_j}}{\lambda\tau_j} - e^{-\lambda\tau_j} \right]$, $F_{i,t}$ is the vector of factors and $\varepsilon_{i,t}$ is the vector of errors for country i at time t.

The generalization of this model for several countries is straightforward:

$$Y_t = \Gamma F_t + \varepsilon_t, \tag{3}$$

where Y_t is the matrix that stacks the yields for all the countries at time t; Γ is a block diagonal matrix of factor loadings that contains three identical submatrices, Γ_i, which in turn contain the factors loadings for the level, slope and curvature for each country. The vector F_t contains the three factors for each of the countries while ε_t is the vector of errors at time t.

To explore the possibility of there being components of the level, slope and curvature that are common across countries, we undertake a preliminary analysis to estimate the NS factor model for the USA, Germany and the UK. The data consist of monthly zero coupon government bond yields for maturities from 1 to 10 years, for the USA, Germany and the UK. The data is from the BIS database and the Bank of England.[5] The data spans August 1997 to May 2010. We demean and standardize the yields.

Estimates of the factors from the NS model for each country are plotted in Fig. 1 while Table 1 documents correlations between the factors.

Figure 1 indicates that there is a similar pattern among the three NS factors for the three countries, suggesting that each of the factors contain a common component across the countries in the sample. Table 1 provides further evidence to suggest that there is a common, cross-country component to each of the factors. Table 1 clearly shows that the same factors are strongly correlated across countries. For example, the correlation between the US and UK level factor is around 0.5 while for the US and Germany it is in the region of 0.9. Such strong correlations are not restricted to the level factor, the correlations for the slope factors ranging from around 0.35–0.7 while for the curvature factor they are in the region of 0.7–0.75. There is also evidence that the factors are correlated although the correlations are typically not nearly as pronounced. The negative correlation between the level and the slope could be due to the fact that increases in the general level of rates reduce the gap between short and long term rates.

In order to examine whether there are common components to the level, slope and curvature factors across the countries, we estimate and extract the first principal component for each factor for the three countries. Table 2 shows the percentage of the variance for each one of the factors explained by the first principal component for each country. The total variance explained for each principal component ranges between 78 and 82% on average. These preliminary results support the idea of an international linkage between yield curves and that yield curves contain common components and suggests that there is something to be gained by considering the

[5]In the case of the data taken from the BIS database, the data is provided to BIS by the central banks of the respective countries.

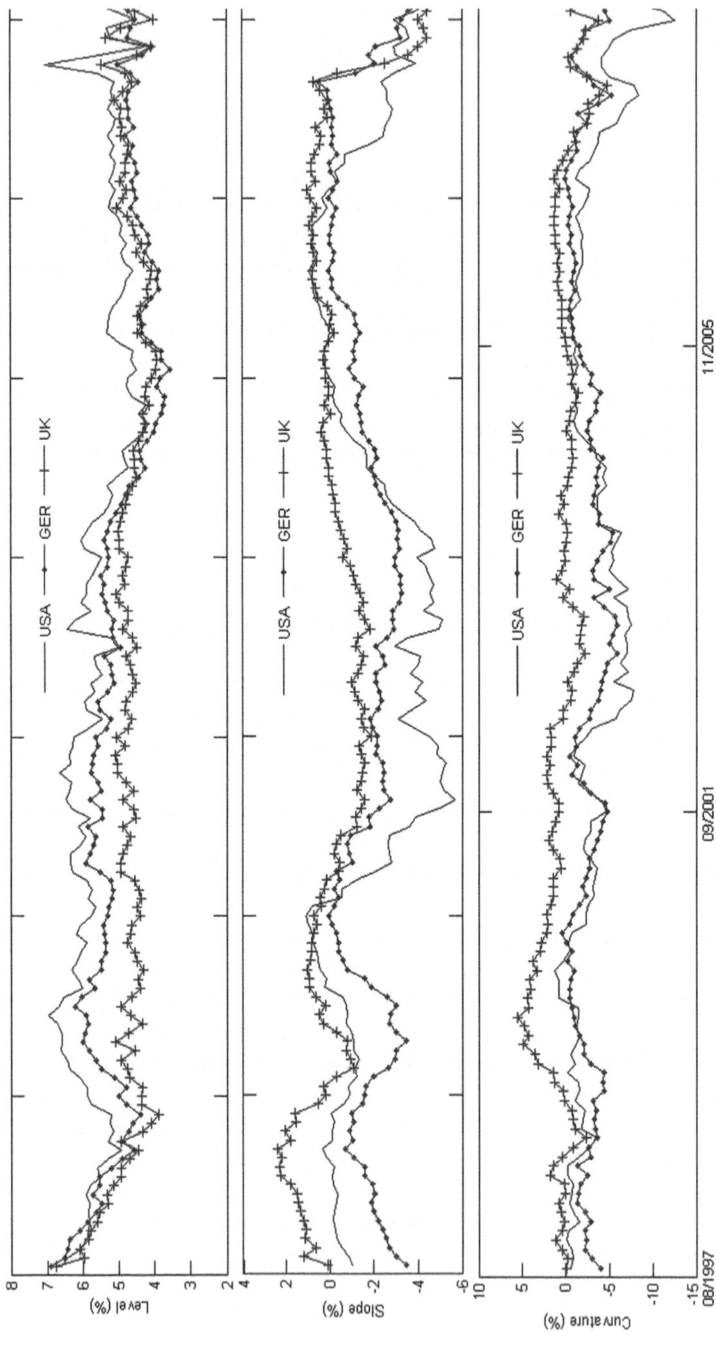

Fig. 1 Nelson and Siegel (1987) factors estimated using the Diebold and Li (2006) reparameterizion of the NS model for the USA, Germany, 1997:08-2010:05

Table 1 Correlation matrix for the NS factors

		Level			Slope			Curvature		
		USA	GER	UK	USA	GER	UK	USA	GER	UK
Level	USA	1.00								
	GER	0.89	1.00							
	UK	0.48	0.59	1.00						
	USA	−0.25	−0.19	−0.28	1.00					
Slope	GER	−0.30	−0.28	−0.36	0.68	1.00				
	UK	0.01	0.14	−0.11	0.75	0.70	1.00			
	USA	0.17	0.23	−0.10	0.66	0.34	0.72	1.00		
Curvature	GER	0.11	0.06	−0.19	0.63	0.55	0.62	0.73	1.00	
	UK	0.52	0.52	−0.15	0.39	0.22	0.55	0.75	0.69	1.00

Table 2 Percentage of variance of the country-specific factors explained by the first principal component (the common factor) for each of the level, slope and curvature factors, 1997:08-2010:05

Country	Level (%)	Slope (%)	Curvature (%)
USA	84	81	84
GER	91	78	80
UK	58	83	81
Mean	78	81	82

effects of global as well as local factors on yield curves. This then allows us to assess the effect of changes in common factors on the yield curve of each country, as well as the effect of changes in the local factors of one country on the yield curve of other countries.

3 The Model

The results in the previous section indicate that there are common components to the level, slope and curvature yield curve factors across countries. Indeed, the results suggest that there are two types of factor: common and local. We can therefore define a factor model for the yield curve that considers both global and local factors. This is the subject matter of this section. We decompose the factors in the factor model given by (3) into the sum of two orthogonal components: the global factors, which we denote by F_t^G, and the local factors denoted by F_t^L. The factor model (3) then becomes

$$F_t = \beta F_t^G + F_t^L \tag{4}$$

where β is a matrix of factor loadings relating to the global factors. Substituting (4) into (3) we have

$$Y_t = \Gamma\left[\beta F_t^G + F_t^L\right] + \varepsilon_t. \tag{5}$$

In this specification of the model, the respective yield curves of each country are defined by the global factors and their loadings, $\Gamma\beta$ and the local factors and their loadings, Γ. Defining $\Gamma^G = \Gamma\beta$ we have

$$Y_t = \Gamma^G F_t^G + \Gamma F_t^L + \varepsilon_t \tag{6}$$

which we can more conveniently write as

$$Y_t = [\Gamma^G \Gamma][F_t^{G'} F_t^{L'}]' + \varepsilon_t, \quad \varepsilon_t \sim N(0, \Sigma) \tag{7}$$

We model the factors as a first order vector autoregressive process:

$$[F_t^{G'} F_t^{L'}]' = \Phi[F_{t-1}^{G'} F_{t-1}^{L'}]' + w_t, \quad w_t \sim N(0, \Omega) \tag{8}$$

where the matrix Φ is a block diagonal matrix of factor loadings and w is the vector of errors. Σ and Ω are the variance-covariance matrices which are independent.

The state space representation of the model described by Eqs. (7) and (8) can be written compactly as

$$Y_t = \Gamma^T F_t^T + \varepsilon_t \tag{9}$$

where Γ^T contains the factor loadings, $\Gamma^T = [\Gamma^G \;\; \Gamma]$ and F_t^T contains the global and local NS factors, $F_t^T = [F_t^{G'} \;\; F_t^{L'}]'$ with F_t^T following the VAR(1) process

$$F_t^T = \Phi F_{t-1}^T + A u_t \tag{10}$$

The reduced form errors are given by $w_t = Au_t$, where the shocks u_t are defined as "primitive" or "fundamental". The shocks are orthogonal and have unit variance. The covariance matrix Σ in (7) is a diagonal matrix while the covariance matrix Ω in (8) is a two-block diagonal matrix. The first block is the variance-covariance matrix of the global factor errors while the second block contains the variance-covariance matrix for the errors of the local factors. The global and local factors in this model are independent of each other both contemporaneously and across time.

It is well known that factor models do not have a unique solution because of the rotational indeterminacy problem whereby different combinations of factors and factor loadings could provide observationally equivalent solutions with the same likelihood but with different financial or economic implications. Hence, in order to uniquely identify the parameters and the unobservable factors in our model, we need to impose restrictions on the factor loadings.

In order to identify β, recall that Eq. (4) decomposes the factor for each country into the loadings, β, the global factors, F_t^G, and the local factors, F_t^L. The loadings β in (4) are not identified so we need to impose some restrictions in order to identify them. First, we constrain β to be block diagonal so that each global factor only loads on to the same global factor (the global level factor only loads onto the global level factor, for example). Second, the factors are restricted to have a variance-covariance matrix equal to the identity matrix.

To estimate the model described by Eqs. (9) and (10), we need to ensure that factor loadings and factors are uniquely identified. Hence, we restrict the factor loadings by imposing the NS factor loading restrictions $\left(1, \frac{1-e^{-\lambda\tau_j}}{\lambda\tau_j} \text{ and } \frac{1-e^{-\lambda\tau_j}}{\lambda\tau_j} - e^{-\lambda\tau_j}\right)$.

In order to disentangle the global and local factors, the estimation of the model proceeds as follows. First, the three NS factors are estimated for each country using the pre-defined factor loadings, assuming that the factors comprise both global and local factors. Second, we estimate the loadings for each country over the global factors $\hat{\beta}$ using quasi maximum likelihood, restricting $\hat{\beta}$ to be block diagonal and standardizing the factors. Third, the latent global and local factors are estimated using the quasi-maximum likelihood approach. We impose orthogonality between the global and local factors. To estimate the global and local factors we use a joint estimation procedure. We initialize the estimates of the global factors using the standardized first principal component for each factor estimated using all the countries. We initialize the estimates of the local factors by subtracting the standardized first principal component from yields.

3.1 Impulse Response Analysis and Variance Decompositions

The impulse response functions IRFs allow us to track the effect of shocks to the factors on the yields of the countries according to the model described by Eqs. (9) and (10). In order to track the effect of shocks to the factors through the system, we write Eq. (10) in its moving average form as follows

$$(I - \Phi L)F_t^T = Au_t \tag{11}$$

$$F_t^T = (I - \Phi L)^{-1}Au_t \tag{12}$$

where L denotes the lag operator and where I is the identity matrix. Substituting (12) into (9) we have

$$Y_t = \Gamma^T(I - \Phi L)^{-1}Au_t + \varepsilon_t \tag{13}$$

Defining $B(L) = \Gamma^T(I - \Phi L)^{-1}A$ we can write

$$Y_t = B(L)u_t + \varepsilon_t \tag{14}$$

which is the moving average representation of the model from which we can estimate the impulse responses. To identify the IRFs, we use the approach proposed by Sims (1980) and Sims (1982) which is based on the Cholesky decomposition. From (8) and (10), we know that $w_t = Au_t$. Since we can estimate w_t, we need to find A and

u_t in order to be able to recover u_t and identify orthogonalized shocks. Using the Cholesky decomposition, $\Omega = AA'$, we define a lower diagonal matrix, A, imposing $K(K-1)/2$ restrictions on the matrix A, where K is the total number of factors. We determine the ordering of the factors in the decomposition as follows. First, in our specification there is zero correlation between the shocks to global factors and the shocks to local factors, so the global and local factors are not contemporaneously correlated. Hence, allocating the global factors first or the local factors first does not change our analysis or our results. Second, the evidence in Sect. 2 suggests that yields are explained primarily by the level, then by the slope and finally by curvature. Therefore, we follow the same order in the hierarchy of both global and local factors for the identification of shocks. Translating this ordering into our model, we have the level influencing both the slope and curvature contemporaneously, and the slope influencing the curvature. Finally, we rank the countries in descending order by GDP. This means that the US factors explain those of Germany and the UK, and the German factors explain those of the UK.

Related to the impulse response analysis is the variance decomposition. The variance decomposition is interested in asking what fraction of the variance of the nth yield is explained by the jth factor. Recalling that from the Cholesky decomposition of Ω, $\Omega = AA'$ and recalling that yields are defined by $Y_t = \Gamma^T F_t^T + \varepsilon_t$, the conditional variance of the yields is equal to $\left[\Gamma^T \left[AIA'\right] \Gamma^{T\prime}\right]$. From this, the variance of nth yield explained by the jth factor is given by the expression $\Theta_n^{-1} \left[\Gamma_n^T \left[AI_jA'\right] \Gamma_n^{T\prime}\right]$, where Θ_n^{-1} is the inverse of the variance of the nth yieldΓ_n^T is the nth row of the matrix Γ^T and I_j is a matrix with one in the jth row and jth column with zeros elsewhere.

4 Results

Table 3 reports estimates of the global factor loadings, β while Table 4 reports the goodness of fit for the model. There are a couple of points of interest from the results in Table 3. First, the loadings of each country for the global slope and curvature

Table 3 Global factor loadings, β

Country	Level	Slope	Curvature
USA	0.60	–	–
GER	0.63	–	–
UK	0.50	–	–
USA	–	0.58	–
GER	–	0.57	–
UK	–	0.59	–
USA	–	–	0.59
GER	–	–	0.57
UK	–	–	0.58

Table 4 Goodness of fit statistics (%) across maturities

	1 yr.	2 yrs.	3 yrs.	4 yrs.	5 yrs.	6 yrs.	7 yrs.	8 yrs.	9 yrs.	10 yrs.	Mean
USA	98.1	99.4	99.4	99.1	98.7	98.2	97.7	97.2	96.6	95.9	98.0
GER	98.7	99.6	99.6	99.5	99.5	99.5	99.5	99.4	99.3	99.1	99.4
UK	98.3	99.3	99.3	99.3	99.2	98.9	98.4	97.7	96.3	94.2	98.1

Table 5 Variance decomposition: fraction of the one step ahead forecast error variance explained by the level (l), slope (s) and curvature (c) factors

	Global			USA			GER			UK		
	l_G (%)	s_G (%)	c_G (%)	l_{USA} (%)	s_{USA} (%)	c_{USA} (%)	l_{GER} (%)	s_{GER} (%)	c_{GER} (%)	l_{UK} (%)	s_{UK} (%)	c_{UK} (%)
USA	43	1	16	23	1	13						
GER	59	5	6	3	2	7	1	1	14			
UK	19	8	10	2	0	4	8	1	4	27	8	2
TOTAL	40	5	10	9	1	8	3	1	6	9	3	1

factors are very similar, suggesting that the slope and curvature factors for each country are important in the composition of the global slope and curvature factors (recall that the factors are standardized). Of more interest is the level factor, where the loading for the UK is somewhat lower than those for the US and Germany. This appears consistent with the results in Table 2, where only 58% of the variance in UK yields is explained by the level factor. The results in Table 4 show that the dynamic model with global and local yield curve factors does a very good job of explaining yields.

Table 5 shows results from the variance decomposition. It is noticeable that the global factors explain roughly 55% of the total variance of yields, and less than 45% is explained by the local factors and the interaction between them. The factor which explains the largest proportion of the total variance of yields is the global level at 40%, followed by the global slope. In terms of individual countries, the global level explains a large proportion of the forecast error variance for the USA and Germany though interestingly not so much for the UK, where the local level is more important. Table 6 looks at the variance decomposition by yield maturity to examine whether there is any noticeable difference in the proportion of the one step ahead forecast error variance explained by global and local factors across maturities. The percentage explained by the global level monotonically increases with maturity for the US and Germany and almost monotonically for the UK. Conversely, the global slope explains more of the forecast error variance of short maturity bonds relative to longer maturity bonds. Global curvature explains more of the forecast variance for maturities between 2 and 3 years.

Results from the impulse response analysis are shown in Figs. 2 through 5. They indicate that, in general, the effects of shocks to the local factors shocks on yields

Table 6 Variance decomposition for different maturities

		Global			USA			GER			UK		
		l_G (%)	s_G (%)	c_G (%)	l_{USA} (%)	s_{USA} (%)	c_{USA} (%)	l_{GER} (%)	s_{GER} (%)	c_{GER} (%)	l_{UK} (%)	s_{UK} (%)	c_{UK} (%)
USA	1 yr.	26	7	18	2	9	16	–	–	–	–	–	–
	2 yrs.	28	3	31	8	3	27	–	–	–	–	–	–
	3 yrs.	31	1	28	15	1	24	–	–	–	–	–	–
	4 yrs.	37	1	22	20	0	19	–	–	–	–	–	–
	5 yrs.	43	0	17	25	0	15	–	–	–	–	–	–
	6 yrs.	48	0	13	28	0	11	–	–	–	–	–	–
	7 yrs.	51	0	9	31	0	8	–	–	–	–	–	–
	8 yrs.	54	0	7	32	0	6	–	–	–	–	–	–
	9 yrs.	55	0	5	33	0	5	–	–	–	–	–	–
	10 yrs.	54	0	4	32	0	4	–	–	–	–	–	–
GER	1 yr.	20	21	6	0	1	18	5	0	14	–	–	–
	2 yrs.	30	11	10	4	5	16	1	1	24	–	–	–
	3 yrs.	39	6	10	5	5	12	0	2	23	–	–	–
	4 yrs.	50	3	8	5	4	9	0	2	19	–	–	–
	5 yrs.	60	2	7	4	3	6	0	2	16	–	–	–
	6 yrs.	69	2	5	3	2	4	1	2	12	–	–	–
	7 yrs.	76	1	4	2	1	3	1	1	10	–	–	–
	8 yrs.	81	1	3	1	1	2	1	1	8	–	–	–
	9 yrs.	84	1	3	1	0	1	1	1	6	–	–	–
	10 yrs.	84	1	2	1	0	1	1	1	5	–	–	–
UK	1 yr.	0	27	7	1	0	2	5	1	3	18	6	2
	2 yrs.	2	19	18	1	1	10	8	2	7	12	15	4
	3 yrs.	6	11	18	1	1	11	9	2	8	12	16	4
	4 yrs.	14	7	16	2	0	9	9	2	7	18	14	3
	5 yrs.	21	4	12	2	0	6	9	1	6	25	11	3
	6 yrs.	26	3	9	2	0	3	9	1	4	32	8	2
	7 yrs.	31	2	7	2	0	2	8	0	3	38	6	1
	8 yrs.	33	1	5	2	0	1	7	0	3	41	4	1
	9 yrs.	31	1	3	2	0	0	6	0	2	40	3	1
	10 yrs.	27	1	2	1	0	0	5	0	1	34	2	0
	Tot.	40	5	10	9	1	8	3	1	6	9	3	1

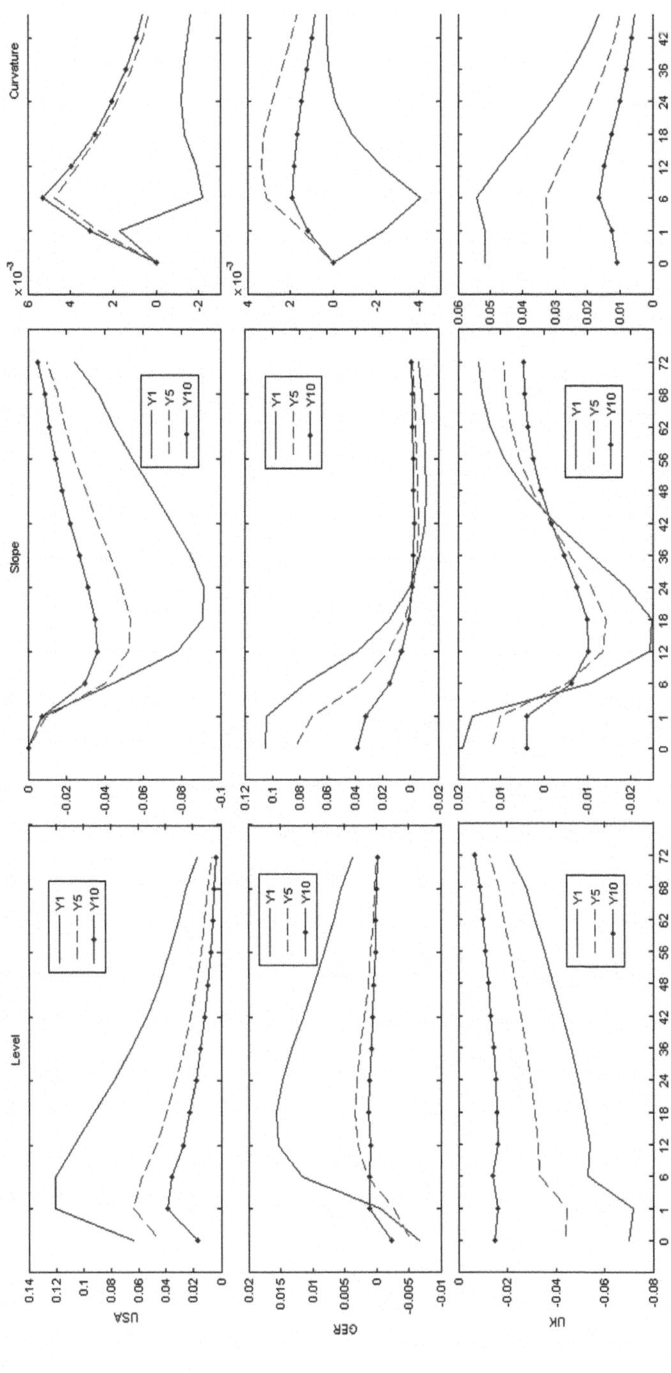

Fig. 2 Effects of a one standard deviation shock to the global factors on yields for the USA, Germany and the UK

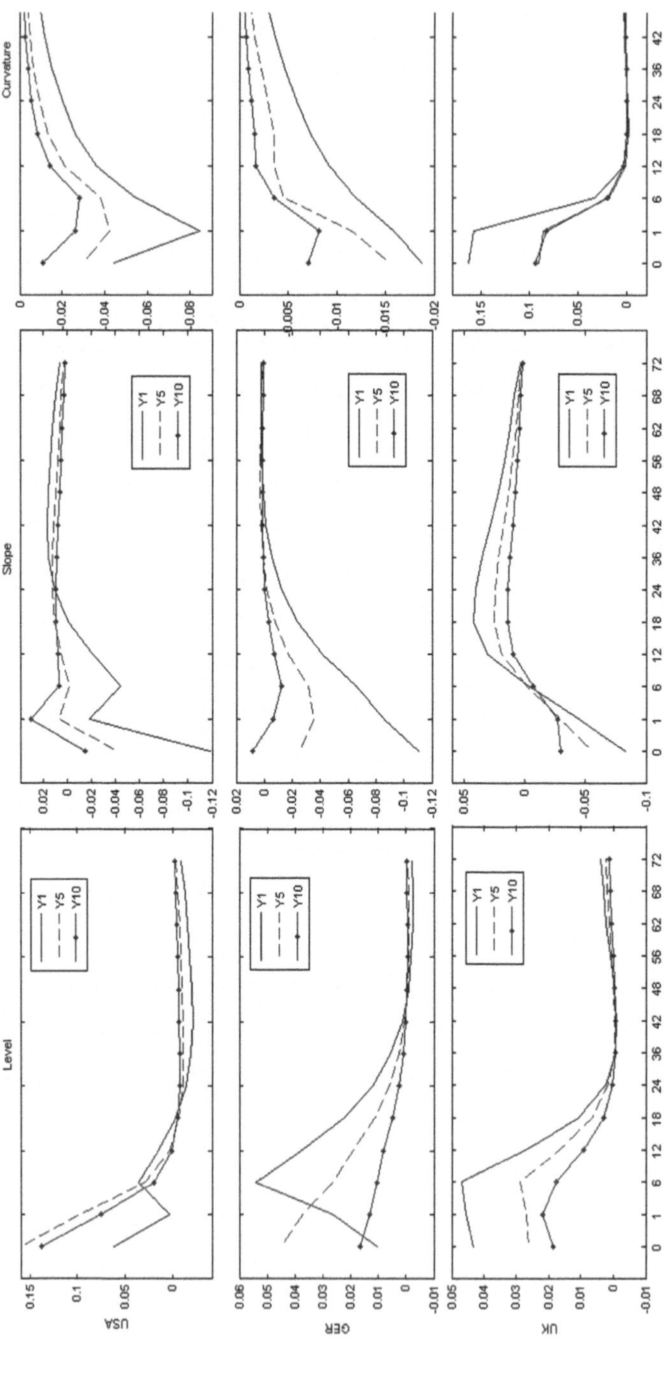

Fig. 3 Effects of a one standard deviation shock to the US local factors on yields for the USA, Germany and the UK

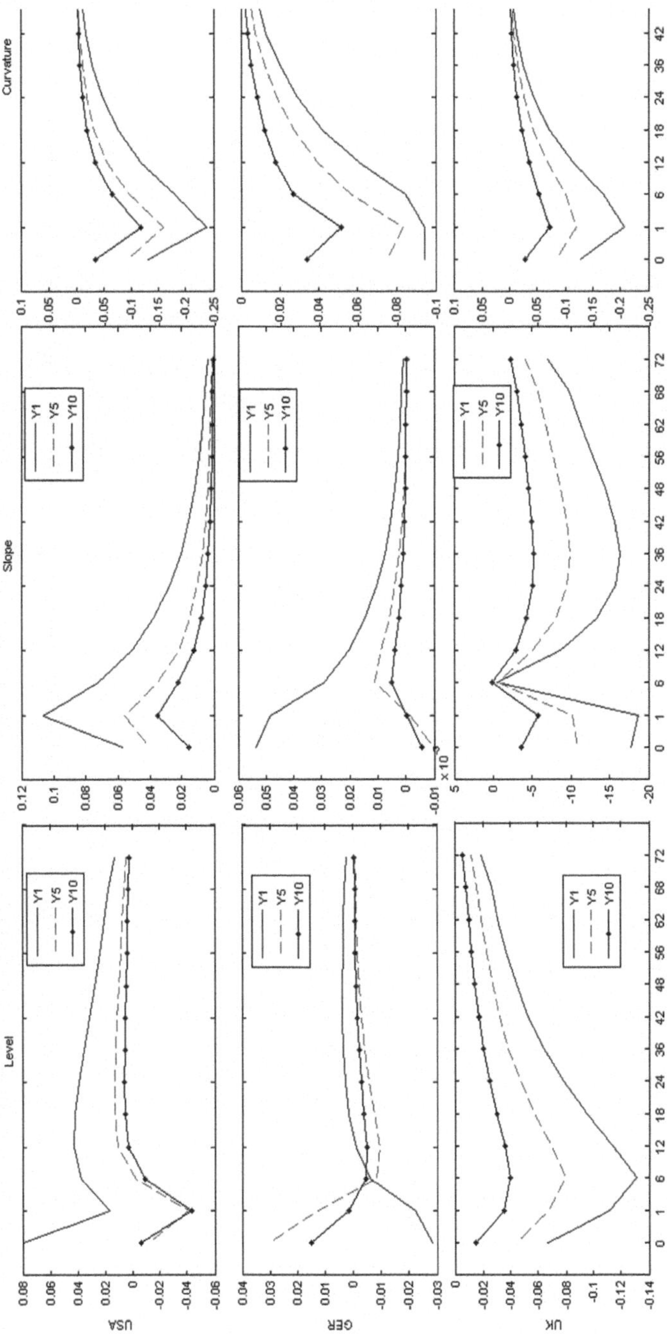

Fig. 4 Effects of a one standard deviation shock to German local factors on yields for the USA, Germany and the UK

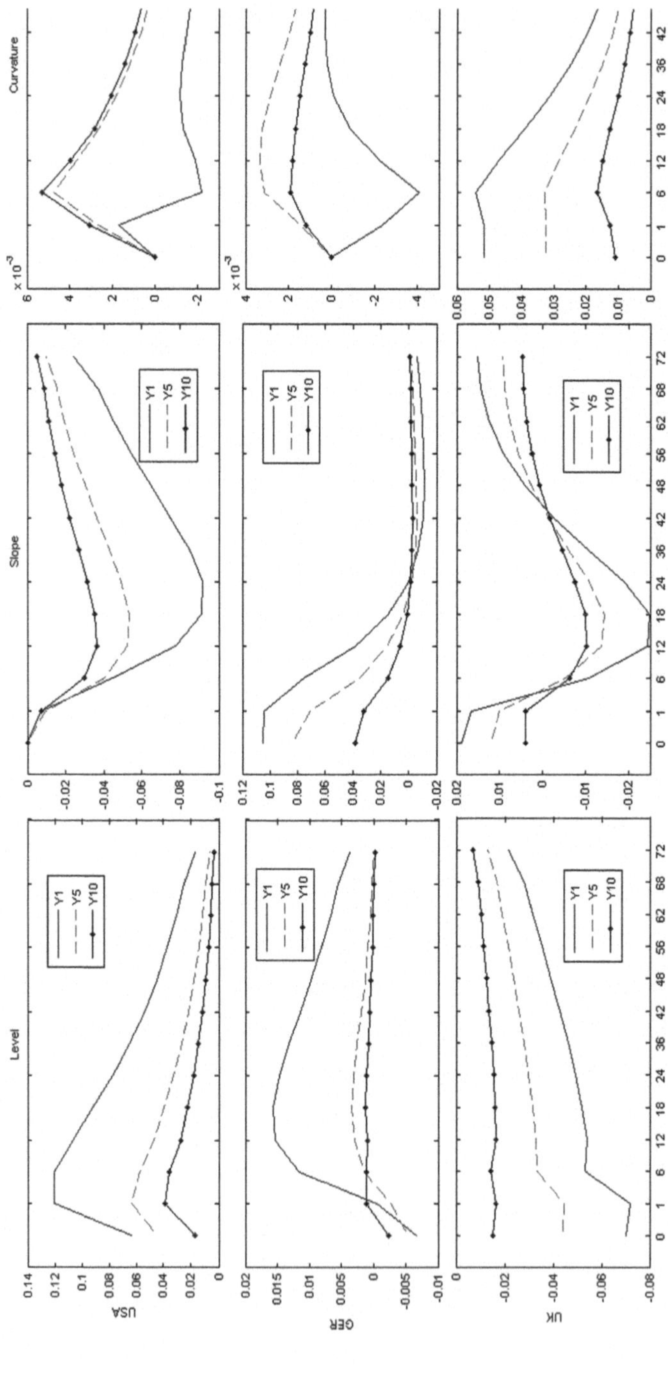

Fig. 5 Effects of a one standard deviation shock to the UK local factors on yields for the USA, Germany and the UK

disappear by no later than after 42 months. The effects of shocks to global factors, on the other hand, disappear slowly and last for about 72 months (see Fig. 2). Over the 72 month horizon that the impulse responses are estimated for, the average response of yields to global factor shocks is 4 basis points (bp). This reaction is larger than the average response of yields to local factor shocks (see Figs. 3, 4 and 5), which is lower than 1 bp, the only exception being the response of yields to shocks to US local factors. The shortest maturities exhibit a larger response than the longest, and this is especially important for global factor shocks. In general, one year yields for all countries are more sensitive than longer maturities to shocks to global and local factors. Our results indicate that there is an important linkage across international yield curves due to the effect of global factors as well as cross-factor dynamic interactions among local factors.

5 Concluding Remarks

We proposed a global and local factor model based on the three NS factors (level, slope and curvature) for the USA, Germany and the UK. We estimated this factor model using monthly government zero coupon yields. Our results indicate that global factors explain on average 55% of the variance of yields, and that the most important factor is the global level, which explains 40% of the variance of yields. Impulse response analysis shows that shocks to local factors dissipate quicker than shocks to global factors, the relevant shocks lasting approximately 42 and 72 months, respectively. Moreover, the size of the effects of global factor shocks are larger than those of local factor shocks. Finally, bonds with shorter maturities are more sensitive to shocks, particularly shocks to the global factors. Therefore, global factors play an important role in explaining yields on bonds of different maturities across different countries. Indeed, the influence of local factors appears to be relatively limited. These results indicate that a yield curve model that considers both global and local factors and the dynamic interaction of these factors can better explain the future evolution of yields.

References

Bank for International Settlements. (2005). Zero-Coupon Yield Curves: Technical Documentation. BIS Paper No. 25. Available at SSRN: https://ssrn.com/abstract=1188514 or http://dx.doi.org/10.2139/ssrn.1188514

Bank for International Settlements. (2009). *79th Annual Report.*

Bank for International Settlements. (2010). *80th Annual Report.*

Christensen, J. H., Diebold, F. X., & Rudebusch, G. D. (2011). The affine arbitrage-free class of Nelson-Siegel term structure models. *Journal of Econometrics*, *164*, 4–20.

Cochrane, J. H., & Piazzesi, M. (2005). Bond risk premia. *American Economic Review*, *95*, 138–160.

Coroneo, L., Nyholm, K., & Vidova-Koleva, R. (2011). How arbitrage-free is the Nelson-Siegel model? *Journal of Empirical Finance*, *18*, 393–407.

Dahlquist, M., & Hasseltoft, H. (2013). International bond risk premia. *Journal of International Economics*, *90*, 17–32.

De Pooter, M. (2007). Examining the Nelson-Siegel class of term structure models. *Tinbergen Institute Discussion Papers*, 07-043/4.

Diebold, F. X., & Li, C. (2006). Forecasting the term structure of government bond yields. *Journal of Econometrics*, *130*, 337–364.

Diebold, F. X., Li, C., & Yue, V. Z. (2008). Global yield curve dynamics and interactions: a dynamic Nelson-Siegel approach. *Journal of Econometrics*, *146*, 351–363.

Doz, C., Giannone, D., & Reichlin, L. (2012). A quasi maximum likelihood approach for large approximate dynamic factor models. *Review of Economics and Statistics*, *94*, 1014–1024.

Ghahramani, Z., & Hinton, G. E. (1996). Parameter estimation for linear dynamical systems. *University of Toronto Technical Report CRG-TR-96-2*.

Modugno, M., & Nikolaou, K. (2009). *The forecasting power of international yield curve linkages*. In: Working Paper Series, European Central Bank, 1044.

Moench, E. (2012). Term structure surprises: the predictive content of curvature, level and slope. *Journal of Applied Econometrics*, *27*, 574–602.

Nelson, C. R., & Siegel, A. F. (1987). Parsimonious modeling of yield curves. *Journal of Business*, *60*, 473–89.

Shumway, R. H., & Stoffer, D. S. (1982). An approach to time series smoothing and forecasting using the em algorithm. *Journal of Time Series Analysis*, *3*, 253–264.

Sims, C. A. (1980). Macroeconomics and Reality. *Econometrica*, *48*, 1–48.

Sims, C. A. (1982). Policy Analysis with Econometric Models. *Brookings Papers on Economics Activity*, *1*, 107–152.

Svensson, L.E. (1994). Estimating and interpreting forward interest rates: Sweden 1992–1994. NBER Working Paper No. w4871. Available at SSRN: https://ssrn.com/abstract=226502

Estimating the No-Negative-Equity Guarantee in Reverse Mortgages: International Sensitivity Analysis

Iván de la Fuente, Eliseo Navarro and Gregorio Serna

Abstract In this paper, we perform a sensitivity analysis to show how the value of the no-negative-equity guarantee (NNEG) embedded in reverse mortgage contracts varies with the value of the mortgage roll-up rate, the rental yield rate, as well as the gender and the age of the borrower. The analysis is performed for four European countries: France, Germany, Spain and the United Kingdom. The results show that the NNEG tends to be higher, and consequently the reverse mortgage provider faces higher risks, for higher roll-up rates, for higher rental yield rates, for the female population, and for relatively young borrowers. Moreover, the country with the highest estimated value of the NNEG is Spain, the country most affected by the 2008 financial crisis.

Keywords Reverse mortgages · Option pricing · No-negative-equity guarantee · Mortality modeling · House price modeling

1 Introduction

In most reverse mortgages, there is a "no-negative-equity guarantee" (NNEG), which ensures that the borrower will never fall into negative equity, so that the redemption amount of the mortgage (principal plus interests) can never be higher than the home's market value. Therefore, at maturity (date T), when the borrower dies, the lender only receives the minimum between the home's market value (S_T) and the amount repayable, principal plus interest (Y): $\min(Y, S_T)$. Given that

I. de la Fuente (✉) · E. Navarro · G. Serna
Facultad de Ciencias Económicas y Empresariales, Universidad de Alcalá,
Plaza de la Victoria, 2, 28802 Alcalá de Henares, Madrid, Spain
e-mail: ivan.fuente@uah.es

E. Navarro
e-mail: eliseo.navarro@uah.es

G. Serna
e-mail: gregorio.serna@uah.es

M. Mili et al. (eds.), *New Methods in Fixed Income Modeling*, Contributions to Management Science, https://doi.org/10.1007/978-3-319-95285-7_13

$\min(Y, S_T) = Y - \max(Y - S_T, 0)$, we can also say that at maturity, the lender receives the amount repayable minus the final value of an European put option owned by the borrower, where the underlying asset is the home's market value and the strike price is the amount repayable at this moment. Thus, this European put option measures the value of the NNEG because if at maturity the sale proceeds of the home are not enough to cover the debt, the borrower will exercise the option and sell the property for the amount repayable at this moment. From the lender's point of view, the NNEG is a source of risk because the higher the value of the NNEG, the lower the amount of money that he receives when the loan is repaid.

Thereby, the value of the NNEG can be estimated using well known option pricing techniques. Nonetheless, the Black-Scholes assumptions do not hold in this case, because the option's underlying asset (the home's market value) exhibits heteroskedasticity and autocorrelation (see Li et al. 2010, among others) and, consequently, it cannot be assumed to follow a constant-volatility geometric Brownian motion. Moreover, the option matures on an uncertain date, when the borrower dies.

In recent times, growing attention has been drawn to the issues of pricing and hedging the NNEG in reverse mortgages. In this way, Li et al. (2010) propose a method for estimating the value of de NNEG in lump-sum reverse mortgages, that assumes an ARMA-EGARCH process for house prices and the Lee and Carter model (1992) for fitting mortality data. These authors derive a closed-form formula for the NNEG that involves valuing the European put option by Monte Carlo simulation. Using data from the United Kingdom, these authors find that the NNEG represents a significant financial burden for reverse mortgage lenders. Within the same line of research, Chen Cox and Wang (2010) propose a model where house prices follow an ARIMA-GARCH process and a Lee and Carter model (1992) allowing for asymmetric jump effects to fit mortality data.

Later, other papers have extended the model by incorporating house price risk together with interest rate risk (Huang et al. 2011), the relationship between house price risk and some macroeconomic factors (Chang et al. 2012), or the joint effect of house price risk, interest rate risk, rental yield risk and the GDP (Alai et al. 2013). Other extensions include Kogure et al. (2014), who take into account house price risk, longevity risk and interest rate risk on a Bayesian framework, Shao et al. (2015), who incorporate idiosyncratic house price risk together with longevity risk, or Wang et al. (2016), who incorporate stochastic interest rates and an instantaneously adjusted loan interest rate.

In this paper, we present an international comparative study, analyzing the differences in the value of the NNEG in lump-sum reverse mortgages within the framework proposed by Li et al. (2010), considering different scenarios for the reverse mortgage interest rate and for the rental yield rate, as well as for the gender and the age of the borrower. Specifically, we calculate the value of the NNEG as a percentage of the amount of money advanced by the reverse mortgage provider, using data from four European countries: France, Germany, Spain and the United Kingdom. This issue is particularly relevant in the current economic context, characterized by low interest rates, relatively high rental yield rates and a high volatility of house prices. Therefore, it is crucial for mortgage providers to analyze the effect of different future scenarios for interest rates and rental yield rates on the risk they are assuming. Moreover, we must keep in mind that the issues related to the risk measuring and regulatory capital requirements calculation are key points in the context of the Basel II and III frameworks.

This paper contributes to the fixed income literature by focusing on the impact of the interest rate that the reverse mortgage provider is charging to the borrower, which is called roll up rate in this context. This is a specific interest rate that has not been studied as much as other interest rates. Specifically, as stated before, this paper tries to analyze the effect on the risk that the mortgage provider is facing under different values of the mortgage roll up rate. Furthermore, this sensitivity analysis can be used as a way to determine the mortgage roll up rate so that the risk assumed by the mortgage provider (i.e. the value of the NNEG) is maintained at acceptable levels.

The results show that the NNEG tends to be higher, and consequently the reverse mortgage provider faces higher risks, for higher roll-up rates, for higher rental yield rates, for the female population, and for relatively young borrowers. More concretely, the reverse mortgage provider faces extremely high risk for borrowers under 80. Moreover, the country with the highest estimated value of the NNEG is Spain, the country most affected by the 2008 financial crisis.

This paper is organized as follows. Section 2 describes the data used in the estimation of the NNEG, as well as the procedure used to fit the house price and mortality data. Section 3 presents the theoretical model used to estimate the NNEG for each country and a sensitivity analysis describing the effect on the value of the NNEG of different scenarios for the reverse mortgage roll-up rate, the rental yield rate, the gender and the age of the borrower. Finally, Sect. 4 concludes with a summary and discussion.

2 Data Set and House Price Model Selection

As explained in Sect. 1, the proposed method for estimating the value of the NNEG in reverse mortgages involves the use of different sources of data concerning risk-free interest rates, rental yield rates, as well as mortality rates and house prices.

Concerning risk-free rates, given that since the 2008 financial crisis many European government bonds are far from been risk-free, it has been decided to use the 10-year zero-coupon German government bond rate as the appropriate proxy for the risk-free rate. The value used in this paper is the 2014 average (2.16%), provided by Eurostat. However, in the case of the United Kingdom, the domestic risk-free rate (as before, the 2014 average of the 10-year zero-coupon government bond) is used (2.14%, also provided by Eurostat).

In the case of the rental yield rate, the proxy used in this paper is the average of the annual rental yield rate in the main cities of each country, provided by the 2014 Global Property Guide (http://www.globalpropertyguide.com). Specifically, the values of the annual rental yield rate for each country used in this paper are:

Table 1 Descriptive statistics

	France	Germany	Spain	UK
Mean	0.0148	0.0053	0.0143	0.0196
Median	0.016	0.0047	0.0144	0.0193
Maximum	0.068	0.0330	0.1493	0.1248
Minimum	−0.0370	−0.0116	−0.1912	−0.0544
Std. Dev.	0.016	0.0080	0.0423	0.0258
Skewness	−0.263	0.3978	−0.6000	0.2680
Kurtosis	3.3635	2.8021	6.9860	4.1954
Jarque-Bera	2.7109	4.4530	114.8025	11.3711
(Probability)	(0.2578)	(0.1079)	(0.0000)	(0.0034)
Observations	159	159	159	159

The table shows the main descriptive statistics of the House Price Index returns for the four countries considered in the study

2.89, 3.34, 3.91 and 3.21% for France, Germany, Spain and the United Kingdom respectively.

Information on mortality rates in all countries considered has been obtained from the Human Mortality Database (http://www.mortality.org/). Based on the information in this database from 1990 to 2102, the probabilities of death and survival have been estimated by means of the Lee and Carter (1992) model.

Regarding house prices, the quarterly House Price Index, from 1975 to 2014, from the International House Price Database, created by the Federal Reserve Bank of Dallas (http://www.dallasfed.org/institute/houseprice/) has been used in his paper. The main descriptive statistics of the Hose Price Index returns are shown in Table 1. This House Price Index is not based on transaction prices, but on appraisal values. The problem with indices based on appraisal values is that appraisers tend to take into accont not only current market information, but also past information when making their appraisal decisions. This is the reason why indices based on appraisal values tend to be smoother and to exhibit more autocorrelation than indices based on transaction prices and consequently tend to exhibit less volatility than transaction-based indices. As stated in Fisher et al. (1994) among others, appraised values used in the construction of the index are relatively stable through time. Therefore, this smoothing effect must be eliminated from the House Price Index. In this paper, it has been decided to use the most common used unsmoothing method suggested by Geltner (1993) and Fisher et al. (1994). This method tries to recover the underlying market value returns from the observed appraisal-based returns, by eliminating the autocorrelation structure present in the latter.

Once the House Price Index has been unsmoothed, a model for house price returns accounting for autocorrelation and heteroscedasticity is needed. In this paper we follow the approach suggested by Li et al. (2010), who propose an ARMA-EGARCH model for house price returns.

Let Y_t represent the unsmoothed log index return for period t. The autocorrelation effect will be capture by means of an ARMA structure for log returns:

$$Y_t = c + \sum_{i=1}^{R} \phi_i Y_{t-i} - \sum_{j=1}^{M} \theta_j a_{t-j} + a_t \tag{1}$$

where c, ϕ_i and θ_j are parameters and a_t is the innovation that is assumed to follow a normal variable with a zero mean and variance h_t.

The heteroscedasticity effect will be capture by means of an EGARCH structure for conditional variance log returns:

$$\ln(h_t) = k + \sum_{i=1}^{P} \alpha_i \ln(h_{t-i}) + \sum_{j=1}^{Q} \beta_j \left[|\tilde{a}_{t-j}| - E(|\tilde{a}_{t-j}|) \right] + \sum_{j=1}^{Q} \gamma_j \tilde{a}_{t-j} \tag{2}$$

where $\tilde{a}_t = a_t / \sqrt{h_t}$ is the standardized innovation, k, α_i, β_j, and γ_j are parameters. This EGARCH structure also account for the well-known leverage effect that can be captured when the conditional variance responds in a different way to positive than to negative shocks.

This ARMA-EGARCH model has been estimated for all four countries considered in this paper. As usual, the values for R, M, P and Q have been selected according to the Ljung-Box statistic and the ACF.

3 Estimating the NNEG for Each Country

In this section, we estimate the value of the NNEG for the four countries under study. Once the best ARMA-EGARCH model has been selected for each country's house price data, the NNEG can be valued using well-known option pricing procedures. The value is the present value of the expectation of the option payoff at expiration under the equivalent risk-neutral probability measure.

However, as stated by Li et al. (2010), the assumption of an ARMA-EGARCH process for the underlying asset (the house price index) implies market incompleteness. The problem with incomplete markets, from the option pricing point of view, is that more than one equivalent probability measure can be found. These authors propose a method to identify an equivalent probability measure that can be used to value the NNEG in an economically consistent and justifiable way. Specifically, the proposed method is an adaptation of the procedure introduced by Siu et al. (2004), which values derivative contracts under GARCH models using the conditional Esscher transform, to a case in which the underlying asset follows an ARMA-EGARCH process under the physical measure.

Let P be the physical probability measure and Φ_t be the set containing all market information available at time t. Then, as stated in the third section, under P, the house price log-returns, Y_t, are assumed to follow an ARMA(R, M)-EGARCH(P, Q) process with standard normal innovations. Therefore, under P, $Y_t|\Phi_{t-1}$ is distributed as a normal variable with mean μ_t and variance h_t, where

$$\mu_t = c + \sum_{i=1}^{R} \phi_i Y_{t-i} - \sum_{j=1}^{M} \theta_j a_{t-j} + a_t \tag{3}$$

and h_t is the conditional variance that is specified in Eq. (2).

Using the results in Buhlmann et al. (1996), Li et al. (2010) prove that an equivalent risk-neutral probability measure, Q, can be found by means of a conditional Esscher transform, so that under Q, $Y_t|\Phi_{t-1}$ is distributed as a normal variable with mean r-g- $h_t/2$ and variance h_t, where r is the risk-free interest rate[1] and g the rental yield.

Therefore, if V_t is the price at time t of a European option maturing at time T, then, under Q,

$$V_t = \mathrm{E}_{\mathrm{Q}}\left[e^{-r(T-t)} V_T|\Phi_t\right] \tag{4}$$

where V_T is the option's payoff at expiration.

Following Li et al. (2010), the value of the NNEG, V_{NNEG}, in a lump-sum reverse mortgage sold to a person aged x at inception is the weighted average of the European put option value, where the weights are the probabilities that a person aged x at inception survives to age $x+k$ and dies during the interval k to $k+1$, with k varying from zero to the maximum lifetime considered for that person. Specifically, the value of the NNEG can be expressed as follows:

$$V_{NNEG} = \sum_{k=0}^{\omega-x-1} {}_k p_x q_{x+k} P(k + 1/2 + \delta, S, X, u, r, g) \tag{5}$$

where ${}_k p_x$ is the probability that a person aged x at inception survives to age $x+k$, q_{x+k} is the probability that a person aged x at inception dies during the time interval k to $k+1$ given that he survived to age $x+k$, and $P(k+½+\delta, S, X, u, r, g)$ is the put option value, assuming that all deaths occur at mid-quarter and that there is a delay of δ (six months) from the home exit until the sale of the property. The value of the put option can be calculated as follows:

[1]Traditionally is has been assumed an inverse relationship between interest rates and home prices. However, during the last years there have been many papers putting into question the existence of a clear relationship between both variables (see for example McQuinn and O'Reilly 2008, and Wong et al. 2003, among others). For this reason, we have decided to assume independence between interest rates and home prices. This is also the approach followed by Li et al. (2010).

$$e^{-r(k+1/2+\delta)}\mathrm{E_Q}[(Xe^{u(k+1/2+\delta)} - S_{k+1/2+\delta})^+] \tag{6}$$

where S is the current price of the property, X is the amount of money advanced by the mortgage provider and u is the roll-up mortgage rate. Under the assumption of an ARMA-EGARCH process for the underlying house price, the expectation in (6) must be computed numerically by Monte Carlo simulation.

Tables 2, 3, 4 and 5 present the results of the valuation of the NNEG for France, Germany, Spain and the United Kingdom respectively. The results shown in these tables are the estimated values of the NNEG, using expression (5), expressed as a percentage of the money advanced by the reverse mortgage lender. As explained in Sect. 2, the German risk-free rate is used for all countries belonging to the Eurozone (the domestic risk-free rate in the case of the United Kingdom), and the probabilities of death and survival have been estimated using the Lee and Carter (1992) model. In every case the results are presented for several values of the roll-up mortgage rate, varying from 3 to 6%,[2] and for several values of the rental yield rate varying from 1 to 5%.

In some countries, as in Spain, since the 2008 financial crisis the reverse mortgage market has decreased sharply, so there is not a reliable market roll-up rate. To solve this problem different values of the roll-up rate have been employed in the estimation of the NNEG shown in the tables.

As stated in the second section, the proxy for the rental yield rate has been obtained from the Global Property Guide 2014, which estimates the annual rental yield rate for each country as the average annual rental yield rate in the main cities of the country. In general, all five countries show relatively high rental yield values (Sect. 2), varying from 2.89% (France) to 3.91% (Spain).

One can argue that it is highly improbable that rental yield rates continue to show these relatively high values over a long time period in a context of low interest rates. Therefore, we have performed a sensitivity analysis to show how the value of the NNEG varies with the value of the rental yield rate. Specifically, we have considered five values for the rental yield rate, from 1 to 5%.

The results in Tables 2, 3, 4, and 5 indicate that, if we take as the appropriate rental yield rate for each country the current market rate described in Sect. 2 (2.89, 3.34, 3.91 and 3.25% for France, Germany, Spain and the United Kingdom respectively[3]), the country with the highest estimated value of the NNEG is Spain, the country most affected by the 2008 financial crisis, followed by France, Germany and the United Kingdom.

Tables 2, 3, 4, and 5 also indicate that, for all countries considered, the estimated NNEG is higher, and consequently the reverse mortgage lender faces higher risk, for the female population, for higher rental yield rates, for higher roll-up rates and for

[2]The estimated values of the NNEG are close to zero for roll-up rates below 3%.

[3]The results for rental yield rates closest to the current market rates are shown in the shaded columns in Tables 2, 3, 4, and 5.

Table 2 Value of the NNEG for different values of the roll-up rate and the rental yield rate. France

		Males					Females				
		Rental Yield					Rental Yield				
Roll-up rate (%)	Age at inception	1%	2%	3%	4%	5%	1%	2%	3%	4%	5%
3	70	0.00	0.00	1.54	8.61	18.05	0.00	0.00	3.75	16.37	30.43
	80	0.00	0.00	0.05	1.42	4.29	0.00	0.00	0.13	3.22	8.41
	90	0.00	0.00	0.00	0.00	0.19	0.00	0.00	0.00	0.00	0.40
3.5	70	0.00	0.13	5.09	15.46	26.87	0.00	0.40	10.74	27.48	43.51
	80	0.00	0.00	0.61	3.03	6.83	0.00	0.00	1.55	6.32	12.75
	90	0.00	0.00	0.00	0.00	0.58	0.00	0.00	0.00	0.01	1.15
4	70	0.00	2.02	11.15	24.56	37.63	0.00	4.95	21.42	41.52	59.10
	80	0.00	0.06	1.83	5.36	10.05	0.00	0.16	4.12	10.50	18.01
	90	0.00	0.00	0.00	0.23	1.13	0.00	0.00	0.00	0.48	2.14
4.5	70	0.25	6.72	19.97	35.98	50.46	0.76	14.23	35.88	58.59	77.39
	80	0.00	0.75	3.81	8.45	13.96	0.00	1.90	7.94	15.76	24.23
	90	0.00	0.00	0.01	0.67	1.86	0.00	0.00	0.02	1.32	3.38
5	70	2.96	14.70	31.65	49.82	65.46	7.20	28.39	54.11	78.84	98.59
	80	0.09	2.23	6.67	12.34	18.61	0.24	5.03	13.05	22.13	31.42
	90	0.00	0.00	0.29	1.28	2.77	0.00	0.00	0.59	2.43	4.88
5.5	70	9.39	26.19	46.23	66.21	82.82	19.85	47.37	76.21	102.50	123.01

(continued)

Table 2 (continued)

Roll-up rate (%)	Age at inception	Males					Females				
		Rental Yield					Rental Yield				
		1%	2%	3%	4%	5%	1%	2%	3%	4%	5%
	80	0.96	4.64	10.42	17.02	24.00	2.42	9.68	19.44	29.58	39.59
	90	0.00	0.03	0.79	2.09	3.87	0.00	0.06	1.55	3.81	6.62
6	70	20.15	41.30	63.90	85.34	102.77	38.91	71.19	102.41	129.89	151.02
	80	2.81	8.06	15.07	22.52	30.15	6.32	15.83	27.09	38.14	48.78
	90	0.00	0.39	1.48	3.10	5.16	0.00	0.78	2.80	5.47	8.61

The table shows the value of the NNEG, as a percentage of the amount of money advanced by the reverse mortgage provider, for the case of France. The results are computed using Germany's risk-free rate and several values for the roll-up rate varying from 3 to 6%, as well as several values for the rental yield rate varying from 1 to 5%. The results for rental yield rates closest to the current market rate are shown in the shaded columns

Table 3 Value of the NNEG for different values of the roll-up rate and the rental yield rate. Germany

		Males					Females				
		Rental yield					Rental yield				
Roll-up rate (%)	Age at inception	1%	2%	3%	4%	5%	1%	2%	3%	4%	5%
3	70	0.00	0.00	0.93	5.51	15.07	0.00	0.00	1.91	9.35	22.93
	80	0.00	0.00	0.01	0.98	3.12	0.00	0.00	0.02	1.87	5.27
	90	0.00	0.00	0.00	0.00	0.13	0.00	0.00	0.00	0.00	0.23
3.5	70	0.00	0.03	3.48	10.75	22.65	0.00	0.06	6.30	17.28	33.55
	80	0.00	0.00	0.30	2.17	5.12	0.00	0.00	0.64	3.85	8.30
	90	0.00	0.00	0.00	0.01	0.40	0.00	0.00	0.00	0.01	0.68
4	70	0.00	1.15	8.16	18.03	31.99	0.00	2.38	13.74	27.87	46.40
	80	0.00	0.05	1.06	3.95	7.70	0.00	0.11	2.04	6.65	12.10
	90	0.00	0.00	0.00	0.17	0.79	0.00	0.00	0.00	0.30	1.32
4.5	70	0.21	4.44	15.32	27.42	43.17	0.50	8.07	24.50	41.20	61.58
	80	0.00	0.50	2.50	6.35	10.89	0.00	1.05	4.43	10.31	16.70
	90	0.00	0.00	0.01	0.48	1.33	0.00	0.00	0.02	0.82	2.15
5	70	2.13	10.47	25.07	39.00	56.30	4.30	17.73	38.64	57.35	79.25
	80	0.04	1.52	4.65	9.42	14.71	0.10	2.90	7.84	14.85	22.08
	90	0.00	0.00	0.21	0.93	2.02	0.00	0.00	0.37	1.55	3.17
5.5	70	6.98	19.68	37.46	52.94	71.54	12.57	31.66	56.22	76.51	99.65

(continued)

Table 3 (continued)

		Males					Females				
		Rental yield					Rental yield				
Roll-up rate (%)	Age at inception	1%	2%	3%	4%	5%	1%	2%	3%	4%	5%
	80	0.56	3.28	7.57	13.20	19.19	1.18	5.83	12.27	20.30	28.30
	90	0.00	0.00	0.56	1.53	2.87	0.00	0.00	0.96	2.47	4.40
6	70	15.36	32.07	52.54	69.34	89.03	25.89	49.80	77.28	98.87	122.99
	80	1.74	5.88	11.27	17.72	24.32	3.35	9.94	17.75	26.69	35.34
	90	0.00	0.19	1.07	2.31	3.87	0.00	0.35	1.78	3.62	5.82

The table shows the value of the NNEG, as a percentage of the amount of money advanced by the reverse mortgage provider, for the case of Germany. The results are computed using Germany's risk-free rate and several values for the roll-up rate varying from 3 to 6%, as well as several values for the rental yield rate varying from 1 to 5%. The results for rental yield rates closest to the current market rate are shown in the shaded columns

Table 4 Value of the NNEG for different values of the roll-up rate and the rental yield rate. Spain

		Males					Females				
		Rental yield					Rental yield				
Roll-up rate (%)	Age at inception	1%	2%	3%	4%	5%	1%	2%	3%	4%	5%
3	70	0.00	0.00	1.14	6.65	15.60	0.00	0.00	2.79	13.28	27.39
	80	0.00	0.00	0.03	1.20	3.71	0.00	0.00	0.07	2.52	6.99
	90	0.00	0.00	0.00	0.00	0.18	0.00	0.00	0.00	0.00	0.31
3.5	70	0.00	0.10	3.91	12.47	23.54	0.00	0.30	8.55	23.23	39.67
	80	0.00	0.00	0.50	2.56	5.95	0.00	0.00	1.14	5.06	10.74
	90	0.00	0.00	0.00	0.00	0.56	0.00	0.00	0.00	0.00	0.94
4	70	0.00	1.60	8.94	20.44	33.33	0.00	3.92	17.90	36.12	54.41
	80	0.00	0.04	1.51	4.55	8.82	0.00	0.11	3.16	8.58	15.37
	90	0.00	0.00	0.00	0.22	1.10	0.00	0.00	0.00	0.38	1.77
4.5	70	0.16	5.35	16.55	30.62	45.06	0.47	11.70	30.96	52.02	71.76
	80	0.00	0.64	3.18	7.24	12.35	0.00	1.45	6.28	13.10	20.89
	90	0.00	0.00	0.01	0.65	1.80	0.00	0.00	0.02	1.08	2.83
5	70	2.25	12.02	26.85	43.10	58.87	5.49	24.12	47.75	71.02	91.91
	80	0.07	1.88	5.59	10.66	16.54	0.18	3.95	10.55	18.64	27.31
	90	0.00	0.00	0.27	1.26	2.67	0.00	0.00	0.48	2.03	4.11
5.5	70	7.33	21.96	39.91	58.01	74.90	16.02	41.28	68.34	93.32	115.11

(continued)

Table 4 (continued)

Roll-up rate (%)	Age at inception	Males					Females				
		Rental yield					Rental yield				
		1%	2%	3%	4%	5%	1%	2%	3%	4%	5%
	80	0.83	3.93	8.82	14.85	21.43	1.88	7.76	15.99	25.22	34.66
	90	0.00	0.02	0.77	2.04	3.71	0.00	0.03	1.28	3.21	5.61
6	70	16.18	35.28	55.86	75.49	93.33	32.54	63.17	92.88	119.18	141.69
	80	2.38	6.85	12.92	19.80	27.03	4.99	12.93	22.64	32.83	42.95
	90	0.00	0.34	1.46	3.01	4.93	0.00	0.59	2.34	4.64	7.34

Table 5 Value of the NNEG for different values of the roll-up rate and the rental yield rate. United Kingdom

		Males					Females				
		Rental yield					Rental yield				
Roll-up rate (%)	Age at inception	1%	2%	3%	4%	5%	1%	2%	3%	4%	5%
3	70	0.00	0.00	0.00	1.02	5.84	0.00	0.00	0.00	1.59	8.06
	80	0.00	0.00	0.00	0.03	1.04	0.00	0.00	0.00	0.06	1.73
	90	0.00	0.00	0.00	0.00	0.00	0.00	0.00	0.00	0.00	0.00
3.5	70	0.00	0.00	0.14	3.46	10.78	0.00	0.00	0.24	5.02	14.37
	80	0.00	0.00	0.00	0.44	2.23	0.00	0.00	0.00	0.77	3.52
	90	0.00	0.00	0.00	0.00	0.00	0.00	0.00	0.00	0.00	0.00
4	70	0.00	0.00	1.51	7.80	17.50	0.00	0.00	2.35	10.78	22.71
	80	0.00	0.00	0.06	1.31	3.99	0.00	0.00	0.11	2.17	6.05
	90	0.00	0.00	0.00	0.00	0.19	0.00	0.00	0.00	0.00	0.32
4.5	70	0.00	0.18	4.84	14.25	26.06	0.00	0.30	7.02	19.04	33.16
	80	0.00	0.00	0.59	2.78	6.35	0.00	0.00	1.03	4.37	9.35
	90	0.00	0.00	0.00	0.00	0.56	0.00	0.00	0.00	0.01	0.89
5	70	0.00	2.05	10.61	22.93	36.52	0.00	3.19	14.69	29.86	45.77
	80	0.00	0.07	1.68	4.91	9.34	0.00	0.12	2.78	7.44	13.43
	90	0.00	0.00	0.00	0.22	1.07	0.00	0.00	0.00	0.36	1.67
5.5	70	0.24	6.59	19.10	33.88	48.99	0.41	9.53	25.57	43.29	60.67

(continued)

Table 5 (continued)

		Males					Females				
		Rental yield					Rental yield				
Roll-up rate (%)	Age at inception	1%	2%	3%	4%	5%	1%	2%	3%	4%	5%
	80	0.00	0.70	3.48	7.76	13.00	0.00	1.24	5.47	11.42	18.32
	90	0.00	0.00	0.01	0.62	1.74	0.00	0.00	0.02	1.00	2.66
6	70	2.69	14.29	30.40	47.21	63.58	4.20	19.79	39.70	59.45	78.04
	80	0.12	2.05	6.05	11.35	17.34	0.23	3.38	9.18	16.33	24.04
	90	0.00	0.00	0.26	1.19	2.58	0.00	0.00	0.42	1.86	3.87

The table shows the value of the NNEG, as a percentage of the amount of money advanced by the reverse mortgage provider, for the U.K. The results are computed using the domestic risk-free rate for the U.K. and several values for the roll-up rate varying from 3 to 6%, as well as several values for the rental yield rate varying from 1 to 5%. The results for rental yield rates closest to the current market rate are shown in the shaded columns younger borrowers. Specifically, the estimated NNEG value is extremely high for borrowers under 80 years old.

Moreover, from Tables 2, 3, 4, and 5, we can see that the value of the NNEG is highly sensitive to changes in the rental yield rate. Specifically, for relatively low values of the roll-up rate, an increase of 1% in the rental yield rate can result in an increase of nearly 50% in the value of the NNEG. For relatively high values of the roll-up rate, the increase in the value of the NNEG is lower but still higher than 20% in most cases. These results show that the reverse mortgage provider can severely underestimate the risk associated with these types of products if the current rental yield rate is relatively low, and it is assumed to be constant over the entire reverse mortgage life.

Finally, it is worth noting that the post-2008 crisis period is characterized by a downtrend in home prices and low interest rates. In the case of Spain the downtrend in home prices is especially pronounced. These two effects, falling home prices and low interest rates, produce a relatively high value of the NNEG for all countries, with Spain showing the highest values. We must take into account that we are valuing a put option, so that the lower the underlying asset (home prices in this case), the higher the option value. Moreover, in the risk-neutral valuation framework, the risk-neutral return of the underlying asset is determined by the risk-free rate (minus the rental yield rate and minus the squared volatility divided by 2). Therefore, the lower the risk-free rate, the lower the final value of the underlying asset and, consequently, the higher the value of the option (the NNEG in this case). This is the reason why the value of the NNEG for all countries considered is higher than the values obtained by Li et al. (2010), who use an older data set (1953–2008). Therefore, the 2008 financial crisis has increased the value of the NNEG, increasing the risk assumed by the reverse mortgage providers.

4 Conclusions

In this paper, we have estimated the value of the NNEG embedded in reverse mortgage contracts using the procedure suggested by Li et al. (2010) for four European countries: France, Germany, Spain and the United Kingdom. The estimation procedure assumes that house price returns follow and ARMA-EGARCH process. The probabilities of death and survival have been estimated by means of the Lee and Carter (1992) model. Using a Monte Carlo simulation procedure we are able to estimate the NNEG for each country as the present value of the corresponding European put option at maturity.

The results show that the highest estimated value for the NNEG is found in the case of Spain, the country most affected by the 2008 financial crisis, followed by France, Germany and the United Kingdom. Moreover, it is found that the NNEG tends to be higher, and consequently the reverse mortgage provider faces higher risks, for higher roll-up rates, for higher rental yield rates, for the female population, and for relatively young borrowers.

Acknowledgements We acknowledge the financial support of Junta de Castilla-La Mancha grant PEII-2014-019-P and of the Ministerio de Educaci6n grant ECO2014-59664-P.

References

Alai, D.H., Chen, H., Cho, D., Hanewald, K., & Sherris, M. (2013). *Devoloping equity relese markets: risk analysis for reverse mortgages and home reversions*. UNSW Australian School of Business Research Paper No. 2013ACTL01.

Buhlmann, H., Delbaen, F., Embrechts, P., & Shiryaev, N. (1996). No-arbitrage, change of measure and conditional esscher transforms. *CWI Quarterly, 9,* 291–317.

Chang, C., Wang, C., & Yang, C. (2012). The effects of macroeconomic factors on pricing mortgage insurance contracts. *Journal of Risk and Insurance, 79,* 867–895.

Chen, H., Cox, S. H., & Wang, S. (2010). Is the home equity conversion mortgage in the United States sustainable? Evidence from pricing mortgage insurance premiums and non-recourse provisions using the conditional Esscher transform. *Insurance: Mathematics and Economics, 46,* 371–384.

Fisher, J. D., Geltner, D. M., & Webb, R. B. (1994). Value indices of commercial real estate: a comparison of index construction methods. *Journal of Real Estate Finance and Economics, 9,* 137–164.

Geltner, D. (1993). Estimating market values from appraised values without assuming and efficient market. *Journal of Real Estate Research, 8,* 325–345.

Huang, H. C., Wang, C. W. & Miao, Y. C. (2011). Securitization of Crossover Risk in Reverse Mortgages. *The Geneva Papers on Risk and Insurance—Issues and practice, 36,* 622–647.

Kogure, A., Li, J., & Kamiya, S. (2014). A bayesian multivariate risk-neutral method for pricing reverse mortgages. *Norh American Actuarial Journal, 18,* 242–257.

Lee, R., & Carter, L. (1992). Modeling and forecasting U.S. mortality. *Journal of the American Statistical Association, 87,* 659–671.

Li, J. S. H., Hardy, M., & Tan, K. S. (2010). On pricing and hedging the No-negative-equity guarantee in equity release mechanisms. *The Journal of Risk and Insurance, 77,* 499–522.

McQuinn, K., & O'Reilly, G. (2008). Assessing the role of income and interest rates in determining house prices. *Economic Modelling, 25,* 377–390.

Shao, A. W., Hanewald, K., & Sherris, M. (2015). Reverse mortgage pricing and risk analysis allowing for idiosyncratic house price risk and longevity risk. *Insurance: Mathematics and Economics, 63,* 76–90.

Siu, T. K., Tong, H., & Yang, H. (2004). On pricing derivatives under GARCH models: a dynamic gerber-shiu approach. *North American Actuarial Journal, 8,* 17–31.

Wang, C.-W., Huang, H.-C., & Lee, Y. T. (2016). On the valuation of reverse mortgage insurance. *Scandinavian Actuarial Journal, 2016,* 293–318.

Wong, T. Y. J., Hui, C. M. E., & Seabrooke, W. (2003). The impact of interest rates upon housing prices: an empirical study of Hong Kong's market. *Property Management, 21,* 153–170.

Institutional Versus Retail Investors' Behavior Around Credit Rating News

Pilar Abad, Antonio Díaz, Ana Escribano and M. Dolores Robles

Abstract This paper investigates the impact of credit rating downgrades on the liquidity and trading behavior of both segments of trading in the U.S. corporate bond market: the institutional- and the retail-sized ones. Using the TRACE dataset, we test if both market segments behave different and what hypotheses explain this potential divergence. We test the regulatory constraints hypothesis (regulatory mandates may force institutional bondholders to sale downgraded bonds), the informed-uninformed traders' hypothesis (information could be responsible for different trading activity patterns before credit rating changes depending the segment of trading), and the usual information hypothesis. We obtain evidence of increased trading activity and price adjustments around downgrades for both segments. Rating-contingent regulation induces larger intensity responses in the institutional segment. Finally, we observe trading anticipation before downgrades that is consistent with the existence of informed institutional investors.

Keywords Credit rating · Institutional · Liquidity · Regulatory constraints
Corporate bonds

This work was supported by the Spanish Ministry of Economy, Industry and Competitiveness (ECO2015-68367-R, ECO2015-67305-P, ECO2017-89715-P), Junta de Comunidades de Castilla-La Mancha (PEII-2014-019-P) and Bank of Spain (PR71/15-20229). Any errors are solely the responsibility of the authors.

P. Abad (✉)
University Rey Juan Carlos of
Madrid, Paseo de los Artilleros s/n, 28032 Vicálvaro, Madrid, Spain
e-mail: pilar.abad@urjc.es

A. Díaz · A. Escribano
University of Castilla-la Mancha, Plaza de la Universidad s/n, 02071 Albacete, Spain
e-mail: antonio.diaz@uclm.es

A. Escribano
e-mail: ana.escribano@uclm.es

M. D. Robles
University Complutense of Madrid and ICAE, Campus de Somosaguas s/n, 28223 Madrid, Spain
e-mail: mdrobles@ccee.ucm.es

M. Mili et al. (eds.), *New Methods in Fixed Income Modeling*, Contributions to Management Science, https://doi.org/10.1007/978-3-319-95285-7_14

JEL Classification G12 · G14 · C34

1 Introduction

Theoretical and empirical literature on corporate bond market provides evidence of both liquidity and credit risk effects on the pricing of these assets. On the one hand, issue credit ratings are commonly used to control for default risk. They are considered as the market's current judgment of the obligor's creditworthiness with respect to a specific financial obligation. On the other hand, yield spreads of corporate bonds are not primarily explained by default risk but are mainly attributable to liquidity and other factors (see, e.g., Sarig and Warga 1989; Elton et al. 2001; Collin-Dufresne et al. 2001; Díaz and Navarro 2002). Most recent papers show how liquidity levels affect bond prices and hence bond yields (see, e.g., Chen et al. 2007; Mahanti et al. 2008; Bao et al. 2011; Dick-Nielsen et al. 2012; Friewald et al. 2012).

Credit rating agencies (CRA) state that they consider insider information when assigning ratings without disclosing specific details to the public at large. According to a long-standing literature, CRA convey new valuable information that have pricing implications in the corporate bond market. However, there is an indirect link between credit rating and liquidity. Many institutional investors face credit rating-based portfolio restrictions. Regulatory and nonregulatory rules contingent on ratings can be found in bank, insurance and broker-dealer capital requirements, suitability requirements in contractual investment mandates and in internal investment procedures, or collateral requirements. The rating of a bond affects the clientele of investors willing to hold it and modifies security price and trading regardless of any fundamental information the rating may convey. Regardless whether a credit rating change discloses relevant information about credit risk, the bond's liquidity should be affected.

The U.S. corporate bond market is mainly dominated by institutional investors. Their trading volume is far greater compared to the trading levels of retail investors, although the number of trades reported by institutional traders is somewhat smaller than for retail-sized traders. The usual practice in the literature is discarding retail-sized trades (trades below \$100,000 in volume) before the analysis (see, e.g. Dick-Nielsen 2009; Dick-Nielsen et al. 2012). In contrast, we explore whether the effects of credit rating news in both the institutional and the retail segments are different, focusing on the liquidity response of both type of traders.

We assume that different fixed income investors, i.e. institutional and retail ones, may behave differently in their trading responses around credit rating downgrades. Several reasons may explain a different behavior. First, they are subject to different rating-specific holding restrictions, rating-contingent regulations and liquidity needs. A downgrade could force or tend most institutional investors to sell the downgraded bond. Second, institutions are relatively more experienced and skillful than retail investors and this could lead to an informational advantage (Barber et al. 2009; Ng and Wu 2007). Third, although corporate bond market is dominated by institutional investors, retail trades represent the 65% of the total number of trades (Ronen and

Zhou 2013). Retail investors are more prone to reach for yields and may act as liquidity providers when institutions need to unwind positions to comply with regulation. Fourth, institutional investors are presumably informed, and the retail ones presumably uninformed. This fact could lead to the existence of information asymmetries among traders, particularly before rating changes. In the corporate bond market, Ronen and Zhou (2013) find evidence of movements in trading and prices consistent with strategic information-based activity. Following traditional market microstructure theory, the presence of investors managing some private information may affect bond trading prices, and hence liquidity (see Easley and O'Hara 2004).

In this paper, we examine the response of these different investors to cuts in the credit quality of corporate bonds. We analyze the potentially different effect of downgrades on bond pricing, liquidity and trading activity for both institutional- and retail-sized trades in the U.S. corporate bond market. We deal with several research questions: Are rating adjustments equally informative for institutional and retail investors? Are institutional investors forced to unwind positions after downgrades due to regulation, overreacting after rating cuts with respect to retails? Does the trading behavior of institutional investors denote the existence of informed trading before downgrades?

We study a comprehensive sample of 2082 U.S. corporate bond rating downgrades involving 1250 straight bonds issued by 245 issuers. We combine bond trading information in the TRACE database, with bond's issue, issuer and rating information provided by the Fixed Income Securities Database (FISD) during the period July 2002 to December 2014. We consider several aspects of liquidity, bond pricing and trading activity, such as the number of trades, the trading volume, the price, the yield spread and the price variation around credit rating announcements. We apply the event study methodology to test whether both types of traders—institutional and retail—respond and/or even anticipate credit rating downgrades. We examine whether the trading behavior of the two segments around credit events go in the same direction, and whether both segments anticipate to a greater or lesser extent credit rating downgrades. Furthermore, our analysis allows us to distinguish different subsamples by specific-downgrade features because downgrades may have different and important implications for bondholders depending on final ratings. We consider subsamples according to some downgrade characteristics such as downgrades within the investment-grade category, within the speculative-grade category and crossing the investment-speculative barrier (the so-called "fallen angel"). We also study the degree of equality in the responses and/or in the anticipation of both types of traders, applying different tests related to the equality in means and distributions of their abnormal trading patterns.

We find increased trading activity and price adjustments around downgrades, particularly around fallen angel bonds, which suggests the induced selling of these bonds. These results are consistent with a forced selling phenomenon for the institutional constrained segment, and a risk-averse selling phenomenon for the retail segment. The different intensity shocks around downgrade announcements support the idea of different motivated-trading. Finally, we find abnormal behavior on trading activity and prices some days before the announcement that are consistent with the existence of informed-based trading.

Our study contributes to the existing literature in several ways. To our knowledge, this is the first study considering separately the two trading segments, i.e. institutional and retail investors. This fact allows us to investigate whether trading activity around rating downgrades is triggered by rating-contingent regulatory constraints. It would be of interest to bond market participants, credit risk managers and regulators. We provide evidence supporting segmentation between institutional and retail traders. Market participants can take advantage if market behavior seems to anticipate downgrade events, especially those bondholders of future downgraded bonds with regulatory implications. Additionally, our findings about the existence of information asymmetry have important implications for corporate bond pricing, as long as prices around credit events may have an embedded information-price premium.

2 Hypotheses of the Effects of Credit Rating Changes

The traditional literature examines bond pricing, returns and trading volume variations around bond rating adjustments. Depending on the information content of credit rating changes, the response of prices, returns and trading volumes should be different. The market reaction should be strong when rating changes convey new and surprising information to the market, whereas the reaction would be lower when the changes disclose information that is already known by the market. Furthermore, these effects might be intensified because of rating-specific constraints, liquidity needs or the crossing of a threshold risk tolerance. Under the efficient markets hypothesis, rating changes convey new relevant information that affect to the pricing process and the trading patterns. This is the widely tested *information content of rating changes hypothesis*.[1] Using the TRACE dataset, several papers find changes in price, volume and trading frequency due to rating downgrades with some persistent effects over time. Da and Gao (2009) document persistent shocks after downgrades for bonds downgraded to "junk" status (from Baa to Ba) during the period −1 to +120 days around the announcement. May (2010) finds statistically significant abnormal bond returns over a two-day and over a one-month event window around downgrades. Jankowitsch et al. (2014) document large trading volumes on the day and the following days of the downgrade for defaulted bonds.[2] According the information hypothesis, rating change should be informative for any investor in the corporate bond market, and this motivates our first hypothesis:

> H1: Market participants react after the disclosure of rating changes as long as they provide new information to the market.

[1]Other studies on the stock market and Credit Default Swaps (CDS) find also evidence of the effects of credit rating changes.

[2]For instance, Norden and Weber (2004) in the stock and CDS markets, Jorion and Zhang (2007) in the CDS market and Hull et al. (2004) in the CDS and bond markets.

The second hypothesis consider that downgrades could have different implications for institutional and retail bondholders. Both kind of investors are subject to different restrictions and regulations affecting their investment decision. Most institutions such as insurance companies, pension funds, investment-grade bond mutual funds or money market funds are generally subject to regulatory constraints that prevent them to invest in speculative-grade bonds.[3] Financial regulation impose institutions restrictions to limit their inherent conflict of interests. Rating-contingent regulation prevents to invest in low credit rating bonds, forcing institutional investors to react to downgrades that cross the investment/speculative barrier by selling them. This forced selling could lead to the "fire sales" phenomenon causing a price pressure effect due to the large price concessions that bond sellers bear to trade downgrades bonds. Da and Gao (2009) find evidence of a persistent price concession of investment-grade funds and insurance companies due to the forced selling of fallen angels. Ellul et al. (2011) find fire sales of fallen angel bonds held by insurance companies, especially when the overall insurance industry is in distress, whereas Ambrose et al. (2008) only find a small selling activity only in a small portion of their overall holdings. Thus, the evidence identified in prior studies indicates that downgraded bonds crossing the investment-speculative barrier will involve stronger effects than any other downgrade due to the intensified response of those investors subject to rating-specific constraints. Regulatory constraints do not affect to retail investors, who should consider these kind of downgrades as any other downgrade. In this sense, we expect to find a weaker response to retails to this kind of rating adjustment. This is the *regulatory constraints hypothesis* that we state as:

> H2: Market participants that are subject to regulatory constraints will overreact after rating changes respect to market participants unaffected by these restrictions.

In addition, literature usually consider the institutional investors more sophisticated and informed than retail investors, that are more prone to chase returns (i.e. Frazzini and Lamont 2008). When private information is linked to a deterioration in the issuer's credit risk that will affect bond prices, informed traders will trade in advance, choosing the moment, size and direction of their transactions to optimize their trades and minimize transaction costs.[4] Ronen and Zhou (2013) find significant institutional trading activity and price movements before earnings announcements consistent with informed trading. Han and Zhou (2014) find that microstructure measures of asymmetric information well explain adverse selection in trading. Information asymmetry also explains the bond yield spreads to a high degree. Chae (2005) analyzes specific corporate events such as unscheduled Moody's rating announcements and finds dramatically increased trading volume before downgrades. We consider that institutional investors possess superior information-gathering and monitoring skills than retail investors, which suffer higher levels of information asymmetry.

[3]Campbell and Taksler (2003) highlight that those institutions that are subject to rating-based restrictions hold more than half of all corporate bonds.

[4]Market microstructure theory suggests that informed traders prefer to trade in large amounts to take advantage of the nature of the pending information they hold (see Easley and O'Hara 1987).

Institutional segment should lead the trading activity prior rating change announcements. We formulate this hypothesis as follows:

> H3: The existence of informed and uninformed traders could be responsible for different trading activity patterns before credit rating changes.

3 Data Description

We study the U.S. corporate bond market. The majority of U.S. corporate bonds are fixed coupon bonds (Becker and Ivashina 2015). We use two sources. The first is the TRACE database by the National Association of Securities Dealers (NASD). The second is the Mergent's FISD that collects complete information on bond characteristics and bond credit rating history. TRACE includes information on prices, yields and volumes. We filter the dataset before usage to avoid errors, applying Dick-Nielsen (2009) algorithms and procedures with minor variations.[5] FISD includes qualitative and quantitative information about bond features as well as bond rating history information.[6] In addition, the FISD data set reports rating information per bond from the main three CRA, i.e. Moody's, Standard & Poor's and Fitch, accounting for event type, rating status and reason.

Matching the TRACE and FISD datasets, we collect trading information for bonds meeting specific trading criteria, particularly we require a minimum trading activity level of the bond involved in the announcement during different periods around the event. The bond should be traded at least once in the 20 working days before the event and once in a similar period after the event.[7] In addition, bonds must be traded on at least 20% of the trading days in the control window. We define the control window as the period starting 41 days prior to the credit announcement and ending 21 days after the announcement. We do this to avoid possible price lead-up preceding the shocks.

Furthermore, we avoid overlapping events meaning that events preceded by other rating announcements in the previous 61 working days are ignored. Table 1 shows the composition of our final sample across rating categories where rating grades are clustered.[8] Doubled or tripled rating announcements by two or more CRA (multiple

[5]Dick-Nielsen (2009) shows that TRACE contains almost 7.7% of the errors among total reports. Edwards et al. (2007) and Dick-Nielsen (2009) show that many errors are due to later corrected or canceled transactions.

[6]We limit the sample to straight corporate bonds, excluding zero coupon bonds, variable coupon bonds, bonds that are part of a unit deal, TIPS, STRIPS, and either perpetual, putable, callable, tendered, preferred, convertible or exchangeable bonds.

[7]This level of trading may seem negligible, but the general liquidity level in the US corporate bond market, the world's largest one, is really low. Mahanti et al. (2008) report that the percentage of the total number of bonds in their sample (2004–2005) that trade at least once a year is between 22 and 34%, each year. Over 40% of bonds do not even trade once a year.

[8]We exclude default bonds because of the reduced information provided by TRACE after defaults. Once a CRA announces the default, TRACE stops reporting yield information.

Table 1 Downgrades by agency across rating categories and summary statistics of bond characteristics

	Final rating category after downgrade									Average		
	AA	A	BBB	BB	B	CCC	CC	C	All	Prior rating	Jump size	Final rating
Fitch	42	257	253	81	26	11	1	0	671	7.10	1.23	8.33
Moody's	79	284	182	214	48	50	5	1	863	7.48	1.54	9.02
S&P	127	254	123	80	43	17	4	1	649	6.17	1.24	7.41
Unique Total	**238**	**719**	**552**	**368**	**116**	**77**	**10**	**2**	**2082**	7.07	1.35	8.42
# issues	213	535	459	287	84	69	11	1	1250			
# issuers	47	137	92	38	23	22	7	1	245			
Issue size	1.65	0.89	0.53	0.17	0.50	0.22	0.44	0.50	0.70			
Coupon	4.38	5.54	5.58	5.83	6.80	6.63	7.53	7.70	5.59			
Age	3.95	4.82	4.38	4.48	6.06	7.63	8.92	5.53	4.74			
Duration	2.87	3.24	2.80	2.43	3.06	2.13	1.36	0.47	2.87			
Time to maturity	4.42	5.81	4.20	3.38	5.20	4.44	3.69	0.48	4.70			

This table shows final ratings after downgrades for different CRA, and for the whole sample across the final rating grade, and summary statistics for rating downgrades. Rating grades are clustered as follows: AA/Aa contain final ratings AA+/Aa1, AA/Aa/Aa2 and AA-/Aa3; A contain A+/A1, A/A2 and A-/A3; and so on. Default downgrade events are excluded. *Issue size* is the value of the amount outstanding in millions of dollars. *Coupon* is the coupon rate for the bond in percentage. *Age* is the age of the bond in years. *Duration* is the Macaulay duration in years. *Time to maturity* is the remaining time until maturity in years. *Prior rating* represents the numeric value of the rating immediately before the downgrade. Rating scale is from AAA = 1 to D = 25 (in the case of S&P). *Jump size* represent the numeric value of the jump measured in rating notches crossed. *Final rating* indicates the final rating after the downgrade. For the final sample, we compute the average numeric values when simultaneously double or triple rated in the same event. The dataset includes 2082 unique downgrade events involving 1250 straight bonds issued by 245 issuers. It covers the period from July 1, 2002 to December 31, 2014, and includes bond's issue, issuer and rating information based on data from the Fixed Income Securities Database (FISD)

rating changes) are considered as one unique event whenever changes are in the same direction and grade category (investment or speculative grade).[9] Default downgrade events are excluded.[10] Hence the "Unique total" row in Table 1 represent final values where events are triggered by only one CRA. The final data sample consists of 2082 unique credit rating downgrade announcements, involving 1250 bonds from 245 issuers. Using the common rating scale from AAA = 1 to D = 25, the average prior rating is 7.07 (around A). Average final rating account for 8.82 (around BBB). Finally, average jump size is equal to 1.34, representing an average jump of one notch, and ranging from 1 to 7 notches. Fallen angel downgrades account for more than 14% in the sample (300 events).

4 Methodology

In order to determine whether the observed responses of institutional and retail investors are statistically different, we first need to identify them from transaction data. According to Edwards et al. (2007), we classify institutional- and retail-sized trades taking as the trading volume cutoff point the amount of \$100,000 (retail-size trades <\$100,000).

We examine the impact of the 2082 downgrades included in our sample on prices, yield spreads and different traditional liquidity proxies. We compute these measures on a day-by-day basis. The number of trades, $NT_{i,t}$, is the sum of the number of transactions of bond i on day t. The cumulative daily trading volume, $TV_{i,t}$, is the sum of the volume traded of bond i on day t.[11] The daily price of bond i on day t, $PR_{i,t}$, is given by the arithmetic mean price. The yield spread, $YS_{i,t}$, is computed as the difference between the average daily yield of bond i on day t and the yield of a Treasury bond with the same time to maturity on the same day t.[12] Finally, the bond's price performance on day t, $PP_{i,t}$, is the difference between the maximum and the minimum prices over the maximum price on that day. We require a minimum of two trades in a day to compute the price performance measure.

The period examined around the rating change day (t=0) includes 41 working days before and 20 working days after.[13] Following Corwin and Lipson (2000), we use the firm-specific past history to compute the expected values in a period of normal

[9]Rating events in the opposite direction are ignored.

[10]Bond prices affected by default events drop around 80% or even 90% of par value, and yields reported by TRACE from defaulted bonds rise 100% on average, biasing price and yield analysis.

[11]It should be noted that TRACE does not report exact volumes when trades are higher than \$5 million for investment-grade bonds and higher than \$1 million for high-yield bonds. Therefore, trading volume values are expected to be higher than reported in this study.

[12]We use Treasury yields on actively traded non-inflation-indexed issues adjusted to constant maturities reported by the Federal Reserve Statistical Release (http://www.federalreserve.gov/releases/h15/data.htm#fn26), and interpolate if necessary.

[13]We discard those re-rated bonds with other confounding rating event in the [−41, 20] window. We also require that the re-rated bond trades on at least 20% of the days in the stable-rating period.

behavior, i.e. in absence of rating event.[14] This control or uncontaminated one-month period is the second month prior the rating change (from $t=-41$ to $t=-21$). We consider a pre-announcement period of one month before the downgrade (from $t=-20$ to $t=-1$) to analyze movements that could be a signal of market anticipation. Finally, the post-announcement period includes one-month period immediately after the downgrade (from $t=0$ to $t=20$). We assume that investors trade downgraded bonds to rebalance their portfolios after the new information and possible regulatory implications. We consider the [0, 5], [6, 10], [11, 20] and [−5, −1] periods, i.e., one week after, the second week after, the second fortnight after and one week before the event.

The expected behavior of the variable X, EX_i, for the credit rating downgrade (CRD) i is computed as the average of the X values on this benchmark window:

$$EX_i = \frac{1}{N}\sum_{t=-41}^{-21} X_{it} \tag{1}$$

where X_{it} represents the price, yield spread, number of trades, volume or price performance of the downgraded bond on day t ∈ (−21, −41). Once we know how the normal behavior of variable X is, we compute their abnormal behavior around the announcement of the credit rating downgrade i, AX_i, as:

$$AX_{i,\tau} = \frac{X_{i,\tau} - EX_i}{EX_i} \tag{2}$$

for $\tau = -20, -19, \ldots, 0, \ldots 20$. We then compute the mean abnormal response of X for day τ as:

$$MAX_\tau = \frac{1}{N}\sum_{i=1}^{N} AX_{i,\tau} \tag{3}$$

where N is the number of downgrades in the sample. We compute expression (3) for all considered variables and for both types of investors, retail and institutional.

To test the statistical significance of the response, we analyze the average abnormal response in the event windows around the CRD date and across subsamples. We compute the daily average of variable MAX_T, for all T in the window of days $s=(t_1, t_2)$ $MAAX_s$. Under the null, the expected value of $MAAX_s$ must be zero. To test the statistical significance of $MAAX_s$ we apply different methods . First, we compute the

[14]Bessembinder et al. (2009) indicate that the use of the firm-specific past history as a benchmark could be less powerful than creating a matching portfolio of stable rating bonds. However, the second approach involves creating an appropriate portfolio with characteristics related to liquidity as similar as possible to the re-rated bond to avoid biased results. To find such a good matching portfolio is very difficult in the case of corporate bonds, due to infrequent trading observed even in the most liquid markets such as the U.S.

well-known t-ratio test, asymptotically normal distributed under the null hypothesis. Second, we compute two non-parametric tests (Fisher sign test and Wilcoxon rank test) that are robust to non-normality, skewness and other statistical characteristics of data that may affect the t-ratio properties. The Fisher sign test equals the number of times abnormal values of the proxies is positive. The Wilcoxon rank test accounts for information of both magnitudes and signs. We compute *p*-values for the asymptotic normal approximation to these tests.

To study to what extent the abnormal response to CDR is statistically the same for both investors, retail vs institutional, we test the equality of means and distributions of their abnormal responses. For the first null hypothesis, we apply the Welch's unequal variances t-test, and for the second one the tie-adjusted Mann-Whitney test (see Sheskin 1997 for statistical details).

5 Results

5.1 Preliminary Graphical Analysis

We first make a graphical analysis of the abnormal trading behavior around credit rating downgrades considering the trade size from day −20 to day 20 in event time. Figure 1 shows the abnormal number of trades, trading volume, price, yield spread and price averaged across CRDs. We observe increasing abnormal number of trades and trading volume around the announcement date for both market segments (upper figures). This preliminary result is consistent with the H1. There is an apparently significant response of bond liquidity to rating changes that could indicate that downgrades convey new information to the market.

The abnormal behavior seems to be different between the retail and institutional segments. The increase in abnormal trading activity is much more pronounced for institutional-sized trades. Their number of trades increases by nearly 50% on average immediately after the event, almost double that in the case of the retail traders. Remarkably, the average abnormal number of institutional-sized trades is smaller than that for retail traders, except from day −2 to day +2. This fact highlights that the increment in the trading activity by the institutional segment around the announcement day is sharply larger than the abnormal response of retails these days. This preliminary evidence supports the H2.

One significant feature is that the trading pattern of the institutional segment in terms of number of trades and trading volume points to some anticipated response 3 days before the CRD announcement. We observe that the retail segment response occurs after the initial reaction of institutional investors, that lead the trading activity prior the CRD. This behavior is consistent to the H3, indicating that institutional segment uses relevant information before the downgrade.

In the second row of Fig. 1, we observe average abnormal prices close to zero with a slightly downward trend for both market segments. There is a clear cut in prices for

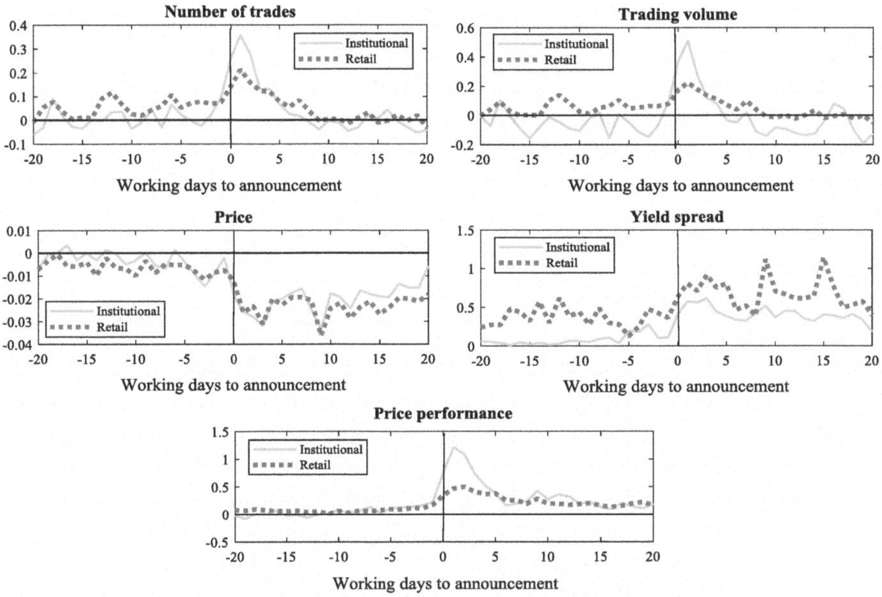

Fig. 1 Abnormal trading behavior around credit rating downgrades by trade size. This figure shows average abnormal values across working days to announcement, where abnormal values are computed as Eq. (2) in a daily basis by bond i, on day τ only for days with non-zero EX_i. Abnormal values are then averaged across days for all N downgrades in the sample. The dataset includes 2082 credit rating downgrades involving 1250 straight bonds issued by 245 issuers. Bond trading information is reported by the TRACE database and rating information is provided by FISD. The dataset covers the period from July 1, 2002 to December 31, 2014

both types of traders the day before the event more evident at $t=0$, when the price in the institutional segment falls more than 3%. This fall seems larger than the one observed in the retail segment. Prices stabilize on the following days, and from day +4 onwards price recover levels similar to one month before the downgrade.

Contrary to prices, the daily average abnormal yield spread is systematically higher for retail than for institutional trades. Prices (yield spreads) in the retail segment seems to react less (more) than the institutional segment. This apparent contradiction may lie in the different weight that retail trades have in the investment and speculative categories. After a downgrade to or inside speculative categories, there are extreme low prices and high yield spreads. However, the dispersion around the mean of yield spreads is much larger than the one of prices, because yield spreads are much more sensitive to the credit risk perception than prices. The impact of these extreme observations on the average yield spread is huge and mainly affects retail transactions.

According to the last row of Fig. 1, the average abnormal price performance around downgrades indicates that the price pressure is larger in the institutional-sized segment. The average abnormal price performance accounts for almost 120% and it is more than twice higher than this value in the retail case.

Overall, this preliminary evidence is consistent with our main hypotheses as we observe (1) that both market segments reacts to downgrades, (2) a stronger institutional response to downgrades and (3) that institutional segment anticipates the response some days before the retail segment.

5.2 Market Reaction

In this section we conduct formal tests to better understand the abnormal response to downgrades and the main differences between institutional and retail-sized investors. We split the sample into three subsamples according to the main characteristics of downgrade: within the speculative-grade, within the investment-grade, and across the investment/speculative boundary (fallen angels). We make a finer partition of the post-event period in three windows to study the timing of effects: one week immediately after the announcement, i.e., the [0, 5] window; the second week after the event, i.e., the [6, 10] window; and the following two weeks, i.e., the [11, 20] window.

Table 2 depicts the results of the three different test statistics for the null hypothesis of absence of abnormal behavior of NT, TV, PR, YS and PP for institutional- and retail-sized trades and the two test statistics for the null hypothesis of equality of effects between institutional- and retail-sized trades. Panel A shows the results for the whole sample (All bonds), Panel B for the downgrades within Investment-grade categories subsample, Panel C for the downgrades within Speculative-grade categories subsample and Panel D for Fallen Angels. Summarizing, we observe statistically significant abnormal behavior for all variables in all subsamples and for both market segments with at least one of the tests, supporting the H1. Downgrades disclose new relevant information that affect retail and institutional investors who alter their trading pattern after the announcement date. We also find evidence on different effects between investors at least with one test, but this depends on the window and downgrade characteristics.

When the whole sample is considered (Panel A of Table 2), we find positive mean abnormal values of the number of trades equal to 0.97 and 0.61 for institutional and retail trades respectively within the first week after the rating change ([0, 5] window).

The event study results for the trading volume, also shows positive mean abnormal values equal to 2.43 and 1.01 for institutional and retail traders respectively in the [0, 5] window. These effects persist over the rest of the period ([6, 10] and [11, 20] windows) although the intensity seems to decrease as time passes by. These results suggest that trading activity increases immediately after downgrades and persists one month after the announcement, although the intensity of the effect decreases over time. The average abnormal prices have negative sign for both segments within the [0, 5] window. It is equal to -1.72 for institutional traders and -2.03 for retail traders. The results are similar for the [6, 10] and [11, 20] windows and are statistically

Table 2 Abnormal response to rating downgrades and their differences

Variable	Window		Panel A: all bonds		Panel B: within investment grade		Panel C: within speculative grade		Panel D: fallen angels	
			Mean	Median	Mean	Median	Mean	Median	Mean	Median
NT	[0, 5]	Ins	0.97^{*}	$0.16^{+\wedge}$	0.91^{*}	$0.14^{+\wedge}$	0.99^{*}	$0.33^{+\wedge}$	1.75^{*}	$0.86^{+\wedge}$
		Ret	0.61^{*}	$0.11^{+\wedge}$	0.72^{*}	$0.13^{+\wedge}$	0.31^{*}	0.00	0.36^{*}	$0.05^{\wedge}$
		Dif	$-2.13^{\dagger}$	$3.18^{\diamond}$	−0.87	0.85	$-3.88^{\dagger}$	$3.58^{\diamond}$	$-2.94^{\dagger}$	$4.30^{\diamond}$
	[6, 10]	Ins	0.55^{*}	$0.03^{\wedge}$	0.46^{*}	$0.00^{\wedge}$	0.83^{*}	$0.14^{\wedge}$	1.21^{*}	$0.64^{+\wedge}$
		Ret	0.31^{*}	$-0.01^{\wedge}$	0.36^{*}	$-0.01^{\wedge}$	0.18^{*}	$0.00^{\wedge}$	0.19^{*}	0.00
		Dif	$-3.73^{\dagger}$	$2.98^{\diamond}$	−1.37	1.04	$-3.15^{\dagger}$	$2.15^{\diamond}$	$-4.45^{\dagger}$	$5.83^{\diamond}$
	[11, 20]	Ins	0.41^{*}	$0.03^{\wedge}$	0.33^{*}	$0.00^{\wedge}$	0.70^{*}	$0.05^{\wedge}$	0.99^{*}	$0.31^{+\wedge}$
		Ret	0.27^{*}	$0.00^{\wedge}$	0.28^{*}	$-0.01^{\wedge}$	0.24^{*}	$0.00^{\wedge}$	0.21^{*}	$0.03^{\wedge}$
		Dif	$-2.73^{\dagger}$	1.21	−0.78	0.67	$-2.30^{\dagger}$	0.53	$-2.83^{\dagger}$	$2.87^{\diamond}$
TV	[0, 5]	Ins	2.43^{*}	$0.04^{\wedge}$	2.55^{*}	$0.01^{\wedge}$	1.46^{*}	$0.19^{+\wedge}$	2.41^{*}	$0.28^{\wedge}$
		Ret	1.01^{*}	$0.08^{+\wedge}$	1.22^{*}	$0.11^{+\wedge}$	0.38^{*}	0.00	0.47^{*}	$0.02^{\wedge}$
		Dif	$-2.81^{\dagger}$	0.04	$-1.97^{\dagger}$	1.55	$-3.96^{\dagger}$	$3.12^{\diamond}$	$-2.80^{\dagger}$	1.36
	[6, 10]	Ins	1.50^{*}	$-0.10^{+\wedge}$	1.46^{*}	$-0.13^{+\wedge}$	1.50^{*}	$-0.06^{\wedge}$	1.99^{*}	$0.40^{+\wedge}$
		Ret	0.39^{*}	$-0.03^{\wedge}$	0.41^{*}	$-0.03^{\wedge}$	0.18^{*}	−0.1	0.44^{*}	0.00
		Dif	$-5.64^{\dagger}$	1.19	$-4.62^{\dagger}$	$2.64^{\diamond}$	$-2.99^{\dagger}$	1.41	$-2.56^{\dagger}$	$3.05^{\diamond}$
	[11, 20]	Ins	1.41^{*}	$-0.05^{\wedge}$	1.21^{*}	$-0.09^{+\wedge}$	1.61^{*}	$0.06^{\wedge}$	3.67^{*}	$0.24^{\wedge}$
		Ret	0.31^{*}	$-0.01^{\wedge}$	0.31^{*}	$-0.02^{\wedge}$	0.33^{*}	$0.07^{\wedge}$	0.28^{*}	$0.00^{\wedge}$
		Dif	$-5.15^{\dagger}$	0.84	$-6.17^{\dagger}$	1.58	$-1.93^{\dagger}$	0.89	−1.36	1.57

(continued)

Table 2 (continued)

Variable	Window		Panel A: all bonds		Panel B: within investment grade		Panel C: within speculative grade		Panel D: fallen angels	
			Mean	Median	Mean	Median	Mean	Median	Mean	Median
PR	[0, 5]	Ins	-1.72^{*}	$-0.25^{+\wedge}$	-1.79^{*}	$-0.25^{+\wedge}$	-1.50^{*}	$-0.31^{\wedge}$	-1.22^{*}	-0.17
		Ret	-2.03^{*}	$-0.41^{+\wedge}$	-2.02^{*}	$-0.44^{+\wedge}$	-2.55^{*}	$-0.33^{+\wedge}$	-1.63^{*}	$-0.27^{+\wedge}$
		Dif	-0.97	$3.14^{\Diamond}$	-0.71	$3.11^{\Diamond}$	-0.67	0.52	-0.49	1.36
	[6, 10]	Ins	-1.47^{*}	$-0.31^{+\wedge}$	-1.36^{*}	$-0.28^{+\wedge}$	-2.00^{*}	$-0.94^{+\wedge}$	-1.88^{*}	-0.44
		Ret	-2.28^{*}	$-0.58^{+\wedge}$	-1.84^{*}	$-0.47^{+\wedge}$	-4.93^{*}	$-1.28^{+\wedge}$	-2.20^{*}	$-0.89^{+\wedge}$
		Dif	$-2.39^{\dagger}$	$4.41^{\Diamond}$	-1.54	$3.35^{\Diamond}$	-1.5	1.36	-0.32	$1.74^{\Diamond}$
	[11, 20]	Ins	-1.72^{*}	$-0.30^{+\wedge}$	-1.72^{*}	$-0.28^{+\wedge}$	-2.62^{*}	$-0.85^{+\wedge}$	-0.23	-0.58^{+}
		Ret	-2.78^{*}	$-0.69^{+\wedge}$	-2.09^{*}	$-0.50^{+\wedge}$	-7.74^{*}	$-2.49^{+\wedge}$	-1.73^{*}	$-1.34^{+\wedge}$
		Dif	$-2.78^{\dagger}$	$5.54^{\Diamond}$	-1.08	$3.50^{\Diamond}$	$-2.45^{\dagger}$	$2.17^{\Diamond}$	-1.64	$3.36^{\Diamond}$
YS	[0, 5]	Ins	0.67^{*}	$0.08^{+\wedge}$	0.71^{*}	$0.08^{+\wedge}$	0.35^{*}	$0.06^{+\wedge}$	0.77^{*}	$0.08^{\wedge}$
		Ret	0.91^{*}	$0.12^{+\wedge}$	1.13^{*}	$0.13^{+\wedge}$	0.33^{*}	$0.13^{+\wedge}$	0.46^{*}	$0.09^{+\wedge}$
		Dif	0.94	$3.39^{\Diamond}$	1.21	$2.99^{\Diamond}$	-0.17	0.8	-1.36	0.78
	[6, 10]	Ins	0.43^{*}	$0.09^{+\wedge}$	0.43^{*}	$0.08^{+\wedge}$	0.29^{*}	$0.14^{+\wedge}$	0.64^{*}	$0.11^{+\wedge}$
		Ret	1.09^{*}	$0.16^{+\wedge}$	1.32^{*}	$0.13^{+\wedge}$	0.59^{*}	$0.26^{+\wedge}$	0.51^{*}	$0.19^{+\wedge}$
		Dif	$2.43^{\dagger}$	$5.15^{\Diamond}$	$2.35^{\dagger}$	$3.51^{\Diamond}$	$2.50^{\dagger}$	$2.51^{\Diamond}$	-0.67	$2.03^{\Diamond}$
	[11, 20]	Ins	0.64^{*}	$0.08^{+\wedge}$	0.70^{*}	$0.07^{+\wedge}$	0.36^{*}	$0.15^{+\wedge}$	0.43^{*}	$0.11^{\wedge}$
		Ret	1.21^{*}	$0.20^{+\wedge}$	1.43^{*}	$0.15^{+\wedge}$	0.93^{*}	$0.42^{+\wedge}$	0.51^{*}	$0.24^{+\wedge}$
		Dif	$2.85^{\dagger}$	$6.59^{\Diamond}$	$2.55^{\dagger}$	$4.09^{\Diamond}$	$4.33^{\dagger}$	$3.82^{\Diamond}$	0.53	$3.37^{\Diamond}$

(continued)

Table 2 (continued)

Variable	Window		Panel A: all bonds		Panel B: within investment grade		Panel C: within speculative grade		Panel D: fallen angels	
			Mean	Median	Mean	Median	Mean	Median	Mean	Median
PP	[0, 5]	Ins	4.56^{*}	$-0.07^{\wedge}$	5.12^{*}	$-0.10^{+\wedge}$	2.14^{*}	$0.09^{\wedge}$	0.87^{*}	$-0.01^{\wedge}$
		Ret	1.64^{*}	$0.08^{+\wedge}$	1.91^{*}	$0.11^{+\wedge}$	0.76^{*}	−0.1	1.09^{*}	$0.11^{+\wedge}$
		Dif	$-1.94^{\dagger}$	$1.71^{\diamond}$	−1.64	$3.83^{\diamond}$	$-2.40^{\dagger}$	$3.27^{\diamond}$	$-1.67^{\dagger}$	$1.77^{\diamond}$
	[6,10]	Ins	2.75^{*}	-0.19^{+}	3.11^{*}	-0.21^{+}	0.94^{*}	−0.1	0.96^{*}	0.00
		Ret	1.29^{*}	$0.00^{\wedge}$	1.49^{*}	$0.01^{\wedge}$	0.86^{*}	0.00	0.66^{*}	0.00
		Dif	$-2.13^{\dagger}$	$3.30^{\diamond}$	$-1.82^{\dagger}$	$4.41^{\diamond}$	−1.52	0.03	$-1.73^{\dagger}$	$1.66^{\diamond}$
	[11, 20]	Ins	1.35^{*}	-0.17^{+}	1.41^{*}	-0.16^{+}	1.42^{*}	-0.18^{+}	0.49^{*}	−0.2
		Ret	1.06^{*}	$0.00^{\wedge}$	1.14^{*}	$-0.01^{\wedge}$	0.98^{*}	$0.06^{\wedge}$	0.75^{*}	$0.10^{\wedge}$
		Dif	$-2.03^{\dagger}$	$4.03^{\diamond}$	−1.2	$3.98^{\diamond}$	−1.49	0.87	−1.12	0.51

This table shows the results from the event study and the differences study in the windows [0, 5], [6, 10] and [11, 20] where t = 0 is the downgrade day. *NT*, *TV*, *PR*, *YS and PP* represent the abnormal value of the number of trades, the trading volume, the price, the yield spread and the price performance variables respectively. *Ins* and *Ret* refers to the results of the event study for the institutional and the retail segments respectively, and *Dif* indicates the results from the study of the differences in the response (Retail vs. Institutional). The dataset includes 2082 unique downgrade events involving 1250 straight bonds issued by 245 issuers. It covers the period from July 1, 2002 to December 31, 2014. Rating information is obtained from the FISD database and trading information is based on data obtained from the TRACE database. *, $^{+}$ and ^ indicate significance at 10% or lower level for the t-ratio, Sign and Rank tests, respectively. And † and $^{\diamond}$ indicate significance at 10% or lower level for the Welch's unequal variances t-test and for the tie-adjusted Mann-Whitney tests respectively significant, suggesting that prices fall after downgrades. The average abnormal yield spread is also statistically significant for both type of traders during the first week after downgrades, with positive values equal to 0.67 and 0.91 for institutional and retail traders respectively. The sign and significance are similar for the other periods, i.e., [6, 10] and [11, 20] windows, indicating larger price impact of downgrades. Finally, the average abnormal price performance is positive and statistically significant for both institutional and retail traders, (4.56 and 1.64 respectively) in the window [0, 5]. The impact in prices is larger immediately after the announcement. We observe similar results within the remaining windows, suggesting that the effect is persistent for some weeks after.

All these prior results are quite similar when considering the different subsamples and windows, for example in the case of downgrades within investment-grade (Panel B) or within speculative-grade (Panel C) subsamples. These results suggest that there is an increase in the trading activity after the release of rating information that is accompanied by a drop in prices, and that the effect persists for at least one month after the downgrade. Downgrades disclose new information and market participants seems to overreact immediately after the announcement day. The medium- and long-term responses indicate that the effects persist almost one-month after this date for both types of traders. These results are consistent with H1 confirming that rating changes add new information to market participants. Besides, these results are in line with those of Da and Gao (2009), May (2010) and Jankowitsch et al. (2014) that document evidence of different price, volume and trading frequency impacts from credit rating downgrades with some persistent effects over time.

Additionally, the results across the investment/speculative threshold (Panel D) are in the same direction and with large intensity. These results are consistent with those in Dao and Gao (2009), that find persistent liquidity shocks in the case of fallen angel bonds. The increase in trading volume after downgrades (presumable much larger than reported), jointly with some price concessions and high price impact, suggest sales at fire sale prices due to rating-specific restrictions faced by institutional investors. This result is consistent with those in Ambrose et al. (2008) and in Ellul et al. (2011).

Most interesting in our research is examining the response differences between bondholders without rating-specific constraints, mainly investors which trade retail-sized transactions, and institutional investors subjected to rating-contingent regulation. We test the two null hypotheses of no differences in mean and no differences in distribution are rejected in most of the cases. We observe statistically significant mean and median differences in abnormal values of the liquidity proxies. Firstly, in the period subsequently after the downgrade, i.e. [0, 5] period, we find negative statistically significant signs in the abnormal number of trades, volume and price performance for large size transactions when we consider the whole sample. In the institutional segment, the average abnormal number of trades and trading volume for the fallen angel bonds are five time higher than in the retail case. This result indicates a strong abnormal response in terms of trading and price impact for institutional investors. The response is higher than for small size orders in all subsamples. One week after, during days [6, 10], mean difference abnormal volume for the large trades shows a negative sign with statistical significance in almost all variables and subsamples. These results indicate that the response remains high in terms of volume for institutional-sized trades. In this period, we also find a high response in the number of trades for large transactions when downgraded bonds are within the investment grade. For the second fortnight following the downgrade, i.e. [11, 20] window, we find a negative sign with statistical significance for abnormal volume traded by institutional-sized trades. This result indicates that abnormal behavior of institutional traders persists being higher than for retail trades. This is particularly true for fallen angel bonds. Furthermore, the abnormal response in prices is also larger for the institutional segment. These results are highly consistent with H2. Institutional investors, subject to rating-specific restrictions that are not mandatory for retail investors seem to be forced to sell downgraded bonds crossing the threshold between investment and speculative grade.

5.3 Informed Trading Before Downgrades

We explore here to what extent there is an anticipated response to downgrades. We analyze the institutional- and retail-segments behavior in the pre-event period [−5, −1]. Table 3 presents the results. In Panel A (All bonds) we observe that the average abnormal values of all variables are statistically significant. The average abnormal number of trades, trading volume, yield spread and price performance are positives and equal to 0.62, 1.77, 0.36 and 2.68 respectively, and the average abnormal price is negative and equal to −0.30. In general, these results suggest that institutional traders increase their trading activity before the downgrade is announced. They seems to trade at prices lower than the prices in normal periods and increased yield spreads. Across subsamples, the results are similar in general, except to the abnormal price in the fallen angel subsample (see Panel D on Table 3). In this case, we find a positive mean abnormal price equal to 1.02 that is statistically significant, suggesting that institutional traders get to trade fallen angel bonds above normal prices than in the control period during these days. We also find significant abnormal trading patterns on the retail-sized trades. In general, the observed responses are seemly weaker than in the case of institutional traders, mainly in the case of fallen angel bonds.

When differences in the response of both type of traders are formally tested, we observe negative statistically significant signs for the difference in the retail vs institutional traders in some subsamples. The abnormal number of trades and trading volumes have negative statistically significant signs for speculative grade category, as well as the price for the fallen angel subsample, and the price performance for the speculative subsample, suggesting that the abnormal response in terms of trading activity and price impact is large for the institutional segment. This result indicates that there are differences in the mean and distribution between the two segments that suggests a clear difference in the response before the event for both large- and small-sized trades. Institutional traders show a strong early response to downgrades, especially for those related to fallen angel bonds.

Table 3 Abnormal trading before rating downgrades and their differences

Variable	Window		Panel A: all bonds		Panel B: within investment grade		Panel C: within speculative grade		Panel D: fallen angels	
			Mean	Median	Mean	Median	Mean	Median	Mean	Median
NT	[−5, −1]	Ins	0.62^{*}	$0.02^{\wedge}$	0.50^{*}	$0.00^{\wedge}$	1.09^{*}	$0.28^{+\wedge}$	1.35^{*}	$-0.01^{\wedge}$
		Ret	0.44^{*}	$-0.04^{+\wedge}$	0.51^{*}	$-0.02^{+\wedge}$	0.22^{*}	-0.10^{+}	0.29^{*}	−0.10
		Dif	−0.93	$2.99^{\diamond}$	0.04	1.17	$-3.97^{\dagger}$	$4.32^{\diamond}$	$-2.51^{\dagger}$	1.64
TV	[−5, −1]	Ins	1.77^{*}	$-0.11^{+\wedge}$	1.73^{*}	-0.17^{+}	1.76^{*}	$0.29^{+\wedge}$	2.20^{*}	0.00
		Ret	0.9	$-0.07^{+\wedge}$	1.10	$-0.06^{+\wedge}$	0.32^{*}	-0.14^{+}	0.37^{*}	0.00
		Dif	−1.31	$2.00^{\diamond}$	−0.71	$3.67^{\diamond}$	$-3.42^{\dagger}$	$3.65^{\diamond}$	−1.54	0.36
PR	[−5, −1]	Ins	-0.30^{*}	$-0.12^{+\wedge}$	-0.34^{*}	$-0.15^{+\wedge}$	-0.73^{*}	0.01	1.02^{*}	0.3
		Ret	-0.74^{*}	$-0.19^{+\wedge}$	-0.56^{*}	$-0.24^{+\wedge}$	-2.15^{*}	$0.10^{\wedge}$	-0.38^{*}	0.19
		Dif	$-1.99^{\dagger}$	1.55	−1.19	1.56	−1.07	0.70	$-1.85^{\dagger}$	$1.83^{\diamond}$
YS	[−5, −1]	Ins	0.36^{*}	$0.05^{+\wedge}$	0.41^{*}	$0.06^{+\wedge}$	0.13^{*}	$0.02^{\wedge}$	0.13^{*}	-0.05^{+}
		Ret	0.35^{*}	$0.05^{+\wedge}$	0.40^{*}	$0.06^{+\wedge}$	0.23^{*}	$0.03^{\wedge}$	0.25^{*}	$-0.02^{\wedge}$
		Dif	0.04	0.92	−0.03	0.51	1.49	0.44	1.31	$1.82^{\diamond}$
PP	[−5, −1]	Ins	2.68^{*}	-0.21^{+}	2.98^{*}	-0.22^{+}	1.71^{*}	$0.12^{\wedge}$	0.29^{*}	-0.31^{+}
		Ret	0.73^{*}	-0.09^{+}	0.74^{*}	-0.09^{+}	0.83^{*}	0.00	0.61^{*}	−0.10
		Dif	−1.61	1.48	−1.31	$2.13^{\diamond}$	$-1.73^{\dagger}$	1.44	−1.29	0.39

This table shows the results from the event study and the differences study in the window [−5, −1] where t = 0 is the downgrade day. *NT*, *TV*, *PR*, *YS and PP* represent the abnormal number of trades, the trading volume, the price, the yield spread and the price performance variables respectively. *Ins* and *Ret* refers to the results of the event study for the institutional and the retail segments respectively, and *Dif* indicates the results from the study of the differences in the response (Retail vs. Institutional). The dataset includes 2082 unique downgrade events involving 1250 straight bonds issued by 245 issuers. It covers the period from July 1, 2002 to December 31, 2014. Rating information is obtained from the FISD database and trading information is based on data obtained from the TRACE database. *, $^{+}$ and ^ indicate significance at 10% or lower level for the t-ratio, Sign and Rank tests, respectively. And † and $^{\diamond}$ indicate significance at 10% or lower level for the Welch's unequal variances t-test and for the tie-adjusted Mann-Whitney tests respectively

To sum up, results in all analyzed subsamples suggest that CDR are partially anticipated by institutional traders before the announcement date to a greater or lesser extent, as they increase trading with reduced price concessions. The modest increase in trading volume of the retail segment suggests a follower role response encouraged by the trading initiated by the institutional segment, confirming the main patterns observed in Fig. 1.

These results would indicate the existence of information asymmetry in the corporate bond market among traders around credit news. Institutional traders, able to exploit their advantages, seems to trade before the disclosure of negative news to anticipate the deterioration of the credit quality of their portfolios mainly in the case of rating cuts with regulatory implications. These results on trading anticipation by institutional traders are essentially consistent with H3, which stands that the existence of informed and uninformed traders is responsible of different trading activity patterns before credit rating changes. This result is also consistent with Grier and Katz (1976), Hite and Warga (1997) or Steiner and Heinke (2001) who provide evidence of abnormal trading activity before credit rating announcements in bond markets. In addition, it is consistent with Chae (2005) that finds evidence of dramatically increased trading volume before downgrade announcements. Finally, it is also in line with Ronen and Zhou (2013) and Kedia and Zhou (2014) that find significant trading activity and price movements before other corporate events consistent with informed trading. The modest increase in trading activity from the retail segment seems to respond to a herding behavior of the trades initiated by institutional investors.

6 Conclusions

This paper investigates the impact of CRD on the trading activity of both the institutional- and the retail-sized segments of the US corporate bond market. Our results show different intensities of liquidity patterns among traders. The increased trading activity and price concessions after downgrades seem to be triggered by rating-specific constraints of the institutional constrained segment. A large price pressure is detected after downgrades, mainly those crossing the investment/speculative threshold. In this case, the trading activity increment in the institutional segment is five time larger than it is in the retail segment. In addition, we observe some degree of trading anticipation before the disclosure of the downgrade. Significant trading activity and liquidity patterns are consistent with the existence of informed-based trading linked to institutional traders. Our results also suggest that the modest increase in trading activity of retail traders before events responds to a follower behavior to trades initiated by the institutional segment.

This study contributes by shedding light onto research into credit rating effects, and specifically into CRD research. The distinction between retail- and institutional-sized transactions allows to obtain more consistent results. It should be of interest to bond market participants, credit risk managers and regulators. Our results suggest that trading patterns after the CRD are mainly leaded by institutions subjected to rating-contingent regulation, and that trading patterns before the CRD are an important source of information for market participants. Trading volume levels before the disclosure of new information can be used to anticipate future bond credit rating deteriorations, especially for those bondholders with rating-sensitive investments in their portfolios. Moreover, retail investors can take advantage of observed patterns in order to make up for their lack of information, particularly when their level of risk-aversion is high.

References

Ambrose, B. W., Cai, N., & Helwege, J. (2008). Forced selling of fallen angels. *Journal of Fixed Income, 18,* 72–85.

Bao, J., Pan, J., & Wang, J. (2011). The illiquidity of corporate bonds. *The Journal of Finance, 66*(3), 911–946.

Barber, B. M., Lee, Y. T., Liu, Y. J., & Odean, T. (2009). Just how much do individual investors lose by trading? *Review of Financial Studies, 22,* 609–632.

Becker, B., & Ivashina, V. (2015). Reaching for yield in the bond market. *The Journal of Finance, 70*(5), 1863–1902.

Bessembinder, H., Kahle, K. M., Maxwell, W. F., & Xu, D. (2009). Measuring abnormal bond performance. *Review of Financial Studies, 22*(10), 4219–4258.

Campbell, J. Y., & Taksler, G. B. (2003). Equity volatility and corporate bond yields. *The Journal of Finance, 58*(6), 2321–2349.

Chae, J. (2005). Trading volume, information asymmetry, and timing information. *The Journal of Finance, 60*(1), 413–442.

Chen, L., Lesmond, D. A., & Wei, J. (2007). Corporate yield spreads and bond liquidity. *The Journal of Finance, 62*(1), 119–149.

Collin-Dufresne, P., Goldstein, R. S., & Martin, J. S. (2001). The determinants of credit spread changes. *The Journal of Finance, 56,* 2177–2207.

Corwin, S. A., & Lipson, M. L. (2000). Order flow and liquidity around NYSE trading halts. *The Journal of Finance, 55*(4), 1771–1805.

Da, Z., & Gao, P. (2009). *Clientele change, persistent liquidity shock, and bond return reversals after rating downgrades*. SSRN.

Díaz, A., & Navarro, E. (2002). La prima de liquidez en la Deuda del Estado española. *Revista de Economía Aplicada, 10*(29), 23–58.

Dick-Nielsen, J. (2009). Liquidity biases in TRACE. *Journal of Fixed Income, 19*(2), 43–55.

Dick-Nielsen, J., Feldhütter, P., & Lando, D. (2012). Corporate bond liquidity before and after the onset of the subprime crisis. *Journal of Financial Economics, 103,* 471–492.

Easley, D., & O'Hara, M. (1987). Price, trade size, and information in securities markets. *Journal of Financial Economics, 19,* 69–90.

Easley, D., & O'hara, M. (2004). Information and the cost of capital. *The Journal of Finance, 59*(4), 1553–1583.

Edwards, A. K., Harris, L. E., & Piwowar, M. S. (2007). Corporate bond market transaction costs and transparency. *Journal of Finance, 62*(3), 1421–1451.

Ellul, A., Jotikasthira, C., & Lundblad, C. T. (2011). Regulatory pressure and fire sales in the corporate bond markets. *Journal of Financial Economics, 101*(3), 596–620.

Elton, E. J., Gruber, M. J., Agrawal, D., & Mann, C. (2001). Explaining the rate spread on corporate bonds. *The Journal of Finance, 56*(1), 247–277.

Frazzini, A., & Lamont, O. A. (2008). Dumb money: Mutual fund flows and the cross-section of stock returns. *Journal of Financial Economics, 88*(2), 299–322.

Friewald, N., Jankowitsch, R., & Subrahmanyam, M. G. (2012). Illiquidity or credit deterioration: A study of liquidity in the US corporate bond market during financial crises. *Journal of Financial Economics, 105*(1), 18–36.

Grier, P., & Katz, S. (1976). The differential effects of bond rating changes among industrial and public utility bonds by maturity. *Journal of Business, 49,* 226–239.

Han, S., & Zhou, X. (2014). Informed bond trading, corporate yield spreads, and corporate default prediction. *Management Science, 60*(3), 675–694.

Hite, G., & Warga, A. (1997). The effect of bond-rating changes on bond price performance. *Financial Analysts Journal, 53*(3), 35–51.

Hull, J., Predescu, M., & White, A. (2004). The relationship between credit default swap spreads, bonds yields, and credit rating announcements. *Journal of Banking & Finance, 28*(11), 2789–2811.

Jankowitsch, R., Nagler, F., & Subrahmanyam, M. G. (2014). The determinants of recovery rates in the US corporate bond market. *Journal of Financial Economics, 114,* 155–177.

Jorion, P., & Zhang, G. (2007). Good and bad credit contagion: Evidence from credit default swaps. *Journal of Financial Economics, 84,* 860–883.

Kedia, S., & Zhou, X. (2014). Informed trading around acquisitions: Evidence from corporate bonds. *Journal of Financial Markets, 18,* 182–205.

Mahanti, S., Nashikkar, A., Subrahmanyam, M. G., Chacko, G., & Mallik, G. (2008). Latent liquidity: A new measure of liquidity, with an application to corporate bonds. *Journal of Financial Economics, 88*(2), 272–298.

May, A. D. (2010). The impact of bond rating changes on corporate bond prices: New evidence from the over-the-counter market. *Journal of Banking & Finance, 34*(11), 2822–2836.

Ng, L., & Wu, F. (2007). The trading behavior of institutions and individuals in Chinese equity markets. *Journal of Banking & Finance, 31*(9), 2695–2710.

Norden, L., & Weber, M. (2004). Informational efficiency of credit default swap and stock markets: The impact of credit rating announcements. *Journal of Banking & Finance, 28,* 2813–2843.

Ronen, T., & Zhou, X. (2013). Trade and information in the corporate bond market. *Journal of Financial Markets, 16,* 61–103.

Sarig, O., & Warga, A. (1989). Bond price data and bond market liquidity. *Journal of Financial and Quantitative Analysis, 24*(3), 367–378.

Sheskin, D. J. (1997). *Handbook of parametric and nonparametric statistical procedures*. New York: CRC Press.

Steiner, M., & Heinke, V. G. (2001). Event study concerning international bond price effects of credit rating actions. *International Journal of Finance and Economics, 6,* 139–157.

The Market and Individual Pricing Kernels Under No Arbitrage Asset Pricing Models

Thomas F. Cosimano and Jun Ma

Abstract This chapter discusses how to use the No Arbitrage Asset Pricing Model (NAAPM) to determine the pricing kernel for both the financial markets and an individual with a given degree of constant risk aversion over her terminal wealth. The existence of the market and individual pricing kernel allows us to value any financial contract whose payoff is dependent on the prices from the NAAPM. Consequently, an individual would raise her lifetime utility by buying the assets which she prices higher than the financial market.

Keywords Pricing kernel · No arbitrage

Journal of Economic Literature Classification Numbers G11 · G12 · G13

1 Introduction

The pricing of financial instruments is accomplished using a function which operates on the payoff of the instrument to determine the current market price. This chapter develops the pricing kernel given a No Arbitrage Asset Pricing Model (NAAPM) and applies the analysis to a standard term structure model. This pricing kernel is expressed as a Guassian function of the current pricing factors to represent the conditional expectation of the pricing kernel, and a log-normal probability distribution for the transitional probability from the current factors to the future random factors. This probability function is found by applying the Forward Kolmogorov Equation (FKE) which leads to a linear partial differential equation (PDE) whose solution is

T. F. Cosimano retired from University of Notre Dame

T. F. Cosimano (✉)
University of Notre Dame, Notre Dame, IN, USA
e-mail: T.F.Cosimano@gmail.com

J. Ma
Department of Economics, Northeastern University, Boston, MA 02115, USA
e-mail: ju.ma@northeastern.edu

M. Mili et al. (eds.), *New Methods in Fixed Income Modeling*, Contributions to Management Science, https://doi.org/10.1007/978-3-319-95285-7_15

the transitional probability function. The solution to this PDE is a Guassian function such that the future factors have a log-normal distribution with mean zero and finite variance-covariance matrix. Thus the pricing kernel under a NAAPM has a conditional expected value, which is a Guassian function of the pricing factors, and a log-normally distributed transitional probability from the current pricing factors to the future value of these factors.

The valuation of a financial instrument by an individual is also developed given the NAAPM. Suppose an individual has a constant relative risk return over wealth with a given investment horizon and a leverage restriction. In addition, the holding period return follows the NAAPM. In this case, the portfolio analysis of Sangvinatsos and Wachter (2005) and Liu (2007) is extended to incorporate the investor's leverage restriction. This analysis yields a portfolio rule which is a linear relation in the expected holding period return under the NAAPM, the leverage restriction, and the elasticity of the lifetime utility with respect to the pricing factors. Given the portfolio rule, the expected lifetime utility of the investor is the solution of a linear PDE. This solution is a Guassian function of the current pricing factors, so that the portfolio rule is linear in only the pricing factors. With the solution for the investor's expected lifetime utility and portfolio rules, Ito's lemma is used to derive the stochastic process for the investor's wealth and the lifetime utility. These stochastic processes have the same functional form as the pricing kernel for the NAAPM. Consequently, the exact same procedure is applied to split these stochastic processes into a Guassian conditional expectation and a normal transitional probability from the current pricing factors to the future value of these factors. Finally, the intertemporal rate of substitution from the current pricing factors to the future factors is derived. This corresponds to a pricing kernel for an individual which is independent of the investor's wealth. In addition, the pricing kernel is the product of a Guassian function of the current pricing factors and a log-normal transitional probability for the investor's lifetime utility. Consequently, any financial contract, which is a function of the assets priced under the NAAPM, can have both a market price and an individual price. If the individual values the asset more (less) than the market, then it would add (deduct) value to the individual's lifetime utility.

The NAAPM was developed by Duffie et al. (2000) with initial application to the term structure by Duffie and Kan (1996), and Duffie and Singleton (1997). Dai and Singleton (2000, 2002) developed the identification strategy for estimating term structure models. A complete survey of this research can be found in Piazzesi (2010). Joslin et al. (2011), and Hamilton and Wu (2012a, 2014a) introduce procedures to improve the estimation of these models. For example Joslin, Singleton and Zhu showed that the factors can be estimated using a VAR model, so that only yield curve and risk premium parameters need to be estimated by maximum likelihood estimation. Adrian et al. (2013) develop a three step regression procedure which focus on matching the holding period return rather than the yield to maturity. The pricing kernel developed in this chapter can be developed for any of these methods as long as the shocks to the yield curve factors are Guassian.[1]

[1] If time varying variance-covariance matrices are introduced then all the PDEs in this chapter will be more complicated in that the coefficients on the second order derivatives will not be constant.

The NAAPM has been used to study: (1) The interconnection among the yield curve and macroeconomic variables[2]; (2) Expected inflation and real rates on treasury securities[3]; (3) Interpretation of monetary policy[4]; (4) The zero lower bond and the yield curve[5]; (5) Bond risk premium[6]; (6) Crude oil future prices[7]; (7) Swap rates and credit quality[8]; (8) Derivatives for fixed income securities.[9] (9) Quantitative easing and the term structure.[10] The pricing kernel for NAAPM developed here can be used to price all these financial instruments and help to interpret their properties.

2 The Market Pricing Kernel

Consider the typical No Arbitrage Asset Pricing Model (NAAPM), which postulates that several latent factors drive all returns on marketable securities, in such a coherent way that no arbitrage is permitted.[11] These underlying latent factors, $X(s)$, are typically assumed to follow a mean-reverting stochastic process. Specifically, the dynamics for the factors under the physical probability distribution are given by

$$dX(s) = \left(\gamma^{\mathcal{P}} - A^{\mathcal{P}} X(s)\right) ds + \Sigma_X d\epsilon_s. \tag{1}$$

where, the N by 1 vector $X(s)$ contains N latent factors, the standard Brownian motion ϵ_s summarizes the uncertainty in the interest rate factors $X(s)$. The vector $\gamma^{\mathcal{P}}$ and the matrix $A^{\mathcal{P}}$ contain model parameters under the physical probability measure. In particular, the stationary mean of the factors is given by $\left(A^{\mathcal{P}}\right)^{-1}\gamma^{\mathcal{P}}$, and $A^{\mathcal{P}}$ determines the speed of mean-reversion. The matrix $\Sigma_X \Sigma_X'$ is the variance-covariance matrix of the shocks, $d\epsilon_s$, to the interest rate factors.

Because the dynamics of the latent interest rate factors are written in a continuous-time process while we observe data at discrete-time intervals, it is helpful to solve (1) for the interest rate factors, over the time interval τ, relative to its stationary value, $\bar{X}$,

$$X(t+\tau) - \bar{X} = e^{-A^{\mathcal{P}}\tau}\left(X - \bar{X}\right) + Y_\tau, \tag{2}$$

---

[2]See Ang and Piazzesi (2003), Joslin et al. (2014), and references in Bauer and Rudebusch (2017).

[3]Ang et al. (2008a), and Chernov and Mueller (2012).

[4]Ang and Piazzesi (2003), and Ang et al. (2008b).

[5]Hamilton and Wu (2012b, 2016), Krippner (2015), Bauer and Rudebusch (2016), and Wu and Xia (2016).

[6]Cochrane and Piazzesi (2005), Adrian et al. (2013), and Greenwood and Vayanos (2014).

[7]Hamilton and Wu (2014b).

[8]Duffie and Huang (1996), and Duffie and Singleton (1997).

[9]Grinblatt and Longstaff (2000), and Longstaff et al. (2001).

[10]See Li and Wei (2013) for a survey.

[11]The NAAPM model is not limited to the yield on zero coupon bonds. Liu (2016), and Durham (2013) use these methods to model both bonds and stocks. Yung (2017), and Durham (2015) use the NAAPM approach to model foreign currency and its forward price.

where

$$Y_\tau = \int_0^\tau e^{-A^{\mathcal{P}}(\tau-s)} \Sigma_X d\epsilon_s. \tag{3}$$

The first term in (2) captures the part of the deviation of the current interest rate factors from its stationary value that is expected to mean-revert as long as all eigenvalues of the matrix $A^{\mathcal{P}}$ are positive. The second term is the random shock to the interest rate factors from time t to $t+\tau$. This random shock can be shown to have a normal probability distribution with mean 0 and variance covariance matrix $\sigma_Y(\tau)$.[12]

For the purpose of this chapter let the assets be zero coupon Treasury securities with yield to maturity $r_{\tau,s}(X(s))$, where s is the time at which the yield is observed and τ is the maturity of the yield. The Treasury yield to maturity is specified as an affine function of the latent factors $X(s)$

$$r_{\tau,s}(X(s)) = A_\tau + B_\tau X(s). \tag{4}$$

where, the matrices of parameters A_τ and B_τ for each yield to maturity are solutions to a set of differential equations in a coherent way so that no arbitrage opportunity is permitted for investors in the financial markets.

To understand the affine structure of NAAPM, note that all yields to maturity conceptually depend on the risk free rate and the risk premium. First, the risk free interest rate $r(s)$ is assumed to be a linear function of the latent factors:

$$r(s) \equiv r(X(s)) = \delta_0 + \delta_1 X(s). \tag{5}$$

Here, the scalar δ_0 and the vector δ_1 are model parameters.

Furthermore, the risk price in the NAAPM is also assumed to be affine in the latent factors.

$$\Lambda(X(s)) = \lambda_0 + \lambda_1 X(s), \tag{6}$$

This specification of the risk price yields the so-called essentially affine model leading to the affine structure in (4).

The model specifications so far imply a risk-neutral probability distribution of the latent factors through a change of measure which accounts for the price of risk. As a result, the dynamics of the process for the factors, $X(s)$, under the risk-neutral distribution, is

$$dX(s) = \left(\gamma^{\mathcal{Q}} - A^{\mathcal{Q}} X(s)\right) ds + \Sigma_X d\epsilon_s^{\mathcal{Q}}. \tag{7}$$

[12]See Arnold (1974) for proof. The variance-covariance matrix is the solution to a Ricatti differential equation. The solution is found by using recursive rules, which are implemented in the lyap subroutine in Matlab with inputs $A^{\mathcal{P}}$ and Σ_X.

Note first that the variance-covariance matrix in this risk-neutral process remains the same as in the physical process, $\Sigma_X \Sigma_X'$. However, the vector $\gamma^{\mathcal{Q}}$ and the matrix $A^{\mathcal{Q}}$ are the risk adjusted parameters of the corresponding parameters in the physical process, through a change of variable using the risk price

$$\gamma^{\mathcal{Q}} = \gamma^{\mathcal{P}} - \Sigma_X \lambda_0 \text{ and } A^{\mathcal{Q}} = A^{\mathcal{P}} + \Sigma_X \lambda_1. \tag{8}$$

The yields to maturity are affine functions of latent factors implies that the bond prices are exponentially affine in latent factors

$$P_{\tau,s} = \exp\left[a_\tau + b_\tau \cdot X(s)\right]. \tag{9}$$

where, $a_\tau = -\tau A_\tau$ and $b_\tau = -\tau B_\tau$.
The holding period return on a zero coupon bond maturing at τ is given by

$$\frac{dP_{\tau,s}}{P_{\tau,s}} = \left[b_\tau \left((\gamma^{\mathcal{P}} - \gamma^{\mathcal{Q}}) - (A^{\mathcal{P}} - A^{\mathcal{Q}})X(s)\right) + r(s)\right] ds + b_\tau \Sigma_X d\epsilon_s. \tag{10}$$

The pricing kernel under NAAPM is given by

$$\begin{aligned} \frac{M_{\tau,t}}{M_{t,t}} &= \exp\left\{-\int_t^{t+\tau}\left[r(X(s)) + \frac{1}{2}\Lambda(X(s))'\Lambda(X(s))\right]ds + \int_t^{t+\tau}\Lambda(X(s))'d\epsilon_s\right\} \\ &= \exp\left\{\int_0^T \left(-\mathcal{M}_1 - \frac{1}{2}\left(X_s'\mathcal{M}_3X_s - 2\mathcal{M}_2X_s\right)\right)\right]ds + \int_t^T (\mathcal{M}_4 + \mathcal{M}_5X_s)\,d\epsilon_s\right\}. \end{aligned} \tag{11}$$

We use the risk free rate, the risk premium and the risk neutral coefficients in this derivation, so that the constants are given by

$$\begin{aligned} \mathcal{M}_1 &\equiv \delta_0 + \frac{1}{2}(\gamma^{\mathcal{P}} - \gamma^{\mathcal{Q}})'\left(\Sigma_X'\Sigma_X\right)^{-1}(\gamma^{\mathcal{P}} - \gamma^{\mathcal{Q}}), \\ \mathcal{M}_2 &\equiv -\left[\delta_1 - (\gamma^{\mathcal{P}} - \gamma^{\mathcal{Q}})'\left(\Sigma_X'\Sigma_X\right)^{-1}(A^{\mathcal{P}} - A^{\mathcal{Q}})\right], \\ \mathcal{M}_3 &\equiv (A^{\mathcal{P}} - A^{\mathcal{Q}})'\left(\Sigma_X'\Sigma_X\right)^{-1}(A^{\mathcal{P}} - A^{\mathcal{Q}}), \\ \mathcal{M}_4 &\equiv (\gamma^{\mathcal{P}} - \gamma^{\mathcal{Q}})'\left(\Sigma_X'\right)^{-1} \text{ and } \mathcal{M}_5 \equiv -(A^{\mathcal{P}} - A^{\mathcal{Q}})'\left(\Sigma_X'\right)^{-1}. \end{aligned}$$

The pricing kernel is a random variable dependent on the solution to the yield curve factors (2). We want to find the probability density function for the pricing kernel (11) using the Forward Kolmogorov Equation (FKE). Represent the transition probability from state X at time t to the state Y at time T by $p(t, X, T, Y)$. For the stochastic process (11) let

$$\phi(t, s) = \exp\left\{-\frac{1}{2}\int_t^s \left[X_v'\mathcal{M}_3(v)X_v - 2\mathcal{M}_2(v)X_v\right]dv\right\}.$$

For fixed (t, X) the function

$$g(\tau, Y) \equiv \phi(t, \tau)p(t, X, \tau, Y) \tag{12}$$

solves the FKE.[13]

$$\frac{\partial g(\tau, Y)}{\partial \tau} = \mathcal{K}_Y^* g(\tau, Y) - \frac{1}{2}\left(Y'\mathcal{M}_3(\tau)Y - 2\mathcal{M}_2(\tau)Y\right) g(\tau, Y). \tag{13}$$

Here, the dual of $\mathcal{K}_X$ given by[14]

$$\begin{aligned}\mathcal{K}_X^* &= -\sum_{i=1}^{N} \frac{\partial}{\partial X_i}\left(\gamma^{\mathcal{P}} - A^{\mathcal{P}}X\right)_i + \frac{1}{2}\sum_{i,j=1}^{N} \frac{\partial^2}{\partial X_i \partial X_j}\Sigma_{ik}\Sigma'_{kj} \\ &= -\gamma^{\mathcal{P}\prime}\frac{\partial}{\partial X} + X'A^{\mathcal{P}\prime}\frac{\partial}{\partial X} + Trace(A^{\mathcal{P}}) + \frac{1}{2}Trace\left(\Sigma\Sigma'\frac{\partial^2}{\partial X \partial X}\right).\end{aligned} \tag{14}$$

Remark Notice that only the distribution of the factors enters (14). The preferences of the investor only influences the discount factor $\phi(t, \tau)$.

To find the initial condition, let the Dirac distribution centered at $X \in \mathbb{R}^N$ be $f(X) = \delta_X$ such that

$$\delta_X(\theta) = \int_{\mathbb{R}^N} \delta_x(Y)\theta(Y)dY = \theta(X).$$

For a given $X_t = X \in \mathbb{R}^N$,

$$g(\tau, X) = \int_{\mathbb{R}^N} \delta_X(Y)\phi(t, \tau)p(t, X, \tau, Y)dY = \phi(t, \tau)p(t, X, \tau, X).$$

Consequently, if the initial condition for the Kolmogorov forward equation (13) is

$$\lim_{\tau \to 0^+} g(\tau, X(\tau)) = \delta_X, \tag{15}$$

then the solution to (13) is $\phi(t, \tau)p(t, X, \tau, Y) = g(\tau, Y)$.

Thus, we have

Theorem 2.1 *The discounted transition probability $\phi(t, \tau)p(t, X, \tau, Y)$ for a given $X_t = X \in \mathbb{R}^N$ is the solution to the Kolmogorov Forward equation* (13) *with* (14) *subject to the initial condition* (15).

Proof See Appendix. ■

---

[13] See Karatzas and Shreve (1988, p. 369) equation (7.24). Also see Theorem 8.7.1. of Calin et al. (2011), and Chirikjian (2009, pp. 118–121).

[14] See Øksendal (2005, p. 169). Also follow the derivation in Chirikjian (2009, p. 121).

The solution to the FKE is difficult to find given the Dirac initial condition (15). To circumvent this problem we use the Fourier transform of the FKE problem, since the Fourier transform of (15) is 1.
Suppose that $f(X) \in \mathcal{S}(\mathbb{R}^N)$, on $\mathbb{R}^N$. This functional space refers to all functions which rapidly decrease, so that $f(X)$ is absolutely integrable over $\mathbb{R}^N$. This allows one to move between Fourier transforms and its inverse. The Fourier transform of $f(X)$ is

$$F[f(X)] = \hat{f}(\xi) = \int_{-\infty}^{\infty} f(X)e^{-i\xi' X}\, dX. \tag{16}$$

Here $\xi \in \mathbb{R}^N$ and $\xi \cdot X \equiv \xi' X = \xi_1 X_1 + \cdots + \xi_N X_N$.
The inverse Fourier transform of $\hat{f}(\xi)$ is

$$F^{-1}[\hat{f}(\xi)] = f(X) = \frac{1}{(2\pi)^N}\int_{-\infty}^{\infty} \hat{f}(\xi)e^{i\xi \cdot X}\, d\xi. \tag{17}$$

In the appendix we apply the Fourier transform to the FKE problem to yield the linear partial differential equation

$$\begin{aligned} &\frac{\partial F\,[g(\tau, Y)]}{\partial \tau} + \frac{1}{2}\xi'\Sigma\Sigma'\xi F[g(\tau, Y)] + i\gamma^{\mathcal{P}\prime}\xi F\,[g(\tau, Y)] \\ &- \left(\frac{\partial F[g(\tau, Y)]}{\partial \xi}\right)' \left(i\mathcal{M}_2(\tau)' - A^{\mathcal{P}\prime}\xi\right) + \frac{1}{2} Trace\left(\mathcal{M}_3(\tau)\frac{\partial^2 \hat{g}(\xi)}{\partial \xi \partial \xi}\right) = 0 \end{aligned} \tag{18}$$

subject to the initial condition

$$F\,[g(0, Y_0)] = 1.$$

We use a guess and verify procedure to find its solution.

$$F\,[g(\tau, Y)] = \exp\left\{ -\frac{1}{2}\left[\xi'\mathcal{G}_3(\tau)\xi - 2i\mathcal{G}_2(\tau)'\xi + \mathcal{G}_1(\tau)\right]\right\}, \tag{19}$$

We do not have to assume the matrix is symmetric, since $\frac{1}{2}\xi'\left(\mathcal{G}_3(\tau) + \mathcal{G}_3(\tau)'\right)\xi = \xi'\mathcal{G}_3(\tau)\xi$.
The coefficients in (19) are the solution to three ordinary differential equations (ODE).

$$\frac{\partial \mathcal{G}_3(\tau)}{\partial \tau} = \mathcal{G}_3(\tau)\mathcal{M}_3(\tau)\mathcal{G}_3(\tau) - 2\mathcal{G}_3(\tau)A^{\mathcal{P}\prime} + \Sigma_X\Sigma_X' \tag{20}$$

subject to

$$\mathcal{G}_3(0) = 0_{N\times N}.$$

$$\frac{\partial \mathcal{G}_2(\tau)}{\partial \tau} = \mathcal{M}_2(\tau)\left(\mathcal{M}_3(\tau)\mathcal{G}_3(\tau) - A^{\mathcal{P}\prime}\right) - \gamma^{\mathcal{P}\prime} - \mathcal{M}_2(\tau)\mathcal{G}_3(\tau) \tag{21}$$

subject to

$$\mathcal{G}_2(0) = 0_N.$$

$$\frac{\partial \mathcal{G}_1(\tau)}{\partial \tau} = 2\mathcal{G}_2(\tau)\mathcal{M}_2(\tau)' - \mathcal{G}_2(\tau)\mathcal{M}_3(\tau)\mathcal{G}_2(\tau)' - Trace\left(\mathcal{M}_3(\tau)\mathcal{G}_3(\tau)\right) \tag{22}$$

subject to

$$\mathcal{G}_1(0) = 0.$$

The solutions to these three ODEs are found using the ODE solver in Matlab. Given the solution to the Fourier transform to the FKE problem, the inverse Fourier transform yields the solution to the FKE problem.

$$\begin{aligned} g(\tau, Y) = & \frac{1}{\sqrt{(2\pi)^N\, det(\mathcal{G}_3(\tau))}} \\ & \times \exp\left\{ -\frac{1}{2}\mathcal{G}_1(\tau) - \frac{1}{2}\left(\mathcal{G}_2(\tau)' - Y\right)' \mathcal{G}_3(\tau)^{-1}\left(\mathcal{G}_2(\tau)' - Y\right)\right\}. \end{aligned} \tag{23}$$

Thus, the discounted transition probability is a Guassian function of the future yield curve factors Y.

The final step in determining both the conditional expectation and probability distribution for the pricing kernel is to use the solution to the random factors (2) in (11) to express the pricing kernel in terms of the current factors X and the random future factors Y_s for $t < s < T$. In the Appendix it is shown that the pricing kernel is given by

$$\frac{M_{T,t}}{M_{t,t}} = \mathcal{M}(\tau, X)\exp\left\{ -\frac{1}{2}\int_t^{t+\tau} Y_s'\mathcal{M}_3 Y_s + \int_t^{t+\tau}\left(\mathfrak{M}_4 + X_t'\mathfrak{M}_5 + Y_s'\mathcal{M}_5\Sigma_X'\right) d\epsilon_s\right\}. \tag{24}$$

Here, the conditional pricing kernel is derived in the Appendix, and is given by

$$\mathcal{M}(\tau, X) \equiv \exp\left\{ -\frac{1}{2}\left(X - \mu_{\mathcal{M}}(\tau)\right)'\sigma_{\mathcal{M}}^{-1}\left(X - \mu_{\mathcal{M}}(\tau)\right) + \frac{1}{2}\mathfrak{M}_2'\mathfrak{M}_3^{-1}\mathfrak{M}_2 + \mathfrak{M}_1\right\} \tag{25}$$

$$
\begin{aligned}
\mu_{\mathcal{M}}(\tau) &\equiv \mathfrak{M}_3^{-1}\mathfrak{M}_2, \qquad \sigma_{\mathcal{M}} \equiv \mathfrak{M}_3^{-1} \\
\mathfrak{M}_1 &\equiv -\mathcal{M}_1(\tau)\tau - \frac{1}{2}\gamma^{\mathcal{P}\prime}\left(A^{\mathcal{P}\prime}\right)^{-1}\mathcal{M}_3\left(A^{\mathcal{P}}\right)^{-1}\gamma^{\mathcal{P}}\tau \\
&\quad + \gamma^{\mathcal{P}\prime}\left(A^{\mathcal{P}\prime}\right)^{-1}\left(A^{\mathcal{P}\prime}\right)^{-1}\left[I - e^{-A^{\mathcal{P}\prime}\tau}\right]\mathcal{M}_3\left(A^{\mathcal{P}}\right)^{-1}\gamma^{\mathcal{P}} \\
&\quad - \frac{1}{2}\gamma^{\mathcal{P}\prime}\left(A^{\mathcal{P}\prime}\right)^{-1}\left[\mathcal{M} - e^{-A^{\mathcal{P}\prime}\tau}\mathcal{M}e^{-A^{\mathcal{P}}\tau}\right]\left(A^{\mathcal{P}}\right)^{-1}\gamma^{\mathcal{P}}, \\
\mathfrak{M}_2 &\equiv \gamma^{\mathcal{P}\prime}\left(A^{\mathcal{P}\prime}\right)^{-1}\left[\mathcal{M} - e^{-A^{\mathcal{P}\prime}\tau}\mathcal{M}e^{-A^{\mathcal{P}}\tau}\right] + \mathcal{M}_2\left(A^{\mathcal{P}}\right)^{-1}\left[I - e^{-A^{\mathcal{P}}(\tau)}\right] \\
&\quad - \gamma^{\mathcal{P}\prime}\left(A^{\mathcal{P}\prime}\right)^{-1}\mathcal{M}_3\left(A^{\mathcal{P}}\right)^{-1}\left[I - e^{-A^{\mathcal{P}}\tau}\right], \\
\mathfrak{M}_3 &\equiv \mathcal{M} - e^{-A^{\mathcal{P}\prime}\tau}\mathcal{M}e^{-A^{\mathcal{P}}\tau}, \\
\mathfrak{M}_4 &\equiv \Sigma_X\mathcal{M}_5'\left[I - e^{-A^{\mathcal{P}}(\tau - t)}\right]\left(A^{\mathcal{P}}\right)^{-1}\gamma^{\mathcal{P}} \text{ and } \mathfrak{M}_5 \equiv e^{-A^{\mathcal{P}\prime}(s-t)}\mathcal{M}_5.
\end{aligned}
$$

In this case the probability distribution of Y_τ is normally distributed with mean 0 and variance covariance matrix $\sigma_Y(\tau)$, where σ_Y is the solution to the Lypunov equation By exercise (1.2.11) of Hijab (1987)

$$
\sigma_Y(\tau) = \sigma_{Y\infty} - e^{-A^{\mathcal{P}}\tau}\sigma_{Y\infty}e^{-A^{\mathcal{P}\prime}\tau}.
$$

Here, the matrix $\sigma_{Y\infty}$ solves the Lyapunov equation

$$
-A^{\mathcal{P}}\sigma_{Y\infty} - \sigma_{Y\infty}A^{\mathcal{P}\prime} = \Sigma_X\Sigma_X'.
$$

As the time horizon tends to infinity, $\sigma_Y(\tau) \to \sigma_{Y\infty}$ given the eigenvalues of $A^{\mathcal{P}}$ are positive. The solution to this equation is a positive definite symmetric matrix, which is easily calculated using lyap.m in Matlab.
For solving the FKE we use $A^{\mathcal{P}} = 0$ and $\gamma^{\mathcal{P}} = 0$, since Y_τ is white noise. In addition, $\mathcal{M}_2 = 0$, since a linear term in Y_τ is not present in (24). In this case, $\mathcal{G}_2(\tau) = 0$ is the solution to (21). It is also the case that $-\mathcal{G}_1(\tau) > 0$ by (22) in this situation, since $\mathfrak{M}_3 > 0$. As a result, the transition probability in (23) has a log-normal distribution with with parameters 0 and σ_M given by the solution to (20). Thus, the conditional probability distribution for the pricing kernel is

$$
\frac{M_{\tau,t}}{M_{t,t}} = \frac{\mathcal{M}(\tau, X)}{\sqrt{(2\pi)^N\, det(\sigma_{\mathcal{M}})}} \exp\left\{-\frac{1}{2}Y'\sigma_M^{-1}Y\right\}. \tag{26}
$$

Here, we include the expected value of this distribution, $e^{-\frac{1}{2}\mathcal{G}_1(\tau)}$, in $\mathfrak{M}_1 > 0$, since they are both constants independent of X and Y.

Thus, the financial market represented by the NAAPM yields a pricing kernel (25) and (26). This pricing kernel is a combination of a Guassian form in the current factors for the conditional expectation and a log-normal probability function for the future factors.

3 Investor's Pricing Kernel

To find the pricing kernel for an investor given holding period returns follows a NAAPM we examine the optimal behavior of an investor. The investor is assumed to be risk averse with a constant relative risk aversion utility (CRRA) with parameter γ. This investor maximizes the expected utility from terminal wealth at a fixed time $\tau = T - t$ given her current wealth, $W(t) = W$ and yield curve factors, $X(t) = X$. The investment horizon of this investor is τ. The investor's conditional expected value is

$$J(W, X, \tau, t) = e^{-\beta\tau} E\left[\frac{(W(\tau))^{1-\gamma}}{1-\gamma}\middle| W(t) = W, X(t) = X\right], \tag{27}$$

where β is the discount rate for the investor.
Suppose the investor trades N marketable securities such that

$$\omega(s)'\iota + \omega_{1\tau}(s) = \xi \tag{28}$$

for $s \in [t, t + \tau]$. We use the following vector notation

$$\omega'(s) = \left(\omega_{2\tau}(s) \cdots \omega_{N\tau}(s)\right), \quad b' = \left(b_{2\tau} \cdots b_{N\tau}\right) \text{ and } \iota' = \left(1 \cdots 1\right). \tag{29}$$

Here, ξ is the leverage ratio, so that $1 - \xi$ represents the amount of wealth $W(s)$ invested in the risk free asset, $\omega_i(s)$ is the percentage wealth invested in assets $i = 1, \cdots N$.
The change in the investor's wealth is

$$\frac{dW(s)}{W(s)} = (1 - \xi)\mu_{1\tau}(s) + \sum_{i=2}^{N} \mu_{i\tau}(s)\omega_i(s) + \sum_{i=1}^{N} \omega_i(s) b'_{i_\tau} \Sigma_X d\epsilon_s. \tag{30}$$

The instantaneous expected excess rates of return on marketable securities, from (10), are

$$\begin{aligned} \mu_{1\tau}(s) - r(s) &\equiv b'_{\tau}\left[(\gamma^{\mathcal{P}} - \gamma^{\mathcal{Q}}) - (A^{\mathcal{P}} - A^{\mathcal{Q}})X(s)\right] \\ \mu_{i\tau}(s) - \mu_{1\tau}(s) &\equiv \left(b'_{i_\tau} - b'_{\tau}\right)\left[(\gamma^{\mathcal{P}} - \gamma^{\mathcal{Q}}) - (A^{\mathcal{P}} - A^{\mathcal{Q}})X(s)\right], i = 2, \cdots, N. \end{aligned} \tag{31}$$

The investor's problem extends the analysis of Sangvinatsos and Wachter (2005) and Liu (2007) to account for the leverage constraint. Chami et al. (2017) uses their procedure to derive the solution to the individual's problem.

$$J(W,X,\tau,t) = \frac{(W(t))^{1-\gamma}}{1-\gamma} h(\tau,X)^{\gamma}, \tag{32}$$
$$\text{where } h(\tau,X) = h(\tau)\exp\left\{-\frac{1}{2}(X-\mu_J(\tau))'(\sigma_J(\tau))^{-1}(X-\mu_J(\tau))\right\}.$$

Given the solution, the individual's portfolio rule is given by

$$\begin{aligned}
\omega(t) &= \omega_1\left\{(b-\iota b_\tau)\left[(\gamma^{\mathcal{P}}-\gamma^{\mathcal{Q}}) - (A^{\mathcal{P}}-A^{\mathcal{Q}})X(t)\right]\right\} + \omega_2\xi \\
&\quad + \omega_3\gamma(\sigma_J(\tau))^{-1}[X-\mu_J(\tau)] \\
\omega_1 &\equiv \left[\gamma\left(b\Sigma_X\Sigma_X'b' + \iota\iota' b_\tau\Sigma_X\Sigma_X'b_\tau' - 2b\Sigma_X\Sigma_X'b_\tau'\iota'\right)\right]^{-1} \text{ with } \iota' = (1,\ldots,1), \\
\omega_2 &\equiv 2\omega_1\left(b\Sigma_X\Sigma_X'b_\tau' - \iota b_\tau\Sigma_X\Sigma_X'b_\tau'\right) \text{ and } \omega_3 \equiv \omega_1(b-\iota b_\tau)\Sigma_X\Sigma_X'. \\
\omega_1(t) &= \xi - \iota'\omega(t).
\end{aligned} \tag{33}$$

Consequently, the portfolio rule is linear in the yield curve factors.
Given the lifetime utility of the investor, the valuation of an investment by an individual can be analyzed. We can find the stochastic process for future lifetime utility by applying Ito's lemma to (32) given the stochastic process from wealth (30), the return on zero coupon bonds (31), the optimal portfolio rule (33) and the stochastic process for the yield curve factors (1).

$$\begin{aligned}
J(W,X(t+\tau),\tau) = \frac{(W)^{1-\gamma}}{1-\gamma} h(0,X)^{\gamma}\exp\Bigg\{&\int_0^{\tau}\left(\mathcal{J}_1 - \frac{1}{2}\left(X(s)'\mathcal{J}_3X(s) - 2\mathcal{J}_2\right)\right)ds \\
&- \int_0^{\tau}\left(\mathcal{J}_4 + X(s)'\mathcal{J}_5\right)d\epsilon_s\Bigg\}.
\end{aligned} \tag{34}$$

The coefficients in this stochastic process are stated in Chami et al. (2017). The stochastic process for lifetime utility (34) also has the same functional form as (24) with different coefficients. Consequently, it can be split into a conditional expectation of lifetime utility as in (24).

$$\begin{aligned}
E_t(J(W,X(t+\tau),\tau)) \equiv \mathcal{J}(W,X,\tau) &= \frac{(W)^{1-\gamma}}{1-\gamma} h(0,X)^{\gamma} \\
&\times \mathcal{J}(\tau)\exp\left\{-\frac{1}{2}\left(X-\mu_{\mathcal{J}}(\tau)\right)'(\sigma_{\mathcal{J}}(\tau))^{-1}\left(X-\mu_{\mathcal{J}}(\tau)\right)\right\},
\end{aligned} \tag{35}$$

and a transitional probability for lifetime utility from the current yield curve factors X at time t to the random yield curve factors Y at time $t+\tau$,

$$p_J(t,X,\tau,Y) = \frac{\exp\left\{-\frac{1}{2}Y'\sigma_J(\tau)^{-1}Y\right\}}{\sqrt{(2\pi)^N\, det(\sigma_J(\tau))}}. \tag{36}$$

We can therefore write the future lifetime utility as

$$J(W, X(t+\tau), \tau) = \mathcal{J}(W, X, \tau) p_J(t, X, \tau, Y), \tag{37}$$

so that the future marginal utility of wealth is

$$\frac{\partial J(W, X(t+\tau), \tau)}{\partial W} = \frac{\partial \mathcal{J}(W, X, \tau)}{\partial W} p_J(t, X, \tau, Y). \tag{38}$$

The current marginal utility of wealth is given by

$$\frac{\partial J(W, X, \tau)}{\partial W} = \frac{(W)^{-\gamma}}{1-\gamma} h(0, X)^{\gamma}, \tag{39}$$

Thus, the intertemporal rate of substitution or pricing kernel for the individual investor is

$$\begin{aligned} \mathcal{P}(t, X, \tau, Y) = {} & \frac{h(\tau, X)^{\gamma}}{h(0, X)^{\gamma}} \mathcal{J}(\tau) \\ & \times \exp\left\{ -\frac{1}{2} \left(X - \mu_{\mathcal{J}}(\tau) \right)' \left(\sigma_{\mathcal{J}}(\tau) \right)^{-1} \left(X - \mu_{\mathcal{J}}(\tau) \right) \right\} p_J(t, X, \tau, Y). \end{aligned} \tag{40}$$

This corresponds to the pricing kernel for an investor with a given degree of risk aversion, γ and leverage ratio ξ, so that any financial payoff can be priced given the characteristics of the investor.

4 Conclusion

This chapter shows how the financial market would price financial assets under the No Arbitrage Asset Pricing Model (NAAPM). The analysis in this chapter can be used to address multiple financial economic problems. An example of this is Chami et al. (2017) which develops a model of a bank holding company (BHC) with an active trading desk. The trading desk invests in marketable securities, so as to maximize the expected lifetime utility subject to a leverage constraint which is imposed by the Chief Operating Officer (COO) of the BHC. The trading desk's wealth is given to her by the COO. Given the trading desk's closed form solution, the COO can solve the optimal decisions of the loan officer.

Chami et al. (2018) develop a model of the treasury market based on the NAAPM pricing kernel discussed here. In this work they analyze the relation between monetary policy and the term structure. It is shown that the impact of monetary policy is dependent on the current yield curve factors. In particular, if these factors are below the mean of the pricing kernel, then an increase in these factors lead to an increase in the pricing kernel rather than the traditional decrease. This property will lead to similar impacts of factors on the pricing kernel for the NAAPM so that problems, such as optimal corporate investment, and the evaluation of funding value adjustments by derivative dealers for swap books, can be influenced by the NAAPM factors.[15] Finally, Yung (2017), and Cosimano and Yung (2018) show how the NAAPM pricing kernel can be used to model and explain exchange rate movements.

Appendix

In this section we find the probability distribution for terms like (34). The yield curve factors follow the Ornstein-Uhlembeck process (3) in the paper.

$$dX(s) = \left(\gamma^{\mathcal{P}} - A^{\mathcal{P}} X(s)\right) ds + \Sigma_X d\epsilon_s. \tag{41}$$

Following Arnold (1974) Theorem 8.2.2, the fundamental solution is

$$\Phi(s) = e^{-A^{\mathcal{P}}(s-t)}.$$

The solution to (41) is

$$X(\tau) = e^{-A^{\mathcal{P}}(\tau-t)} X(t) + \left(I - e^{-A^{\mathcal{P}}(\tau-t)}\right) \left(A^{\mathcal{P}}\right)^{-1} \gamma^{\mathcal{P}} + \int_t^{t+\tau} e^{-A^{\mathcal{P}}(\tau-\upsilon)} \Sigma_X d\epsilon_\upsilon. \tag{42}$$

Here $\tau > t$.

Following Arnold (1974) Theorem 8.2.12 the integral

$$Y_\tau = \int_t^{t+\tau} e^{-A^{\mathcal{P}}(\tau-\upsilon)} \Sigma_X d\epsilon_\upsilon \sim N(Y; 0, K(\tau)). \tag{43}$$

Here, $N(Y; 0, K(\tau))$ represents a normal distribution with mean zero.

Its variance-covariance matrix is given by

$$K(\tau) = \int_t^{t+\tau} e^{-A^{\mathcal{P}}(\tau-\upsilon)} \Sigma_X \Sigma_X' e^{-A^{\mathcal{P}\prime}(\tau-\upsilon)} d\upsilon.$$

By exercise (1.2.11) of Hijab (1987)

$$K(\tau) = K_\infty - e^{-A^{\mathcal{P}}\tau} K_\infty e^{-A^{\mathcal{P}\prime}\tau}.$$

[15] See Tevlin and Whelan (2003), and Kothari et al. (2017) for the first type of study and Anderson et al. (2018) for the second problem.

Here, the matrix K_∞ solves the Lyapunov equation

$$-A^{\mathcal{P}} K_\infty - K_\infty A^{\mathcal{P}\prime} = \Sigma_X \Sigma_X'.$$

As the time horizon tends to infinity, $K(\tau) \to K_\infty$. The solution to this equation is a positive definite symmetric matrix, which is easily calculated using lyap.m in Matlab.

We have encountered several stochastic processes that have the form

$$Z(X, \tau) = \exp\left\{ -\frac{1}{2} \int_t^T \left[X_s' \mathcal{D}_3(s) X_s - 2\mathcal{D}_2(s) X_s \right] ds + \int_t^T \left(\mathcal{D}_4(s) + X_s' \mathcal{D}_5(s) \right) d\epsilon_s \right\}. \tag{44}$$

In particular, see (34) in which $\mathcal{D}_i(s)$ are replaced by $\mathcal{J}_i(s)$ for $i = 1, 2, 3, 4, 5$. We use the notation X_s rather than $X(s)$, used in the text, to indicate that X is a stochastic process. In addition, the calculations are for a given terminal time T or time horizon τ.

We want $Z(X, \tau)$ to be a uniformly integrable martingale. We recognize that it is a stochastic exponential (Doléans-Dade exponential). See Protter (2005, pp. 84–89). In our case, we have a continuous stochastic process for the factor. As a result, we have

$$\mathcal{E}(X_t) = exp\left\{ X_t - \frac{1}{2} [X, X]_t \right\},$$

where $[X, X]_t$ is the quadratic variation of $Z(X, \tau)$.

Theorem 45 of Protter (2005, p. 141) demonstrates $Z(X, \tau)$ to be a uniformly integrable martingale as long as

$$E\left[exp\left\{ \frac{1}{2} [X, X]_t \right\} \right] < \infty.$$

In this case, the quadratic variation includes all the terms associated with the variance-covariance matrix $\Sigma_X \Sigma_X'$. In this case the quadratic variation is

$$E\left\{ \exp\left[\left(\mathcal{D}_4(0) + X(s)'\mathcal{D}_5(0) \right)' \left(\mathcal{D}_4(0) + X(s)'\mathcal{D}_5(0) \right) \right] \right\} < \infty. \tag{45}$$

This is called the Novikov's Criterion. Below we show these expectations are bounded for the investor's problem.

If this is true, then the stochastic process is given by

$$Z(X, \tau) = Z(X, 0) E_t \left[\exp\left\{ \int_0^T \left(\mathcal{D}_1(0) - \frac{1}{2}\left(X(s)'\mathcal{D}_3(0) X(s) - 2\mathcal{D}_2(0) \right) \right) ds \right\} \right]. \tag{46}$$

For this stochastic process to have a solution, the Novikov condition (45) must be satisfied. In this case, the quadratic variation is dependent on the convergence of the stochastic process for X_s. Its solution is given by (42). The deterministic part of this solution is convergent, as long as $A^{\mathcal{P}}$ has all positive roots. The stochastic part Y includes all the terms associated with the variance-covariance matrix which is bounded by

$$K(\tau) = K_\infty - e^{-A^{\mathcal{P}}\tau} K e^{-A^{\mathcal{P}\prime}\tau} \leq K \text{ with } \tau = T - t.$$

This together with the convergence of the solution X_s (42) assures the quadratic variation (45) exists.

We will now explain how the Backward and Forward Kolmogorov Equations apply to our problem. We then find the solution to these Kolmogorov equations.

The Backward Kolmogorov Equation

To solve for the expectation of the stochastic process (44) we use the backward Kolmogorov equation. We represent the transition probability from state X at time t to the state Y at time T by $p(t, X, T, Y)$. Subsequently, we will derive the transition probability using the forward Kolmogorov equation. In the text X is the vector of interest rate factors at the current time and Y is the random component of these factors at time T given by (43).

We now consider the conditional expectation of (44). As long as the Novikov's Criterion (45) holds, the conditional expectation of (44) is

$$\begin{aligned} f(t, X) = \int_{\mathbb{R}^N} \exp\left\{ -\frac{1}{2}\int_t^T \left[X_s'\mathcal{D}_3(s)X_s - 2\mathcal{D}_2(s)X_s \right] ds \right\} \\ \times f(T, Y)p(t, X, T, Y)dY. \end{aligned} \tag{47}$$

We will show $f(t, X)$ for any $t \in [0, T]$ is the solution to the backward Kolmogorov equation

$$\begin{aligned} &\frac{\partial f(t, X)}{\partial t} - \frac{1}{2}\left(X'\mathcal{D}_3(t)X - 2\mathcal{D}_2(t)X\right) f(t, X) \\ &+ \left(\frac{\partial f(t, X)}{\partial X}\right)' \left(\gamma^{\mathcal{P}} - A^{\mathcal{P}}X\right) + \frac{1}{2} Trace\left(\Sigma_X \Sigma_X' \frac{\partial^2 f(t, X)}{\partial X \partial X}\right) = 0 \end{aligned} \tag{48}$$

under the stochastic process (41).[16] We will be using in the subsequent argument the operator $\mathcal{K}_X$ defined by

$$\mathcal{K}_X \equiv \left(\frac{\partial}{\partial X}\right)' \left(\gamma^{\mathcal{P}} - A^{\mathcal{P}}X\right) + \frac{1}{2} Trace\left(\Sigma_X \Sigma_X' \frac{\partial^2}{\partial X \partial X}\right) \tag{49}$$

[16]This is a variation on the argument for Theorem 8.4.1 of Calin et al. (2011). Also see Duffie (2001) Appendix E, and Karatzas and Shreve (1988, pp. 366–369).

so that

$$\frac{\partial f(t,X)}{\partial t} - \frac{1}{2}\left(X'\mathcal{D}_3(t)X - 2\mathcal{D}_2(t)X\right)f(t,X) + \mathcal{K}_X f(t,X). \tag{50}$$

The Kolmogorov backward PDE is solved subject to the terminal condition

$$\lim_{t\uparrow T} f(t,X) = f(X), \quad X \in \mathbb{R}^N. \tag{51}$$

Proof Define the integrating factor

$$\phi(t,s) = \exp\left\{-\frac{1}{2}\int_t^s \left[X_v'\mathcal{D}_3(v)X_v - 2\mathcal{D}_2(v)X_v\right]dv\right\}.$$

Let

$$Y_s = \phi(t,s)f(s,X_s) \quad s \in [t,T]$$

which is a function of the solution to the stochastic differential equation for X (42). As a result, we can apply Theorem 6.3.1 of Shreve (2006). For a Borel measurable function $h(y)$ on $t \in [0,T]$, we have

$$E\left[h(X(T)) \mid \mathcal{F}(t)\right] = g(t, X(t)).$$

Under these conditions, Lemma 6.4.2 of Shreve (2006), the stochastic process $g(t,X(t))$ is a martingale. Now introduce the discount process

$$D(t) = \phi(0,t).$$

Define

$$Y(t,X) - E\left[\phi(t,T)h(X(T)) \mid \mathcal{F}(t)\right],$$

then

$$Y(t,X) = \phi(0,t)f(t,X)$$

is a martingale and satisfies the PDE (50). However, $f(t,X)$ is not a martingale.

To see the reason for the PDE (50), apply Ito's lemma to Y_s under the stochastic process (41) to yield

$$\begin{aligned}
dY_s = &-\frac{1}{2}\left[X_s'\mathcal{D}_3(s)X_s - 2\mathcal{D}_2(s)X_s\right]\phi(t,s)f(s,X_s)ds + \phi(t,s)\frac{\partial f(s,X_s)}{\partial s}ds \\
&+ \phi(t,s)\left(\frac{\partial f(s,X_s)}{\partial X}\right)'\left(\gamma^{\mathcal{P}} - A^{\mathcal{P}}X_S\right)ds + \frac{1}{2}\phi(t,s)Trace\left(\Sigma_X\Sigma_X'\frac{\partial^2 f(s,X_S)}{\partial X\partial X}\right)ds \\
&+ \phi(t,s)\left(\frac{\partial f(s,X_S)}{\partial X}\right)'\Sigma_X d\epsilon_s.
\end{aligned}$$

For Y_s to be a martingale the drift term must be zero. This property is satisfied by the PDE (50).

Since Y_s is a martingale we can integrate from t to T

$$\phi(t,T)f(T,X_T) - \phi(t,t)f(t,X_t) = \int_t^T \phi(t,s)\Big[\frac{\partial f(s,X_s)}{\partial s} - \frac{1}{2}\left(X_s'\mathcal{D}_3(s)X_s - 2\mathcal{D}_2(s)X_s\right)$$
$$\times f(s,X_s) + \mathcal{K}_{X_s} f(s,X_s)\Big]ds + \int_t^T \phi(t,s)\left(\frac{\partial f(s,X_s)}{\partial X_s}\right)' \Sigma_X d\epsilon_s.$$

We impose (48) subject to the terminal condition (51). In addition we can use the martingale property to take expectations, since Novikov's Criterion (45) is true.

$$f(t,X(t)) = E_t\Big[\phi(t,T)f(Y)\Big] + E_t\Big[\phi(t,s)\left(\frac{\partial f(s,X_S)}{\partial X}\right)' \Sigma_X d\epsilon_s\Big].$$

The second term is zero which leads to the result: Thus, solving the backward Kolmogorov equation (48) for $f(t,X)$ yields the expectation (47).

Solving the Backward Kolmogorov Equation

We set the terminal condition for the backward Kolmogorov equation

$$f(X) = \exp\left\{\frac{1}{2}X'\mathcal{D}_3X + \mathcal{D}_2X + \mathcal{D}_1\right\},$$

where $\mathcal{D}_i$ are constants for the terminal condition.

Guess the solution of (48) has the form

$$f(t,X) = \exp\left\{-\frac{1}{2}\Big[X'\mathcal{F}_3(t)X - 2\mathcal{F}_2(t)X + \mathcal{F}_1(t)\Big]\right\}, \tag{52}$$

$$\frac{\partial f(t,X)}{\partial X} = f(t,X)\left[-\mathcal{F}_3(t)X + \mathcal{F}_2(\tau)'\right].$$

$$\frac{\partial^2 f(t,X)}{\partial X \partial X} = f(t,X)\Big(\mathcal{F}_3(t)XX'\mathcal{F}_3(t) - 2\mathcal{F}_3(\tau)X\mathcal{F}_2(t) + \mathcal{F}_2(t)'\mathcal{F}_2(\tau) - \mathcal{F}_3(\tau)\Big).$$

$$\frac{\partial f(t,X)}{\partial t} = f(\tau,X)\left[-\frac{1}{2}X'\frac{\partial \mathcal{F}_3(t)}{\partial t}X + \frac{\partial \mathcal{F}_2(t)}{\partial t}X - \frac{1}{2}\frac{\partial \mathcal{F}_1(t)}{\partial t}\right].$$

Now substitute these results into the Kolmogorov backward equation (48).

$$\left[-\frac{1}{2}X'\frac{\partial \mathcal{F}_3(t)}{\partial t}X+\frac{\partial \mathcal{F}_2(t)}{\partial t}X-\frac{1}{2}\frac{\partial \mathcal{F}_1(t)}{\partial t}\right]-\frac{1}{2}\left(X'\mathcal{D}_3(t)X-2\mathcal{D}_2(t)X\right)$$
$$+\left[-X'\mathcal{F}_3(t)+\mathcal{F}_2(\tau)\right]\left(\gamma^{\mathcal{P}}-A^{\mathcal{P}}X\right)$$
$$+\frac{1}{2}Trace\left(\Sigma_X\,\Sigma_X'\left(\mathcal{F}_3(t)XX'\mathcal{F}_3(t)-2\mathcal{F}_3(\tau)X\mathcal{F}_2(t)+\mathcal{F}_2(t)'\mathcal{F}_2(\tau)-\mathcal{F}_3(\tau)\right)\right)=0.$$

$$\left[-\frac{1}{2}X'\frac{\partial \mathcal{F}_3(t)}{\partial t}X+\frac{\partial \mathcal{F}_2(t)}{\partial t}X-\frac{1}{2}\frac{\partial \mathcal{F}_1(t)}{\partial t}\right]-\frac{1}{2}X'\mathcal{D}_3(t)X+\mathcal{D}_2(t)X$$
$$-X'\mathcal{F}_3(t)\gamma^{\mathcal{P}}+X'\mathcal{F}_3(t)A^{\mathcal{P}}X+\mathcal{F}_2(\tau)\gamma^{\mathcal{P}}-\mathcal{F}_2(\tau)A^{\mathcal{P}}X+\frac{1}{2}X'\mathcal{F}_3(t)\Sigma_X\,\Sigma_X'\mathcal{F}_3(t)X$$
$$-\mathcal{F}_2(t)\Sigma_X\,\Sigma_X'\mathcal{F}_3(t)X+\frac{1}{2}\mathcal{F}_2(t)\Sigma_X\,\Sigma_X'\mathcal{F}_2(t)'-\frac{1}{2}Trace\left(\Sigma_X\,\Sigma_X'\mathcal{F}_3(t)\right)=0.$$

Now equate quadratic, linear, and constant terms to obtain three ODEs.

$$\frac{\partial \mathcal{F}_3(t)}{\partial t}=\mathcal{F}_3(t)\Sigma_X\,\Sigma_X'\mathcal{F}_3(t)-\mathcal{D}_3(t)+2\mathcal{F}_3(t)A^{\mathcal{P}} \tag{53}$$

subject to

$$\mathcal{F}_3(0)=\mathcal{D}_3.$$

This is the Lyapunov equation.

$$\frac{\partial \mathcal{F}_2(t)}{\partial t}=\mathcal{F}_2(t)\left(\Sigma_X\,\Sigma_X'\mathcal{F}_3(t)+A^{\mathcal{P}}\right)-\mathcal{D}_2(t)+\gamma^{\mathcal{P}\prime}\mathcal{F}_3(t) \tag{54}$$

subject to

$$\mathcal{F}_2(0)=\mathcal{D}_2.$$

This ODE is linear so that we can use integrating factor to solve for $\mathcal{F}_2(t)$. The Final ODE is

$$\frac{\partial \mathcal{F}_1(t)}{\partial t}=2\mathcal{F}_2(\tau)\gamma^{\mathcal{P}}+\mathcal{F}_2(t)\Sigma_X\,\Sigma_X'\mathcal{F}_2(t)'-Trace\left(\Sigma_X\,\Sigma_X'\mathcal{F}_3(t)\right) \tag{55}$$

subject to

$$\mathcal{F}_1(0)=\mathcal{D}_1.$$

This initial value problem is the simplest since everything on the right hand side of the ODE is known.

The Forward Kolmogorov Equation

Following Karatzas and Shreve (1988) the solution to the backward Kolmogorov equation (48) $f(t,X)$ for fixed (T,Y) is

$$f(t,X)\equiv p(t,X,T,Y). \tag{56}$$

In addition, for fixed (t, X) the function

$$g(\tau, Y) \equiv \phi(t, \tau)p(t, X, \tau, Y) \tag{57}$$

solves the forward Kolmogorov equation.[17]

$$\frac{\partial g(\tau, Y)}{\partial \tau} = \mathcal{K}_Y^* g(\tau, Y) - \frac{1}{2}\left(Y'\mathcal{D}_3(\tau)Y - 2\mathcal{D}_2(\tau)Y\right) g(\tau, Y). \tag{58}$$

Here, the dual of $\mathcal{K}_X$ given by[18]

$$\begin{aligned}\mathcal{K}_X^* &= -\sum_{i=1}^{N} \frac{\partial}{\partial X_i}\left(\gamma^{\mathcal{P}} - A^{\mathcal{P}}X\right)_i + \frac{1}{2}\sum_{i,j=1}^{N} \frac{\partial^2}{\partial X_i \partial X_j}\Sigma_{ik}\Sigma'_{kj} \\ &= -\gamma^{\mathcal{P}\prime}\frac{\partial}{\partial X} + X'A^{\mathcal{P}\prime}\frac{\partial}{\partial X} + Trace(A^{\mathcal{P}}) + \frac{1}{2}Trace\left(\Sigma\Sigma'\frac{\partial^2}{\partial X \partial X}\right).\end{aligned} \tag{59}$$

To find the initial condition, let the Dirac distribution centered at $X \in \mathbb{R}^N$ be $f(X) = \delta_X$ such that

$$\delta_X(\theta) = \int_{\mathbb{R}^N} \delta_x(Y)\theta(Y)dY = \theta(X).$$

For a given $X_t = X \in \mathbb{R}^N$,

$$g(\tau, X) = \int_{\mathbb{R}^N} \delta_X(Y)\phi(t, \tau)p(t, X, \tau, Y)dY = \phi(t, \tau)p(t, X, \tau, X).$$

Consequently, if the initial condition for the Kolmogorov forward equation (13) is

$$\lim_{\tau \to 0^+} g(\tau, X(\tau)) = \delta_X, \tag{60}$$

then the solution to (13) is $\phi(t, \tau)p(t, X, \tau, Y) = g(\tau, Y)$.

Thus, we have the proof of Theorem 2.1.

Proof We will use the property of the dual for the Kolmogorov operator, $\mathcal{K}_Y$ given by

$$\int_{\mathbb{R}^N} \mathcal{K}_Y f(Y)g(Y)dY = \int_{\mathbb{R}^N} f(Y)\mathcal{K}_Y^* g(Y)dY. \tag{61}$$

---

[17] See Karatzas and Shreve (1988, p. 369) equation (7.24). Also see Theorem 8.7.1. of Calin et. al (2011), and Chirikjian (2009, pp. 118–121).

[18] See Øksendal (2005, p. 169). Also follow the derivation in Chirikjian (2009, p. 121).

We know from (47) that

$$
\begin{aligned}
f(t,X) &= \int_{\mathbb{R}^N} \exp\left\{-\frac{1}{2}\int_t^T \left[X_s'\mathcal{D}_3(s)X_s - 2\mathcal{D}_2(s)X_s\right]ds\right\} \\
&\quad \times f(Y)p(t,X,T,Y)dY \\
&= \int_{\mathbb{R}^N}\int_{\mathbb{R}^N} \exp\left\{-\frac{1}{2}\int_t^{\tau} \left[X_s'\mathcal{D}_3(s)X_s - 2\mathcal{D}_2(s)X_s\right]ds\right\} \\
&\quad \times \exp\left\{-\frac{1}{2}\int_{\tau}^T \left[X_s'\mathcal{D}_3(s)X_s - 2\mathcal{D}_2(s)X_s\right]ds\right\} \\
&\quad \times f(Y)p(t,X,\tau,Z)p(\tau,Z,T,Y)dZdY \\
&= \int_{\mathbb{R}^N} \phi(t,\tau)f(\tau,Z)p(t,X,\tau,Z)dZ.
\end{aligned}
$$

The next to last step uses the Chapman-Kolmogorov equation for a Markov process[19] and the last step uses the definition of $f(t,X)$. As a result, we know for any $t < \tau \leq T$

$$
f(t,X) = \int_{\mathbb{R}^N} f(\tau,Y)\phi(t,\tau)p(t,X,\tau,Y)dY. \tag{62}
$$

Next differentiate in τ

$$
\begin{aligned}
0 = \frac{\partial f(t,X)}{\partial \tau} &= \int_{\mathbb{R}^N} \frac{\partial f(\tau,Y)}{\partial \tau}\phi(t,\tau)p(t,X,\tau,Y)dY \\
&\quad + \int_{\mathbb{R}^N} f(\tau,Y)\frac{\partial \phi(t,\tau)p(t,X,\tau,Y)}{\partial \tau}dY \\
&= \int_{\mathbb{R}^N} f(\tau,Y)\frac{\partial \phi(t,\tau)p(t,X,\tau,Y)}{\partial \tau}dY \\
&\quad - \int_{\mathbb{R}^N} \mathcal{K}_Y f(\tau,Y)\phi(t,\tau)p(t,X,\tau,Y)dY \\
&\quad + \frac{1}{2}\int_{\mathbb{R}^N} \big(Y'\mathcal{D}_3(\tau)Y - 2\mathcal{D}_2(\tau)Y\big) f(\tau,X)\phi(t,\tau)p(t,X,\tau,Y)dY.
\end{aligned} \tag{63}
$$

The second step uses the backward Kolmogorov equation (48).

Now apply the property (61) to find

$$
\begin{aligned}
0 = \int_{\mathbb{R}^N} f(\tau,Y)\Bigg[&\frac{\partial \phi(t,\tau)p(t,X,\tau,Y)}{\partial \tau} - \mathcal{K}_Y^*(\phi(t,\tau)p(t,X,\tau,Y)) \\
&+ \frac{1}{2}\big(Y'\mathcal{D}_3(\tau)Y - 2\mathcal{D}_2(\tau)Y\big)\,\phi(t,\tau)p(t,X,\tau,Y)\Bigg]dY.
\end{aligned}
$$

This means we want to define $g(\tau,Y) = \phi(t,\tau)p(t,X,\tau,Y)$ for (13) to hold.

---

[19] See Chirikjian (2009, p. 108) equation (4.16).

Solving the Forward Kolmogorov Equation

It is difficult to impose the initial condition (15), since there is no explicit form for it. However, the Fourier transform of δ_X is 1. As a result, we will take the Fourier transform of the Kolmogorov equation (13) and find its solution. We will then apply the inverse Fourier transform to find the solution to the Kolmogorov forward equation given the initial condition.

If the Fourier transforms of $f(X)$ (16) exists, then

$$F_X\left[\frac{\partial f(X)}{\partial X_j}\right] = i\xi_j F_X[f(X)] \Rightarrow F_X\left[\frac{\partial f(X)}{\partial X}\right] = i\xi F_X[f(X)].$$
$$F_X\left[\frac{\partial^2 f(X)}{\partial X_j \partial X_k}\right] = -\,\xi_j\xi_k F_X\left[f(X)\right] \Rightarrow F_X\left[\frac{\partial^2 f(X)}{\partial X\,\partial X}\right] = -\xi\xi' F_X[f(X)]. \quad (64)$$

The subscript X is added to keep track of the integration over X not t.

$$F_X[-iXf(X)] = \frac{\partial \hat{f}(\xi)}{\partial \xi} \Rightarrow F_X[Xf(X)] = i\frac{\partial \hat{f}(\xi)}{\partial \xi}. \quad (65)$$

Proof

$$\begin{aligned}
\frac{\partial \hat{f}(\xi)}{\partial \xi_j} &= \frac{\partial}{\partial \xi_j}\int_{-\infty}^{\infty} f(X)e^{-i\xi\cdot X}\,dX \\
&= \int_{-\infty}^{\infty} -iX_j f(X)e^{-i\xi\cdot X}\,dX = F_X[-iX_j f(X)]. \\
&\Rightarrow F_X[-iXf(X)] = \frac{\partial F_X\left[f(X)\right]}{\partial \xi}.
\end{aligned}$$

$$\begin{aligned}
F_X\left[\left(\frac{\partial f}{\partial X}\right)' A^{\mathcal{P}} X\right] &= Trace\left(A^{\mathcal{P}} F_X\left[X\left(\frac{\partial f}{\partial X}\right)'\right]\right) = iTrace\left(A^{\mathcal{P}}\frac{\partial F_X\left[\left(\frac{\partial f}{\partial X}\right)'\right]}{\partial \xi}\right) \\
&= i^2 Trace\left(A^{\mathcal{P}}\frac{\partial \xi' F_X[f(X)]}{\partial \xi}\right) \\
&= -Trace\left(A^{\mathcal{P}}\frac{\partial F_X[f(X)]}{\partial \xi}\xi' + A^{\mathcal{P}} F_X[f(X)]\right).
\end{aligned}$$

The first result applies the Trace to a quadratic form. The second step uses (65) for the function $\left(\frac{\partial f}{\partial X}\right)'$. In the third equality we use the first result in (64). Finally, we use the product rule of differentiation and $i^2 = -1$.

We also have to consider $F_X[X'Xf(X)]$.

$$F_X[X'Xf(X)] = \frac{\partial^2 \hat{f}(\xi)}{\partial\xi\partial\xi}.$$

Proof

$$\begin{aligned}\frac{\partial \hat{f}(\xi)}{\partial\xi_j\partial\xi_k} &= \frac{\partial}{\partial\xi_k}\int_{-\infty}^{\infty} -iX_j f(X)e^{-i\xi\cdot X}\,dX \\ &= \int_{-\infty}^{\infty} iX_k iX_j f(X)e^{-i\xi\cdot X}\,dX = F_X[-X_k X_j f(X)]. \\ &\Rightarrow F_X[-XX'f(X)] = \frac{\partial^2 F_X\left[f(X)\right]}{\partial\xi\partial\xi}.\end{aligned}$$

Notice

$$\begin{aligned}F_X\left[X'\mathcal{D}_3(\tau)Xf(\tau,X)\right] &= F_X\left[Trace\left(X'\mathcal{D}_3(\tau)X\right)f(\tau,X)\right] \\ &= F_X\left[Trace\left(\mathcal{D}_3(\tau)XX'f(\tau,X)\right)\right] \\ &= Trace\left(F_X\left[\mathcal{D}_3(\tau)XX'f(\tau,X)\right]\right) \\ &= Trace\left(\mathcal{D}_3(\tau)\frac{\partial^2\hat{f}(\xi)}{\partial\xi\partial\xi}\right).\end{aligned}$$

The first step is true since $X'\mathcal{D}_3(\tau)X \in \mathbb{R}$. The second step uses the property $Trace\,(ABC) = Trace\,(BCA)$. The third step takes advantage of the trace being a linear operator so that the additive property of integrals can be used. Since $X'X$ is symmetric the last step uses the last property of Fourier transforms.

Recall the Kolmogorov forward equation

$$\begin{aligned}\frac{\partial g(\tau,Y)}{\partial t} =& -\gamma^{\mathcal{P}\prime}\frac{\partial g(\tau,Y)}{\partial Y} + \left(\frac{\partial g(\tau,Y)}{\partial Y}\right)' A^{\mathcal{P}}Y + Trace\left(A^{\mathcal{P}}\right)g(\tau,Y) \\ &+ \frac{1}{2}Trace\left(\Sigma\Sigma'\frac{\partial^2 g(\tau,Y)}{\partial Y\partial Y}\right) - \frac{1}{2}\left(Y'\mathcal{D}_3(\tau)Y - 2\mathcal{D}_2(\tau)Y\right)g(\tau,Y).\end{aligned} \tag{66}$$

subject to the initial condition

$$g(0,Y_0) = \delta_Y.$$

Apply the Fourier transform to the forward Kolmogorov equation.

$$\begin{aligned}\frac{\partial F_Y\left[g(\tau,Y)\right]}{\partial\tau} &= -\gamma^{\mathcal{P}\prime}F_Y\left[\frac{\partial g(\tau,Y)}{\partial Y}\right]+F_Y\left[\left(\frac{\partial g(\tau,Y)}{\partial Y}\right)'A^{\mathcal{P}}Y\right]\\ &+Trace(A^{\mathcal{P}})F_Y\left[g(\tau,Y)\right]+\frac{1}{2}Trace\left(\Sigma\Sigma'F_Y\left[\frac{\partial^2 g(\tau,Y)}{\partial Y\partial Y}\right]\right)\\ &-\frac{1}{2}F_Y\left[\left(Y'\mathcal{D}_3(\tau)Y-2\mathcal{D}_2(\tau)Y\right)g(\tau,Y)\right]\end{aligned} \tag{67}$$

subject to the initial condition

$$F_Y\left[g(0,Y_0)\right]=1.$$

Next use the rules for Fourier transform to obtain

$$\begin{aligned}\frac{\partial F_Y\left[g(\tau,Y)\right]}{\partial\tau} &= -i\gamma^{\mathcal{P}\prime}\xi F_Y\left[g(\tau,Y)\right]-Trace\left(A^{\mathcal{P}}\frac{\partial F_Y[g(\tau,Y)]}{\partial\xi}\xi'+A^{\mathcal{P}}F_Y[g(\tau,Y)]\right)\\ &+Trace(A^{\mathcal{P}})F_Y\left[g(\tau,Y)\right]-\frac{1}{2}Trace\left(\Sigma\Sigma'\xi\xi'F_Y[g(\tau,Y)]\right)-\frac{1}{2}Trace\left(\mathcal{D}_3(\tau)\frac{\partial^2\hat{g}(\xi)}{\partial\xi\partial\xi}\right)\\ &+i\left(\frac{\partial F_Y[g(\tau,Y)]}{\partial\xi}\right)'\mathcal{D}_2(t,\tau)'\end{aligned}$$

$$\begin{aligned}&\Rightarrow\frac{\partial F_Y\left[g(\tau,Y)\right]}{\partial\tau}+\frac{1}{2}\xi'\Sigma\Sigma'\xi F_Y[g(\tau,Y)]+i\gamma^{\mathcal{P}\prime}\xi F_Y\left[g(\tau,Y)\right]\\ &-\left(\frac{\partial F_Y[g(\tau,Y)]}{\partial\xi}\right)'\left(i\mathcal{D}_2(\tau)'-A^{\mathcal{P}\prime}\xi\right)+\frac{1}{2}Trace\left(\mathcal{D}_3(\tau)\frac{\partial^2\hat{g}(\xi)}{\partial\xi\partial\xi}\right)=0.\end{aligned} \tag{68}$$

subject to the initial condition

$$F_Y\left[g(0,Y_0)\right]=1.$$

Now that the initial value problem is defined we can use a guess and verify procedure to find its solution.

$$F_Y\left[g(\tau,Y)\right]=\exp\left\{-\frac{1}{2}\left[\xi'\mathcal{G}_3(\tau)\xi-2i\mathcal{G}_2(\tau)'\xi+\mathcal{G}_1(\tau)\right]\right\}, \tag{69}$$

We do not assume the matrix is symmetric, since $\frac{1}{2}\xi'\left(\mathcal{G}_3(\tau)+\mathcal{G}_3(\tau)'\right)\xi=\xi'\mathcal{G}_3(\tau)\xi$.

$$\frac{\partial F_Y\left[g(\tau,Y)\right]}{\partial\xi}=F_Y\left[g(\tau,Y)\right]\left[-\mathcal{G}_3(\tau)\xi-\mathcal{G}_3(\tau)'\xi+i\mathcal{G}_2(\tau)\right].$$

$$\frac{\partial^2 F_Y[g(\tau,Y)]}{\partial\xi\partial\xi} = F_Y[g(\tau,Y)]\Big(-\left[\mathcal{G}_3(\tau)+\mathcal{G}_3(\tau)'\right]\xi\xi'\left[\mathcal{G}_3(\tau)+\mathcal{G}_3(\tau)'\right]$$
$$-2i\left[\mathcal{G}_3(\tau)+\mathcal{G}_3(\tau)'\right]\xi\mathcal{G}_2(\tau)'-\mathcal{G}_2(\tau)\mathcal{G}_2(\tau)'-\left[\mathcal{G}_3(\tau)+\mathcal{G}_3(\tau)'\right]\Big).$$

$$\frac{\partial F_Y[g(\tau,Y)]}{\partial\tau} = F_Y[g(\tau,Y)]\left[-\frac{1}{2}\xi'\frac{\partial\mathcal{G}_3(\tau)}{\partial\tau}\xi+i\frac{\partial\mathcal{G}_2(\tau)}{\partial\tau}\xi-\frac{1}{2}\frac{\partial\mathcal{G}_1(\tau)}{\partial\tau}\right].$$

Now substitute these results into the Fourier transform (68) of the forward Kolmogorov equation (13).

$$\left[-\frac{1}{2}\xi'\frac{\partial\mathcal{G}_3(\tau)}{\partial\tau}\xi+i\frac{\partial\mathcal{G}_2(\tau)}{\partial\tau}\xi-\frac{1}{2}\frac{\partial\mathcal{G}_1(\tau)}{\partial\tau}\right]+\frac{1}{2}\xi'\Sigma_X\Sigma_X'\xi+i\xi'\gamma^{\mathcal{P}}$$
$$-\left(-\xi'\left[\mathcal{G}_3(\tau)+\mathcal{G}_3(\tau)'\right]+i\mathcal{G}_2(\tau)\right)\left(i\mathcal{D}_2(\tau)'-A^{\mathcal{P}\prime}\xi\right)$$
$$+\frac{1}{2}Trace\Big(\mathcal{D}_3(\tau)\Big(\left[\mathcal{G}_3(\tau)+\mathcal{G}_3(\tau)'\right]\xi\xi'\left[\mathcal{G}_3(\tau)+\mathcal{G}_3(\tau)'\right]$$
$$-2i\left[\mathcal{G}_3(\tau)+\mathcal{G}_3(\tau)'\right]\xi\mathcal{G}_2(\tau)-\mathcal{G}_2(\tau)'\mathcal{G}_2(\tau)-\left[\mathcal{G}_3(\tau)+\mathcal{G}_3(\tau)'\right]\Big)\Big)=0$$

$$\Rightarrow\left[-\frac{1}{2}\xi'\frac{\partial\mathcal{G}_3(\tau)}{\partial\tau}\xi+i\frac{\partial\mathcal{G}_2(\tau)}{\partial\tau}\xi-\frac{1}{2}\frac{\partial\mathcal{G}_1(\tau)}{\partial\tau}\right]+\frac{1}{2}\xi'\Sigma_X\Sigma_X'\xi+i\gamma^{\mathcal{P}\prime}\xi$$
$$+\mathcal{D}_2(\tau)\mathcal{G}_3(\tau)i\xi-\xi'\mathcal{G}_3(\tau)A^{\mathcal{P}\prime}\xi$$
$$+\mathcal{G}_2(\tau)\mathcal{D}_2(\tau)'+\mathcal{G}_2(\tau)A^{\mathcal{P}\prime}i\xi+\frac{1}{2}\xi'\mathcal{G}_3(\tau)\mathcal{D}_3(\tau)\mathcal{G}_3(\tau)\xi$$
$$-\mathcal{G}_2(\tau)\mathcal{D}_3(\tau)\mathcal{G}_3(\tau)i\xi-\frac{1}{2}\mathcal{G}_2(\tau)\mathcal{D}_3(\tau)\mathcal{G}_2(\tau)'-\frac{1}{2}Trace\left(\mathcal{D}_3(\tau)\mathcal{G}_3(\tau)\right)=0.$$

Now equate quadratic, linear ($i\xi$), and constant terms to obtain three ODEs.

$$\frac{\partial\mathcal{G}_3(\tau)}{\partial\tau}=\mathcal{G}_3(\tau)\mathcal{D}_3(\tau)\mathcal{G}_3(\tau)-2\mathcal{G}_3(\tau)A^{\mathcal{P}\prime}+\Sigma_X\Sigma_X' \tag{70}$$

subject to

$$\mathcal{G}_3(0)=0_{N\times N}.$$

Again this is the Lyapunov equation.

$$\frac{\partial\mathcal{G}_2(\tau)}{\partial\tau}=\mathcal{G}_2(\tau)\left(\mathcal{D}_3(\tau)\mathcal{G}_3(\tau)-A^{\mathcal{P}\prime}\right)-\gamma^{\mathcal{P}\prime}-\mathcal{D}_2(\tau)\mathcal{G}_3(\tau) \tag{71}$$

subject to

$$\mathcal{G}_2(0) = 0_N.$$

This ODE is linear so that we can use integrating factor to solve for $\mathcal{G}_2(\tau)$. The integrating factor is

$$int = e^{-(\mathcal{D}_3(\tau)\mathcal{G}_3(\tau)-A^{\mathcal{P}'})\tau}.$$

Consequently,

$$\frac{\partial e^{-(\mathcal{D}_3(s)\mathcal{G}_3(s)-A^{\mathcal{P}'})s}\mathcal{G}_2(s)}{\partial s}ds = -e^{-(\mathcal{D}_3(s)\mathcal{G}_3(s)-A^{\mathcal{P}'})s}\left(\gamma^{\mathcal{P}'} - \mathcal{D}_2(s,X)\mathcal{G}_3(s)\right)ds.$$

Now integrate from τ to 0

$$\mathcal{G}_2(\tau, X) = e^{\left(\mathcal{D}_3(\tau)\mathcal{G}_3(\tau)-A^{\mathcal{P}'}\right)\tau}\mathcal{G}_2(0) - \int_0^{\tau} e^{-\left(\mathcal{D}_3(s)\mathcal{G}_3(s)-A^{\mathcal{P}'}\right)(s-\tau)}\left(\gamma^{\mathcal{P}'} - \mathcal{D}_2(s,X)\mathcal{G}_3(s)\right)ds.$$

The Final ODE is

$$\frac{\partial \mathcal{G}_1(\tau)}{\partial \tau} = 2\mathcal{G}_2(\tau)\mathcal{D}_2(\tau)' - \mathcal{G}_2(\tau)\mathcal{D}_3(\tau)\mathcal{G}_2(\tau)' - Trace\left(\mathcal{D}_3(\tau)\mathcal{G}_3(\tau)\right) \tag{72}$$

subject to

$$\mathcal{G}_1(0) = 0.$$

This initial value problem is the simplest since everything on the right hand side of the ODE is known.

Solving these three ODEs leads to the solution (19) to the Fourier transform of the Kolmogorov equation (68). The final step is to take the inverse Fourier transform to (19)

$$g(\tau, Y) = \frac{1}{(2\pi)^N}\int_{-\infty}^{\infty} \exp\left\{-\frac{1}{2}\left[\xi'\mathcal{G}_3(\tau)\xi - 2\left(\mathcal{G}_2(\tau) - Y'\right)i\xi + \mathcal{G}_1(\tau)\right]\right\}d\xi. \tag{73}$$

To calculate this integral we use the following Lemma following Strauss (2008, p. 345) and Strichartz (2008, pp. 41–43).

Lemma 4.1 *Let α be a positive number and let x_0 and y_0 be real numbers.*

$$\int_{-\infty}^{\infty} e^{-\alpha(x+x_0+iy_0)^2}\,dx = \sqrt{\frac{\pi}{\alpha}}. \tag{74}$$

We also need the multiple dimension version of Lemma 4.1.

Lemma 4.2 *Let A be a $N \times N$ symmetric matrix with all positive eigenvalues and let $Z \in \mathbb{C}^N$.*

$$\int_{\mathbb{R}^N} e^{-\frac{1}{2}(X+A^{-1}Z)'A(X+A^{-1}Z)}dX = \sqrt{\frac{(2\pi)^N}{detA}}. \tag{75}$$

To apply the Lemma 4.2 to the inverse Fourier transform (73) we have to multiply out the quadratic exponent

$$\left(X + A^{-1}Z\right)' A \left(X + A^{-1}Z\right) = X'AX + 2Z'X + Z'(A^{-1})Z. \tag{76}$$

Now match up the coefficients in (73) to yield

$$A = \mathcal{G}_3(\tau) \text{ and } Z = \left(\mathcal{G}_2(\tau)' - X\right) i. \tag{77}$$

As a result, we can complete the square in the exponent of (73) to find

$$\begin{aligned}
g(\tau, Y) &= \frac{1}{(2\pi)^N} \int_{-\infty}^{\infty} \exp\left\{ -\frac{1}{2}\Big[\xi'\mathcal{G}_3(\tau)\xi - 2\left(\mathcal{G}_2(\tau) - Y'\right) i\xi + \mathcal{G}_1(\tau)\Big]\right\} d\xi \\
&= \exp\left\{ -\frac{1}{2}\mathcal{G}_1(\tau) - \frac{1}{2}\left(\mathcal{G}_2(\tau)' - Y\right)' \mathcal{G}_3(\tau)^{-1} \left(\mathcal{G}_2(\tau)' - Y\right) \right\} \\
&\times \frac{1}{(2\pi)^N} \int_{-\infty}^{\infty} \exp\left\{ -\frac{1}{2}\left(Y + \mathcal{G}_3(\tau)^{-1}\left(\mathcal{G}_2(\tau)' - Y\right) i\right)' \mathcal{G}_3(\tau) \right. \\
&\times \left. \left(Y + \mathcal{G}_3(\tau)^{-1}\left(\mathcal{G}_2(\tau)' - Y\right) i\right) \right\} d\xi \\
&= \frac{1}{\sqrt{(2\pi)^N\, det(\mathcal{G}_3(\tau))}} \exp\left\{ -\frac{1}{2}\mathcal{G}_1(\tau) - \frac{1}{2}\left(\mathcal{G}_2(\tau)' - Y\right)' \mathcal{G}_3(\tau)^{-1} \left(\mathcal{G}_2(\tau)' - Y\right) \right\}.
\end{aligned} \tag{78}$$

By applying this solution to the forward Kolmogorov equation for the stochastic process (44), we can find the probability distribution for the investor's lifetime utility (34).

These random terms are probability densities of a normal distribution. We denote these probabilities densities by

$$\mathcal{N}(x; \mu, \Sigma) \equiv \frac{1}{\sqrt{(2\pi)^N\, det(\Sigma)}} \exp\left\{ -\frac{1}{2}(x-\mu)'\Sigma^{-1}(x-\mu)\right\} \tag{79}$$

for $x \in \mathbf{R}^n$.

By (23) the discounted transition probability can be written as

$$\phi(t, \tau)p(t, X, \tau, Y) = \exp\left\{ -\frac{1}{2}\mathcal{G}_1(\tau)\right\}\mathcal{N}\left(Y; \mathcal{G}_2(\tau)', \mathcal{G}_3(\tau)\right). \tag{80}$$

Note that

$$\phi(t,s) = \exp\left\{ -\frac{1}{2}\int_t^s \left[X_v' \mathcal{D}_3(v) X_v - 2\mathcal{D}_2(v) X_v \right] dv \right\}$$

does not include the constant term

$$\mathcal{D}_0(\tau) = \exp\left\{ -\frac{1}{2}\mathcal{D}_1(\tau)\tau \right\}$$

so it has to be added back in. The same is true for the backward Kolmogorov equation (48).

In the analysis of option values and VaR we will use various rules for Gaussian probability distributions which we recall from Petersen and Pedersen (2008). First we use the rule for the product of two normal distributions.

$$\begin{aligned} &\mathcal{N}(x;\mu_1,\Sigma_1) \times \mathcal{N}(x;\mu_1,\Sigma_1) = \vartheta \mathcal{N}(x;\mu_c,\Sigma_c) \\ &\text{where } \vartheta \equiv \frac{1}{\sqrt{(2\pi)^N \, det\,(\Sigma_1+\Sigma_2)}} \exp\left\{ -\frac{1}{2}(\mu_1-\mu_2)'(\Sigma_1+\Sigma_2)^{-1}(\mu_1-\mu_2) \right\}, \\ &\mu_c \equiv \left(\Sigma_1^{-1}+\Sigma_2^{-1}\right)^{-1}\left(\Sigma_1^{-1}\mu_1+\Sigma_2^{-1}\mu_2\right), \\ &\text{and } \Sigma_c = \left(\Sigma_1^{-1}+\Sigma_2^{-1}\right)^{-1}. \end{aligned} \tag{81}$$

We also use the linear rule[20]

$$Ax \sim \mathcal{N}(x, A\mu, \Sigma A'), \tag{82}$$

Finally, we convert to a standard normal using the rule

$$x = \sigma Z + \mu \text{ such that } Z \sim \mathcal{N}(0_N, I_N). \tag{83}$$

Here, $\Sigma = \sigma\sigma'$ is the Cholesky decomposition of the variance covariance matrix. By following these basic rules for a normal distribution we are able to represent the probability distribution for the investor's wealth and her lifetime utility.

Stochastic Discount Factor

We now have all the tools necessary to break a stochastic process like (44) into expected and random components. First, we apply the argument to the stochastic discount factor. The other stochastic processes will be solved using the same technique.

[20] See Petersen and Pedersen (2008, 8.1.4, p. 41).

$$\frac{M_{\mathcal{T},t}}{M_{t,t}} = \exp\left\{-\int_t^{t+\mathcal{T}}\left[r(X(s)) + \frac{1}{2}\Lambda(X(s))'\Lambda(X(s))\right]ds + \int_t^{t+\mathcal{T}}\Lambda(X(s))'d\epsilon_s\right\}$$

$$= \exp\left\{-\int_t^{t+\mathcal{T}}\left[r(X(s)) + \frac{1}{2}\left(\gamma^{\mathcal{P}} - \gamma^{\mathcal{Q}} - (A^{\mathcal{P}} - A^{\mathcal{Q}})X(s)\right)'\left(\Sigma_X'\Sigma_X\right)^{-1}\right.\right.$$

$$\left.\times\left(\gamma^{\mathcal{P}} - \gamma^{\mathcal{Q}} - (A^{\mathcal{P}} - A^{\mathcal{Q}})X(s)\right)\right]ds$$

$$\left.+\int_t^{t+\mathcal{T}}\left(\gamma^{\mathcal{P}} - \gamma^{\mathcal{Q}} - (A^{\mathcal{P}} - A^{\mathcal{Q}})X(s)\right)'\left(\Sigma_X'\right)^{-1}d\epsilon_s\right\}$$

$$= \exp\left\{-\int_t^{t+\mathcal{T}}\left[\delta_0 + \frac{1}{2}(\gamma^{\mathcal{P}} - \gamma^{\mathcal{Q}})'(\Sigma_X'\Sigma_X)^{-1}(\gamma^{\mathcal{P}} - \gamma^{\mathcal{Q}})\right.\right.$$

$$+\left(\delta_1 - (\gamma^{\mathcal{P}} - \gamma^{\mathcal{Q}})'\left(\Sigma_X'\Sigma_X\right)^{-1}(A^{\mathcal{P}} - A^{\mathcal{Q}})\right)X(s)$$

$$\left.+\frac{1}{2}X(s)'(A^{\mathcal{P}} - A^{\mathcal{Q}})'\left(\Sigma_X'\Sigma_X\right)^{-1}(A^{\mathcal{P}} - A^{\mathcal{Q}})X(s)\right]ds$$

$$\left.+\int_t^{t+\mathcal{T}}\left((\gamma^{\mathcal{P}} - \gamma^{\mathcal{Q}})'\left(\Sigma_X'\right)^{-1} - X(s)'(A^{\mathcal{P}} - A^{\mathcal{Q}})'\left(\Sigma_X'\right)^{-1}\right)d\epsilon_s\right\}$$

$$= \exp\left\{\int_0^{\mathcal{T}}\left(-\mathcal{M}_1(0) - \frac{1}{2}\left(X_s'\mathcal{M}_3(0)X_s - 2\mathcal{M}_2(0)X_s\right)\right\}\right]ds$$

$$\left.+\int_t^T(\mathcal{M}_4 + \mathcal{M}_5X_s)\,d\epsilon_s\right\}.$$

We use the risk free rate, the risk premium and the risk neutral coefficients in this derivation.

The constants are given by

$$\begin{aligned}
\mathcal{M}_1 &\equiv \delta_0 + \frac{1}{2}(\gamma^{\mathcal{P}} - \gamma^{\mathcal{Q}})'\left(\Sigma_X'\Sigma_X\right)^{-1}(\gamma^{\mathcal{P}} - \gamma^{\mathcal{Q}}), \\
\mathcal{M}_2 &\equiv -\left[\delta_1 - (\gamma^{\mathcal{P}} - \gamma^{\mathcal{Q}})'\left(\Sigma_X'\Sigma_X\right)^{-1}(A^{\mathcal{P}} - A^{\mathcal{Q}})\right], \\
\mathcal{M}_3 &\equiv (A^{\mathcal{P}} - A^{\mathcal{Q}})'\left(\Sigma_X'\Sigma_X\right)^{-1}(A^{\mathcal{P}} - A^{\mathcal{Q}}), \\
\mathcal{M}_4 &\equiv (\gamma^{\mathcal{P}} - \gamma^{\mathcal{Q}})'\left(\Sigma_X'\right)^{-1} \text{ and } \mathcal{M}_5 \equiv -(A^{\mathcal{P}} - A^{\mathcal{Q}})'\left(\Sigma_X'\right)^{-1}.
\end{aligned} \tag{84}$$

As a result, the stochastic process for the pricing kernel is

$$\frac{M_{\mathcal{T},t}}{M_{t,t}} = \exp\left\{\int_0^{\mathcal{T}}\left(-\mathcal{M}_1(0) - \frac{1}{2}\left(X_s'\mathcal{M}_3(0)X_s - 2\mathcal{M}_2(0)X_s\right)\right)ds\right.$$

$$\left.+\int_t^{t+\mathcal{T}}(\mathcal{M}_4 + X_s'\mathcal{M}_5)\,\Sigma_X'd\epsilon_s\right\}. \tag{85}$$

We need the probability distribution for the pricing kernel in solving this stochastic process. Before applying the forward Kolmogorov results, we factor out all the deterministic terms from (85). We have from (42)

$$X(\tau) = A_0(\tau) + e^{-A^{\mathcal{P}}(\tau-t)}X(t) + Y_\tau, \tag{86}$$

where

$$A_0(\tau) = \left(I - e^{-A^{\mathcal{P}}(\tau-t)}\right)\left(A^{\mathcal{P}}\right)^{-1}\gamma^{\mathcal{P}}.$$

We also will use

$$\int_t^{t+\tau} e^{-A^{\mathcal{P}}(s-t)}ds = \left(A^{\mathcal{P}}\right)^{-1}\left[I - e^{-A^{\mathcal{P}}\tau}\right].$$

Now factor the square term to find

$$\begin{aligned}
X(\tau)'\mathcal{M}_3X(\tau) &= \left(A_0(\tau) + e^{-A^{\mathcal{P}}(\tau-t)}X(t) + Y_\tau\right)'\mathcal{M}_3\left(A_0(\tau) + e^{-A^{\mathcal{P}}(\tau-t)}X(t) + Y_\tau\right)\\
&= \left(\gamma^{\mathcal{P}\prime}\left(A^{\mathcal{P}\prime}\right)^{-1}\left(I - e^{-A^{\mathcal{P}\prime}(\tau-t)}\right) + X(t)'e^{-A^{\mathcal{P}\prime}}\right)\mathcal{M}_3\\
&\left(\left(I - e^{-A^{\mathcal{P}}(\tau-t)}\right)\left(A^{\mathcal{P}}\right)^{-1}\gamma^{\mathcal{P}} + e^{-A^{\mathcal{P}}(\tau-t)}X(t)\right)\\
&+2\left(\gamma^{\mathcal{P}\prime}\left(A^{\mathcal{P}\prime}\right)^{-1}\left(I - e^{-A^{\mathcal{P}\prime}(\tau-t)}\right) + X(t)'e^{-A^{\mathcal{P}\prime}(\tau-t)}\right)\mathcal{M}_3Y_\tau + Y_\tau'\mathcal{M}_3Y_\tau\\
&= \gamma^{\mathcal{P}\prime}\left(A^{\mathcal{P}\prime}\right)^{-1}\left(I - e^{-A^{\mathcal{P}\prime}(\tau-t)}\right)\mathcal{M}_3\left(I - e^{-A^{\mathcal{P}}(\tau-t)}\right)\left(A^{\mathcal{P}}\right)^{-1}\gamma^{\mathcal{P}}\\
&+2\gamma^{\mathcal{P}\prime}\left(A^{\mathcal{P}\prime}\right)^{-1}\left(I - e^{-A^{\mathcal{P}\prime}(\tau-t)}\right)\mathcal{M}_3e^{-A^{\mathcal{P}}(\tau-t)}X(t) + X(t)'e^{-A^{\mathcal{P}\prime}}\mathcal{M}_3e^{-A^{\mathcal{P}}(\tau-t)}X(t)\\
&+2\left(\gamma^{\mathcal{P}\prime}\left(A^{\mathcal{P}\prime}\right)^{-1}\left(I - e^{-A^{\mathcal{P}\prime}(\tau-t)}\right) + X(t)'e^{-A^{\mathcal{P}\prime}(\tau-t)}\right)\mathcal{M}_3Y_\tau + Y_\tau'\mathcal{M}_3Y_\tau.
\end{aligned}$$

Now integrate the first term over the time horizon τ given $X(t) = X$.

$$\begin{aligned}
&-\frac{1}{2}\int_t^{t+\tau} X(s)'\mathcal{M}_3X(s)ds =\\
&-\frac{1}{2}\int_t^{t+\tau}\gamma^{\mathcal{P}\prime}\left(A^{\mathcal{P}\prime}\right)^{-1}\left(I - e^{-A^{\mathcal{P}\prime}(s-t)}\right)\mathcal{M}_3\left(I - e^{-A^{\mathcal{P}}(s-t)}\right)ds\left(A^{\mathcal{P}}\right)^{-1}\gamma^{\mathcal{P}}\\
&-\gamma^{\mathcal{P}\prime}\left(A^{\mathcal{P}\prime}\right)^{-1}\int_t^{t+\tau}\left(I - e^{-A^{\mathcal{P}\prime}(s-t)}\right)\mathcal{M}_3e^{-A^{\mathcal{P}}(s-t)}dsX(t)\\
&-\frac{1}{2}X(t)'\int_t^{t+\tau} e^{-A^{\mathcal{P}\prime}(s-t)}\mathcal{M}_3e^{-A^{\mathcal{P}}(s-t)}dsX(t)\\
&-\left(\gamma^{\mathcal{P}\prime}\left(A^{\mathcal{P}\prime}\right)^{-1}\int_t^{t+\tau}\left(I - e^{-A^{\mathcal{P}\prime}(s-t)}\right)\mathcal{M}_3Y_sds\right.\\
&\left.+X(t)'\int_t^{t+\tau} e^{-A^{\mathcal{P}\prime}(s-t)}\mathcal{M}_3Y_sds\right) - \frac{1}{2}\int_t^{t+\tau} Y_s'\mathcal{M}_3Y_s
\end{aligned}$$

$$
\begin{aligned}
&= -\frac{1}{2}\gamma^{\mathcal{P}\prime}\left(A^{\mathcal{P}\prime}\right)^{-1}\mathcal{M}_3\left(A^{\mathcal{P}}\right)^{-1}\gamma^{\mathcal{P}}\tau \\
&+\gamma^{\mathcal{P}\prime}\left(A^{\mathcal{P}\prime}\right)^{-1}\int_t^{t+\tau} e^{-A^{\mathcal{P}\prime}(s-t)}ds\mathcal{M}_3\left(A^{\mathcal{P}}\right)^{-1}\gamma^{\mathcal{P}} \\
&-\frac{1}{2}\gamma^{\mathcal{P}\prime}\left(A^{\mathcal{P}\prime}\right)^{-1}\int_t^{t+\tau} e^{-A^{\mathcal{P}\prime}(s-t)}\mathcal{M}_3e^{-A^{\mathcal{P}}(s-t)}ds\left(A^{\mathcal{P}}\right)^{-1}\gamma^{\mathcal{P}} \\
&-\gamma^{\mathcal{P}\prime}\left(A^{\mathcal{P}\prime}\right)^{-1}\mathcal{M}_3\int_t^{t+\tau} e^{-A^{\mathcal{P}}(s-t)}dsX(t) \\
&+\gamma^{\mathcal{P}\prime}\left(A^{\mathcal{P}\prime}\right)^{-1}\int_t^{t+\tau} e^{-A^{\mathcal{P}\prime}(s-t)}\mathcal{M}_3e^{-A^{\mathcal{P}}(s-t)}dsX(t) \\
&-\frac{1}{2}X(t)'\int_t^{t+\tau} e^{-A^{\mathcal{P}\prime}(s-t)}\mathcal{M}_3e^{-A^{\mathcal{P}}(s-t)}dsX(t)-\gamma^{\mathcal{P}\prime}\left(A^{\mathcal{P}\prime}\right)^{-1}\mathcal{M}_3\int_t^{t+\tau}Y_sds \\
&+\left(\gamma^{\mathcal{P}\prime}\left(A^{\mathcal{P}\prime}\right)^{-1}\int_t^{t+\tau} e^{-A^{\mathcal{P}\prime}(s-t)}\mathcal{M}_3Y_sds-X(t)'\int_t^{t+\tau} e^{-A^{\mathcal{P}\prime}(s-t)}\mathcal{M}_3Y_sds\right) \\
&-\frac{1}{2}\int_t^{t+\tau} Y_s'\mathcal{M}_3Y_s.
\end{aligned}
$$

If we use the definition of Y_s, we have

$$
\begin{aligned}
&-\gamma^{\mathcal{P}\prime}\left(A^{\mathcal{P}\prime}\right)^{-1}\mathcal{M}_3\int_t^{t+\tau}Y_sds+\left(\gamma^{\mathcal{P}\prime}\left(A^{\mathcal{P}\prime}\right)^{-1}\int_t^{t+\tau}e^{-A^{\mathcal{P}\prime}(s-t)}\mathcal{M}_3\int_t^s e^{-A^{\mathcal{P}}(s-v)}\Sigma_X d\epsilon_v ds\right. \\
&\left.-X(t)'\int_t^{t+\tau}e^{-A^{\mathcal{P}\prime}(s-t)}\mathcal{M}_3\int_t^s e^{-A^{\mathcal{P}}(s-v)}\Sigma_X d\epsilon_v ds\right)=0,
\end{aligned}
\tag{87}
$$

since $d\epsilon_v ds = 0$ by Ito's Rule.

We need the result

$$
\int_t^{t+\tau} e^{-A^{\mathcal{P}\prime}(s-t)}\mathcal{M}_3e^{-A^{\mathcal{P}}(s-t)}ds=\left[\mathcal{M}-e^{-A^{\mathcal{P}\prime}\tau}\mathcal{M}e^{-A^{\mathcal{P}}\tau}\right],
$$

where the matrix $\mathcal{M}$ solves the Lyapunov equation

$$
-A^{\mathcal{P}}\mathcal{M}-\mathcal{M}A^{\mathcal{P}\prime}=\mathcal{M}_3. \tag{88}
$$

The solution to this equation is a positive definite symmetric matrix, which is easily calculated using lyap.m in Matlab.

$$
\begin{aligned}
&-\frac{1}{2}\int_t^{t+\tau} X(\tau)'\mathcal{M}_3 X(\tau)ds = -\frac{1}{2}\gamma^{\mathcal{P}\prime}\left(A^{\mathcal{P}\prime}\right)^{-1}\mathcal{M}_3\left(A^{\mathcal{P}}\right)^{-1}\gamma^{\mathcal{P}}\tau \\
&+\gamma^{\mathcal{P}\prime}\left(A^{\mathcal{P}\prime}\right)^{-1}\left(A^{\mathcal{P}\prime}\right)^{-1}\left[I-e^{-A^{\mathcal{P}\prime}\tau}\right]\mathcal{M}_3\left(A^{\mathcal{P}}\right)^{-1}\gamma^{\mathcal{P}} \\
&-\frac{1}{2}\gamma^{\mathcal{P}\prime}\left(A^{\mathcal{P}\prime}\right)^{-1}\left[\mathcal{M}-e^{-A^{\mathcal{P}\prime}\tau}\mathcal{M}e^{-A^{\mathcal{P}}\tau}\right]\left(A^{\mathcal{P}}\right)^{-1}\gamma^{\mathcal{P}} \\
&-\gamma^{\mathcal{P}\prime}\left(A^{\mathcal{P}\prime}\right)^{-1}\mathcal{M}_3\left(A^{\mathcal{P}}\right)^{-1}\left[I-e^{-A^{\mathcal{P}}\tau}\right]X(t)+\gamma^{\mathcal{P}\prime}\left(A^{\mathcal{P}\prime}\right)^{-1}\left[\mathcal{M}-e^{-A^{\mathcal{P}\prime}\tau}\mathcal{M}e^{-A^{\mathcal{P}}\tau}\right]X(t) \\
&-\frac{1}{2}X(t)'\left[\mathcal{M}-e^{-A^{\mathcal{P}\prime}\tau}\mathcal{M}e^{-A^{\mathcal{P}}\tau}\right]X(t)-\frac{1}{2}\int_t^{t+\tau} Y_s\mathcal{M}_3 Y_s
\end{aligned}
$$

We also need

$$
\begin{aligned}
\int_t^{t+\tau}\mathcal{M}_2 X_s ds =&\mathcal{M}_2\int_t^{t+\tau}\left(A^{\mathcal{P}}\right)^{-1}\gamma^{\mathcal{P}}ds-\mathcal{M}_2\int_t^{t+\tau}e^{-A^{\mathcal{P}}(\tau-t)}ds\left(A^{\mathcal{P}}\right)^{-1}\gamma^{\mathcal{P}}ds \\
&+\mathcal{M}_2\int_t^{t+\tau}e^{-A^{\mathcal{P}}(s-t)}dsX(t)+\int_t^{t+\tau}\mathcal{M}_2 Y_s ds \\
=&\mathcal{M}_2\left(A^{\mathcal{P}}\right)^{-1}\gamma^{\mathcal{P}}\tau-\mathcal{M}_2\left(A^{\mathcal{P}}\right)^{-1}\left[I-e^{-A^{\mathcal{P}}(\tau)}\right]\left(A^{\mathcal{P}}\right)^{-1}\gamma^{\mathcal{P}} \\
&+\mathcal{M}_2\left(A^{\mathcal{P}\prime}\right)^{-1}\left[I-e^{-A^{\mathcal{P}}(\tau)}\right]X(t)+\int_t^{t+\tau}\mathcal{M}_2\int_t^{s}e^{-A^{\mathcal{P}}(s-v)}\Sigma_X d\epsilon_v ds \\
=&\mathcal{M}_2\left(A^{\mathcal{P}}\right)^{-1}\gamma^{\mathcal{P}}\tau-\mathcal{M}_2\left(A^{\mathcal{P}}\right)^{-1}\left[I-e^{-A^{\mathcal{P}}(\tau)}\right]\left(A^{\mathcal{P}}\right)^{-1}\gamma^{\mathcal{P}} \\
&+\mathcal{M}_2\left(A^{\mathcal{P}}\right)^{-1}\left[I-e^{-A^{\mathcal{P}}(\tau)}\right]X(t).
\end{aligned}
$$

The last step uses the rule $d\epsilon_v dt = 0$

We also need

$$
\begin{aligned}
\int_t^{t+\tau}d\epsilon_s'\Sigma_X\mathcal{M}_5'X_s =&\int_t^{t+\tau}d\epsilon_s'\Sigma_X\mathcal{M}_5'\left(A^{\mathcal{P}}\right)^{-1}\gamma^{\mathcal{P}}-\int_t^{t+\tau}d\epsilon_s'\Sigma_X\mathcal{M}_5'e^{-A^{\mathcal{P}}(\tau-t)}\left(A^{\mathcal{P}}\right)^{-1}\gamma^{\mathcal{P}} \\
&+\int_t^{t+\tau}d\epsilon_s'\Sigma_X\mathcal{M}_5'e^{-A^{\mathcal{P}}(s-t)}X(t)+\int_t^{t+\tau}d\epsilon_s'\Sigma_X\mathcal{M}_5'Y_s \\
=&\int_t^{t+\tau}d\epsilon_s'\Sigma_X\Big[\mathcal{M}_5'\left(A^{\mathcal{P}}\right)^{-1}\gamma^{\mathcal{P}}-\mathcal{M}_5'e^{-A^{\mathcal{P}}(\tau-t)}\left(A^{\mathcal{P}}\right)^{-1}\gamma^{\mathcal{P}} \\
&+\mathcal{M}_5'e^{-A^{\mathcal{P}}(s-t)}X(t)\Big]+\int_t^{t+\tau}d\epsilon_s'\Sigma_X\mathcal{M}_5'Y_s \\
=&\int_t^{t+\tau}\left(\mathbb{M}_4+X_t'\mathbb{M}_5+Y_s'\mathcal{M}_5\Sigma_X'\right)d\epsilon_s.
\end{aligned}
$$

We now bring all these calculations into the stochastic process for the pricing kernel.

$$
\begin{aligned}
\frac{M_{\tau,t}}{M_{t,t}} = \exp\Bigg\{ & -\mathcal{M}_1(\tau)\tau - \frac{1}{2}\gamma^{\mathcal{P}\prime}\left(A^{\mathcal{P}\prime}\right)^{-1}\mathcal{M}_3\left(A^{\mathcal{P}}\right)^{-1}\gamma^{\mathcal{P}}\tau \\
& + \gamma^{\mathcal{P}\prime}\left(A^{\mathcal{P}\prime}\right)^{-1}\left(A^{\mathcal{P}\prime}\right)^{-1}\left[I - e^{-A^{\mathcal{P}\prime}\tau}\right]\mathcal{M}_3\left(A^{\mathcal{P}}\right)^{-1}\gamma^{\mathcal{P}} \\
& - \frac{1}{2}\gamma^{\mathcal{P}\prime}\left(A^{\mathcal{P}\prime}\right)^{-1}\left[\mathcal{M} - e^{-A^{\mathcal{P}\prime}\tau}\mathcal{M}e^{-A^{\mathcal{P}}\tau}\right]\left(A^{\mathcal{P}}\right)^{-1}\gamma^{\mathcal{P}} \\
& - \gamma^{\mathcal{P}\prime}\left(A^{\mathcal{P}\prime}\right)^{-1}\mathcal{M}_3\left(A^{\mathcal{P}}\right)^{-1}\left[I - e^{-A^{\mathcal{P}}\tau}\right]X(t) \\
& + \gamma^{\mathcal{P}\prime}\left(A^{\mathcal{P}\prime}\right)^{-1}\left[\mathcal{M} - e^{-A^{\mathcal{P}\prime}\tau}\mathcal{M}e^{-A^{\mathcal{P}}\tau}\right]X(t) \\
& - \frac{1}{2}X(t)'\left[\mathcal{M} - e^{-A^{\mathcal{P}\prime}\tau}\mathcal{M}e^{-A^{\mathcal{P}}\tau}\right]X(t) + \mathcal{M}_2\left(A^{\mathcal{P}}\right)^{-1}\gamma^{\mathcal{P}}\tau \\
& - \mathcal{M}_2\left(A^{\mathcal{P}}\right)^{-1}\left[I - e^{-A^{\mathcal{P}}(\tau)}\right]\left(A^{\mathcal{P}}\right)^{-1}\gamma^{\mathcal{P}} \\
& + \mathcal{M}_2\left(A^{\mathcal{P}}\right)^{-1}\left[I - e^{-A^{\mathcal{P}}(\tau)}\right]X(t) - \frac{1}{2}\int_t^{t+\tau} Y_s'\mathcal{M}_3 Y_s \\
& + \int_t^{t+\tau}\left(\mathbb{M}_4 + X_t'\mathbb{M}_5 + Y_s'\mathcal{M}_5\Sigma_X'\right)d\epsilon_s\Bigg\}.
\end{aligned} \tag{89}
$$

Define

$$
\begin{aligned}
\mathcal{M}(\tau, X) \equiv \exp\Bigg\{ & -\mathcal{M}_1(\tau)\tau - \frac{1}{2}\gamma^{\mathcal{P}\prime}\left(A^{\mathcal{P}\prime}\right)^{-1}\mathcal{M}_3\left(A^{\mathcal{P}}\right)^{-1}\gamma^{\mathcal{P}}\tau \\
& + \gamma^{\mathcal{P}\prime}\left(A^{\mathcal{P}\prime}\right)^{-1}\left(A^{\mathcal{P}\prime}\right)^{-1}\left[I - e^{-A^{\mathcal{P}\prime}\tau}\right]\mathcal{M}_3\left(A^{\mathcal{P}}\right)^{-1}\gamma^{\mathcal{P}} \\
& - \frac{1}{2}\gamma^{\mathcal{P}\prime}\left(A^{\mathcal{P}\prime}\right)^{-1}\left[\mathcal{M} - e^{-A^{\mathcal{P}\prime}\tau}\mathcal{M}e^{-A^{\mathcal{P}}\tau}\right]\left(A^{\mathcal{P}}\right)^{-1}\gamma^{\mathcal{P}} \\
& + \mathcal{M}_2\left(A^{\mathcal{P}}\right)^{-1}\gamma^{\mathcal{P}}\tau - \mathcal{M}_2\left(A^{\mathcal{P}}\right)^{-1}\left[I - e^{-A^{\mathcal{P}}(\tau)}\right]\left(A^{\mathcal{P}}\right)^{-1}\gamma^{\mathcal{P}} \\
& + \Big[\gamma^{\mathcal{P}\prime}\left(A^{\mathcal{P}\prime}\right)^{-1}\left[\mathcal{M} - e^{-A^{\mathcal{P}\prime}\tau}\mathcal{M}e^{-A^{\mathcal{P}}\tau}\right] + \mathcal{M}_2\left(A^{\mathcal{P}}\right)^{-1}\left[I - e^{-A^{\mathcal{P}}(\tau)}\right] \\
& - \gamma^{\mathcal{P}\prime}\left(A^{\mathcal{P}\prime}\right)^{-1}\mathcal{M}_3\left(A^{\mathcal{P}}\right)^{-1}\left[I - e^{-A^{\mathcal{P}}\tau}\right]\Big]X(t) \\
& - \frac{1}{2}X(t)'\left[\mathcal{M} - e^{-A^{\mathcal{P}\prime}\tau}\mathcal{M}e^{-A^{\mathcal{P}}\tau}\right]X(t)\Bigg\} \\
= \exp\Bigg\{ & -\frac{1}{2}\left(X - \mathfrak{M}_3^{-1}\mathfrak{M}_2\right)'\mathfrak{M}_3\left(X - \mathfrak{M}_3^{-1}\mathfrak{M}_2\right) + \frac{1}{2}\mathfrak{M}_2'\mathfrak{M}_3^{-1}\mathfrak{M}_2 + \mathfrak{M}_1\Bigg\}.
\end{aligned} \tag{90}
$$

This result can be used to separate the portion of the pricing kernel dependent on the current factors X from future random changes in these factors Y_s for $s > t$. We substitute the known part (25) into the pricing kernel (24) so that

$$\frac{1}{\mathcal{M}(\tau,X)}\frac{M_{T,t}}{M_{t,t}} = \exp\left\{-\frac{1}{2}\int_t^{t+T} Y_s'\mathcal{M}_3 Y_s ds + \int_t^{t+T} \left(\mathbb{M}_4 + X_t'\mathbb{M}_5 + Y_s'\mathcal{M}_5\Sigma_X'\right) d\epsilon_s\right\}. \tag{91}$$

This relation is an example of the stochastic process (44) so that its probability distribution is the solution to the forward Kolmogorov equation (13). Notice (91) is dependent on the current X through $\mathbb{M}_5$. This means that $\mathcal{D}_4 \equiv \mathbb{M}_4 + X_t'\mathbb{M}_5$ and $\mathcal{D}_5 = \mathcal{M}_5\Sigma_X'$. These terms do not influence the forward Kolmogorov equation, since this error term has mean zero.

The solution to the forward Kolmogorov equation yields the probability distribution for the pricing kernel.

$$\frac{1}{\mathcal{M}(\tau,X)}\frac{M_{T,t}}{M_{t,t}} \sim \frac{1}{\sqrt{(2\pi)^N\, det(\mathcal{A}_3(\tau,X))}} \exp\left\{-\frac{1}{2}\mathcal{A}_1(\tau,X) - \frac{1}{2}Y'\mathcal{A}_3(\tau,X)^{-1}Y\right\}$$

which has the same form as (44) with the appropriate definitions of the coefficients $\mathcal{D}'s$.

Thus the probability distribution for the pricing kernel is given by

$$\frac{M_{T,t}}{M_{t,t}} \sim \exp\left\{-\frac{1}{2}\left(X - \mathfrak{M}_3^{-1}\mathfrak{M}_2\right)'\mathfrak{M}_3\left(X - \mathfrak{M}_3^{-1}\mathfrak{M}_2\right) + \frac{1}{2}\mathfrak{M}_2'\mathfrak{M}_3^{-1}\mathfrak{M}_2 + \mathfrak{M}_1 - \frac{1}{2}\mathcal{A}_1(\tau)\right\}$$
$$\times \frac{1}{\sqrt{(2\pi)^N\, det(\mathcal{A}_3(\tau))}} \exp\left\{-\frac{1}{2}Y'\mathcal{A}_3(\tau)^{-1}Y\right\}.$$

This leads to equation (25) and (26) in the text with $\sigma_M \equiv \mathcal{A}_3(\tau)$.

$$E_t\left[\frac{M_{T,t}}{M_{t,t}}\right] = \exp\left\{-\frac{1}{2}\left(X - \mathfrak{M}_3^{-1}\mathfrak{M}_2\right)'\mathfrak{M}_3\left(X - \mathfrak{M}_3^{-1}\mathfrak{M}_2\right)\right.$$
$$\left. + \frac{1}{2}\mathfrak{M}_2'\mathfrak{M}_3^{-1}\mathfrak{M}_2 + \mathfrak{M}_1 - \frac{1}{2}\mathcal{A}_1(\tau)\right\}$$
$$\times \frac{1}{\sqrt{(2\pi)^N\, det(\mathcal{A}_3(\tau))}} \int_{-\infty}^{\infty} \exp\left\{-\frac{1}{2}Y'\mathcal{A}_3(\tau)^{-1}Y\right\}dY$$
$$= \exp\left\{-\frac{1}{2}\left(X - \mathfrak{M}_3^{-1}\mathfrak{M}_2\right)'\mathfrak{M}_3\left(X - \mathfrak{M}_3^{-1}\mathfrak{M}_2\right) + \frac{1}{2}\mathfrak{M}_2'\mathfrak{M}_3^{-1}\mathfrak{M}_2 + \mathfrak{M}_1 - \frac{1}{2}\mathcal{A}_1(\tau)\right\}. \tag{92}$$

This corresponds to equation (25) in the text with

$$(\sigma_{\mathcal{M}}(\tau))^{-1} \equiv \mathfrak{M}_3$$
$$\mathcal{M}(\tau) \equiv \exp\left\{\frac{1}{2}\mathfrak{M}_2'\mathfrak{M}_3^{-1}\mathfrak{M}_2 + \mathfrak{M}_1 - \frac{1}{2}\mathcal{A}_1(\tau)\right\}. \tag{93}$$

References

Adrian, T., Crump, R. K., & Moench, E. (2013). Pricing the term structure with linear regressions. *Journal of Financial Economics*, *110*, 110–138.

Anderson, L., Duffie, D., & Song, Y. (2018). Funding value adjustments. forthcoming *Journal of Finance* and working paper Stanford Business School.

Ang, A., & Pizzesi, M. (2003). A no-arbitrage vector autoregression of term structure dynamics with macroeconomic and latent variables. *Journal of Monetary Economics*, *50*, 745–787.

Ang, A., Behaert, G., & Wei, M. (2008a). The term structure of real rates and expected inflation. *Journal of Finance*, *63*, 797–849.

Ang, A., Boivin, J., Dong, S., & Loo-Kung, R. (2008b). Monetary policy shifts and the term structure. *Review of Economic Studies*, *78*, 429–457.

Arnold, L. (1974). *Stochastic differential equations: theory and applications*. New York: Wiley.

Bauer, M. D., & Rudebusch, G. D. (2016). Monetary policy expectations at the zero lower bound. *Journal of Money, Credit and Banking*, *48*, 1439–1465.

Bauer, M. D., & Rudebusch, G. D. (2017). Resolving the spanning puzzle in macro-finance term structure models. *Review of Finance*, *21*, 511–553.

Calin, O., Chang, D.-C., Furutani, K., & Iwasaki, C. (2011). *Heat kernels for elliptic and sub-elliptic operators*. Basel: Birkhäuser.

Chami, R., Cosimano, T. F., Ma, J.., & Rochon, C. (2017). What's Different about Bank Holding Companies? IMF Working Paper 17/26.

Chami, R., Cosimano, T. F., Rochon, C., & Yung, J. (2018). Monetary Policy and the Term Structure. Working Paper.

Chernov, Ml, & Mueller, P. (2012). The term structure of expected inflation. *Journal of Financial Economics*, *106*, 367–394.

Chirikjian, G. S. (2009). *Stochastic models, information theory and Lie groups*. Basel: Birkhäuser.

Cochrane, J., & Piazzesi, M. (2005). Bond risk premia. *American Economic Review*, *95*, 138–160.

Cosimano, T. F., & Yung, J. (2018). Foreign Exchange and the Term Structure. Working paper.

Dai, Q., & Singleton, K. (2000). Specification analysis of affine term structure models. *Journal of Finance*, *55*, 1943–1978.

Dai, Q., & Singleton, K. (2002). Expectations puzzles, time-varying risk premia, and affine models of the term structure. *Journal of Financial Economics*, *63*, 415–441.

Duffie, D. (2001). *Dynamic asset pricing theory*. Princeton: Princeton University Press.

Duffie, D., & Huang, M. (1996). Swap rates and credit quality. *Journal of Finance*, *51*, 921–949.

Duffie, D., & Kan, R. (1996). A yield-factor model of interest rates. *Mathematical Finance*, *6*, 379–406.

Duffie, D., Pan, J., & Singleton, K. (2000). Transform analysis and asset pricing for affine jump-diffusions. *Econometrica*, *68*, 1343–1376.

Duffie, D., & Singleton, K. (1997). An econometric model of the term structure of interest rate swap yields. *Journal of Finance*, *52*, 1287–1323.

Durham, J. B. (2013). Arbitrage-Free Models of Stocks and Bonds. Working paper Federal Reserve Bank of New York.

Durham, J. B. (2015). Arbitrage-Free Affine Models of the Forward Price of Foreign Currency. Working paper Federal Reserve Bank of New York.

Greenwood, R., & Vayanos, D. (2014). Bond supply and excess bond return. *Review of Financial Studies*, *27*, 663–713.

Grinblatt, M., & Longstaff, F. (2000). Financial inovation and the role of derivative securities: An empirical analysis of the treasury STRIPS program. *Journal of Finance*, *55*, 1415–1436.

Hamilton, J., & Wu, J. C. (2012a). Identification and estimation of gaussian affine term structure models. *Journal of Econometrics*, *168*, 315–331.

Hamilton, J., & Wu, J. C. (2012b). The effectiveness of alternative monetary policy tools in a zero lower bound environment. *Journal of Money, Credit and Banking*, *44*, 3–46.

Hamilton, J., & Wu, J. C. (2014a). Testable implications of affine term structure models. *Journal of Econometrics*, *178*, 231–242.
Hamilton, J., & Wu, J. C. (2014b). Risk premia in crude oil futures prices. *Journal of International Money and Finance*, *42*, 9–37.
Hamilton, J., & Wu, J. C. (2016). Measuring the macroeconomic impact of monetary policy at the zero lower bond. *Journal of Money, Credit and Banking*, *48*, 253–291.
Hijab, O. (1987). *Stabilization of control systems*. Berlin: Springer.
Joslin, S., Singleton, K. J., & Zhu, H. (2011). A new perspective on gaussian dynamic term structure models. *Review of Financial Economics*, *24*, 2184–2227.
Joslin, S., Priebsch, M., & Singleton, K. J. (2014). Risk premiums in dynamic term structure models with unspanned macro risks. *Journal of Finance*, *69*, 1197–1233.
Karatzas, I., & Shreve, S. E. (1988). *Brownian motion and stochastic calculus*. Berlin: Springer.
Kothari, S. P., Lewellen, J., & Warner, J. B. (2017). The Behavior of Aggregate Corporate Investment working paper Sloan School of Management MIT.
Krippner, L. (2015). *Zero lower bound term structure modeling: A practitioner's guide*. Basingstoke: Palgrave Macmillan.
Li, C., & Wei, M. (2013). Term structure modeling with supply factors and the federal reserve's large-scale asset purchases programs. *International Journal of Central Banking*, *14*, 3–40.
Liu, J. (2007). Portfolio selection in stochastic environments. *Review of Financial Studies*, *20*, 1–39.
Liu, L. (2016). Essays on Asset Returns and Portfolio Allocation. Chapter 2 of Ph.D. Dissertation, University of Notre Dame.
Longstaff, F., Santa-Clara, P., & Swartz, E. (2001). The relative valuation of caps and swaaptions theory and empirical evidence. *Journal of Finance*, *56*, 2067–2109.
Øksendal, B. (2005). *Stochastic differential equations: An introduction with applications*. Berlin: Springer.
Petersen, K. B., & Pedersen, M. S. (November 14, 2008). The Matrix Cookbook at http://matrixcookbook.com.
Piazzesi, M. (2010). Affine term structure models. In Y. Aït-Sahalia & L. H. Hansen (Eds.), *Handbook of financial econometrics* (Vol. 12, pp. 691–766). New York: North Holland.
Protter, P. E. (2005). *Stochastic integration and differential equations* (2nd ed.). Berlin: Springer.
Sangvinatsos, A., & Wachter, J. A. (2005). Does the failure of the expectations hypothesis matter for long-term investors? *The Journal of Finance*, *60*, 179–230.
Shreve, S. E. (2006). *Stochastic calculus for finance II: Continuous-time models*. Berlin: Springer.
Strauss, W. A. (2008). *Partial differential equations: An introduction*. New York: Wiley.
Strichartz, R. S. (2008). *A guide to distribution theory and fourier transforms*. Singapore: World Scientific.
Tevlin, S., & Whelan, K. (2003). Explaining the investment boom of the 1990s. *Journal of Money, Credit and Banking*, *35*, 1–22.
Wu, J. C., & Xia, F. D. (2016). Measuring the macroeconomic impact of monetary policy at the zero lower bound. *Journal of Money, Credit and Banking*, *48*, 253–291.
Yung, J. (2017). Can Interest Rate Factors Explain Exchange Rate Fluctuations? working paper Bates College.

The manufacturer's authorised representative in the EU is Springer Nature Customer Service Centre GmbH, Europaplatz 3, 69115 Heidelberg, Germany. If you have any concerns regarding our products, please contact ProductSafety@springernature.com

Printed and bound by CPI Group (UK) Ltd, Croydon, CR0 4YY
05/10/2026
02228135-0012